D1236422

Inorganic Chemistry in Biology and Medicine

Arthur E. Martell, Editor
Texas A&M University

Based on a symposium sponsored by the Division of Inorganic Chemistry at the 178th Meeting of the American Chemical Society, Washington, D.C., September 10–11, 1979.

ACS SYMPOSIUM SERIES **140**

AMERICAN CHEMICAL SOCIETY
WASHINGTON, D. C. 1980

Library of Congress CIP Data

Inorganic chemistry in biology and medicine.
(ACS symposium series; 140 ISSN 0097-6156)

Includes bibliographies and index.

1. Metals in the body—Congresses. 2. Metals—Therapeutic use—Congresses. 3. Cancer—Chemotherapy—Congresses. 4. Chelation therapy—Congresses. 5. Chemistry, Inorganic—Congresses.
I. Martell, Arthur Earl, 1916– II. American Chemical Society. Division of Inorganic Chemistry. III. Series. IV. Series: American Chemical Society. ACS symposium series; 140.

QP532.I56 616 80-23248
ISBN 0-8412-0588-4 ACSMC8 140 1-436 1980

PRINTED IN THE UNITED STATES OF AMERICA

ACS Symposium Series

M. Joan Comstock, *Series Editor*

FOREWORD

The ACS SYMPOSIUM SERIES was founded in 1974 to provide a medium for publishing symposia quickly in book form. The format of the Series parallels that of the continuing ADVANCES IN CHEMISTRY SERIES except that in order to save time the papers are not typeset but are reproduced as they are submitted by the authors in camera-ready form. Papers are reviewed under the supervision of the Editors with the assistance of the Series Advisory Board and are selected to maintain the integrity of the symposia; however, verbatim reproductions of previously published papers are not accepted. Both reviews and reports of research are acceptable since symposia may embrace both types of presentation.

CONTENTS

PREFACE

At its inception, the original plan for this symposium was to emphasize the medical aspects of inorganic chemistry, rather than to go over once more new developments in bioinorganic chemistry, important as the subject is, since the latter topic has been treated many times in recent symposia reviews and monographs. The objectives of this symposium were to review and interpret the remarkable advances that have occurred recently in medical inorganic chemistry and to stimulate interest on the part of inorganic chemists to become involved in the developing research problems in this area. The interactions of metal ions with biomolecules and the functions of metal ions in physiological systems are very complex, and the precise nature of these interactions and processes are, for the most part, unknown. In addition to the applications of metal ions and complexes for medical purposes, extensive fundamental studies are needed to understand the basis of these applications and thereby make it possible to carry out systematic improvements in current methods as well as to develop new approaches in this interesting field.

Of the approximately eighty metallic elements, a considerable number have been identified as essential to life; many others have been indicated as possibly essential, while a large number of metals are of concern because of toxic effects that result when they are introduced into the body accidentally or through environmental influences. Major metal ions such as Na^{+}, K^{+}, Mg^{2+}, and Ca^{2+} are important in maintaining electrolyte concentration in body fluids or as skeletal constituents. Many of the transition metal ions are essential in trace amounts for the activation of enzyme systems. In many cases, these essential metal ions become toxic or even carcinogenic when present at sufficient levels to overwhelm the natural ligands and macromolecules that function as carriers for these ions, and thus more than saturate the normal physiological processes for their control. Under such conditions, they may function, as do many unnatural toxic metals, by reacting with other biomolecules, distorting or blocking their essential functions. In many cases, the differences between the essential and toxic levels are surprisingly narrow. This duality of behavior between natural and toxic levels constitutes the basis of threshold concentrations for several carcinogenic metals—below which these metals exist as essential and noncarcinogenic compounds. It also provides a strong refutation of the validity of the linear extrapolation method still in active use for the interpretation of carcinogenicity of compounds observed at high concentration levels in test animals.

The topics covered in this symposium were selected so as to provide examples of current and potential medical applications of metal compounds. The emphasis and amount of attention given were in many cases not in proportion to the importance or activity levels of these applications, for a number of reasons. The use of platinum complexes for the treatment of cancer is perhaps under-represented because several symposia, some of which have been published, have been held on this subject in recent years. Similarly, iron nutrition, although very important, has been omitted because it is well covered by periodic and continuing conferences and conference proceedings devoted entirely to this field of research. New developments of ionophores and on the use of chelating agents for the removal of radioactive metals from the body were not given the attention that they deserve in this symposium because these subjects were treated in separate symposia at the same American Chemical Society Meeting.

Because of the large number and complexity of the functions of metal ions in physiological systems, the applications of complexes of both essential and unnatural metal ions for medical purposes are expected to expand dramatically in the next decade. It is hoped that this book will help to attract more inorganic chemists to this field, to provide the expertise in coordination chemistry needed for the achievement of significant new developments in this potentially important area of medicine.

The Editor wishes to express his appreciation for the many helpful suggestions received from professional colleagues during the formative stages of this symposium. Special thanks are due to L. G. Marzilli for assistance with subject matter planning, and to J. H. Timmons for valuable editorial assistance.

Texas A&M University
College Station, Texas

A. E. MARTELL

August 7, 1980

METAL COMPLEXES
IN NUTRITION AND METABOLISM

1

Molecular and Biological Properties of Ionophores

BERTON C. PRESSMAN, GEORGE PAINTER, and MOHAMMAD FAHIM

Department of Pharmacology, University of Miami, Miami, FL 33101

The ionophores are a group of natural and synthetic compounds which form lipid-soluble cation complexes which can transport cations across low polarity barriers such as organic solvents and lipids (1). From a biological standpoint, the most important low polarity barrier is the lipid bilayer which lies within biological membranes; ionophores possess unique and potent biological properties which derive from their ability to perturb transmembrane ion gradients and electrical potentials. Each ionophore has its own characteristic ion selectivity pattern arising from the interaction between the conformational options of the host ionophore and the effective atomic radius and charge density of the guest cation. The ability of ionophores to complex and transport cations has an ever growing list of applications in experimental biology and technology and may ultimately provide the basis for novel cardiovascular drugs. Ionophores are also intriguing intellectually as objects for study of chemical and physical complexation processes at the molecular level and as challenges to the state of the art of chirally selective organic synthesis (2). Several reviews are available for expanding the description of ionophores provided here (3,4,5).

General Structural Features of Ionophores

Several of the general structural features of ionophores are illustrated in Figure 1. All ionophores deploy an array of liganding oxygen atoms about a cavity in space into which the complexed cation fits. X-ray crystallography reveals that the principal bonding energy is provided by induced dipolar interaction between the complexed cation and those specific oxygens which are filled in.

Valinomycin consists of alternating residues of hydroxyacids and aminoacids constituting a cyclic dodecadepsipeptide. In space the ring undulates defining a bracelet 4 Å high and 10 Å in diameter. The liganding oxygens, the ester carbonyls, form a three

0-8412-0588-4/80/47-140-003$05.00/0

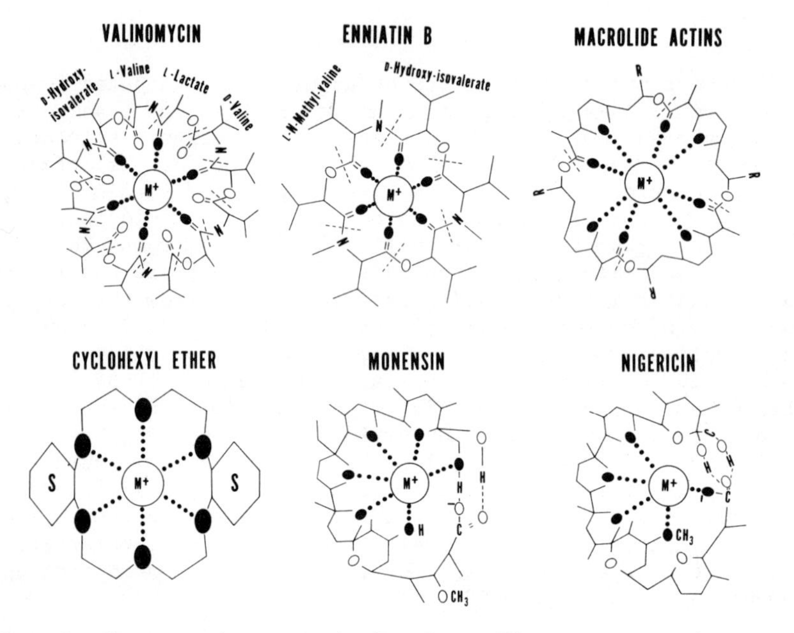

Figure 1. Structures of representative ionophores. The oxygen atoms that x-ray crystallography indicates to be primarily involved in liganding to cations are filled in.

dimensional cage which accommodates K^+ (r = 1.33 Å) much more snugly than Na^+ (r = 0.95 Å) resulting in a $K^+:Na^+$ preference of 10,000:1 (4).

Enniatin B is a cyclic hexadepsipeptide; the smaller ring results in a relatively planer array of liganding oxygen atoms; the more open and more flexible cage results in a $K^+:Na^+$ discrimination of only 3:1 (6).

A new feature appears in the cyclic tetraesters, the macrolide nactins. In addition to the ester carbonyls, four heterocyclic ether oxygens participate in complexation; the oxygens are arranged at the apices of a cubic cage. Five variant nactins are known depending whether 0-4 of the R groups are methyls (nonactin) or ethyls (monactin, dinactin, trinactin, tetranactin)(7).

While the aforementioned ionophores are Streptomyces metabolites, the crown polyethers, the depicted prototype of which is dicyclohexyl-18-crown-6, are synthetic (8). Although they lack the intricate conformations of the natural ionophores arising from multiple asymmetric carbon atoms, their molecular liganding properties are analogous. While they are less efficient ion carriers, their lack of labile linkages confers increased chemical stability; they find extensive use in organic synthesis for solubilizing electrolytes, e.g. K^+ enolates, in nonpolar solvents thereby providing reactive naked anions (9).

The ionophores thus far described lack ionizable groups and are collectively classified as neutral ionophores; their complexes acquire the net charge of whatever ion is complexed. We shall now examine two representatives of the carboxylic subclass of ionophores. Only the anionic form of these ionophores complex cations, hence they form electrically neutral zwitterionic complexes. This distinction is fundamental for explaining the profound differences in biological behavior of the ionophore subclasses, hence we prefer carboxylic ionophore to the term polyether antibiotic used by Westley (5). The latter term, furthermore, leads to functional ambiguity with the ethereal macrolide nactins and crown polyethers which are neutral ionophores.

The naturally occurring carboxylic ionophores, typified by monensin, lack the structural redundancy of the neutral ionophores. Monensin consists of a formally linear array of heterocyclic ether-containing rings, however the molecular chirality arising from the rings and asymmetric carbons favors the molecule assuming a quasi-cyclic configuration. Additional stabilization of the ring is conferred by head-to-tail hydrogen bonding. In addition to its liganding ether oxygens, monensin has a pair of liganding hydroxyl oxygens (10).

The tail portion of nigericin closely resembles monensin, however, an additional tetrahydropyranol ring thrusts the head carboxyl group into the complexation sphere. Thus, in addition to the induced dipole ion bonds previously described, nigericin complexes feature a true ionic bond. Despite major similarities in structure, nigericin prefers K^+ over Na^+ by a factor of 100 while monensin prefers Na^+ over K^+ by a factor of 10 (11).

Dynamics of Ionophore-Mediated Transport

Neutral Ionophores. The relationship between equilibrium ionophore affinities and dynamic biological transmembrane transport is detailed in Figure 2. The transport cycle catalyzed by neutral ionophores is given on the left. Ionophore added to a biological membrane partitions predominately into the membrane. A portion of the ionophore diffuses to the membrane interface where it encounters a hydrated cation. A loose encounter complex is formed followed by replacement of the cationic hydration sphere by engulfment of the cation by the ionophore. The dehydrated complex is lipid-soluble and hence can diffuse across the membrane. The cation is then rehydrated, released, and the uncomplexed ionophore freed to return to its initial state within the membrane. The net reaction catalyzed is the movement of an ion with its charge across the membrane.

Two independent factors determine the thermodynamic gradient governing net transport by neutral ionophores: the membrane potential, i.e. ΔE_{AB}, and the concentration gradient, $[M^+]_A/[M^+]_B$. At equilibrium, the electrochemical potential (a combined function of electrical and concentration terms) of M^+ on side A becomes equal to the electrochemical potential of M^+ on side B, i.e. $\tilde{\mu}_{M_A^+} = \tilde{\mu}_{M_B^+}$. In terms of experimentally measurable parameters, the relationship

$$\Delta E_{AB} = -59 \text{ mV} \log [M^+]_A/[M^+]_B$$

applies. This signifies that if the electrical term, ΔE_{AB}, exceeds the concentration term, 59 mV log $[M_A^+/M_B^+]$, the ion will flow down the potential gradient and dissipate it (electrophoretic transport mode). If the concentration term exceeds the pre-existing potential term, the movement of M^+ down its concentration term will increase ΔE_{AB} (electrogenic transport). The relevant significance of this transport mode is that neutral ionophores perturb not only the transmembrane ion gradients of biological systems but also their transmembrane electrical potentials. Since the latter are so important in biological control, it is not surprising that the neutral ionophores are exceedingly toxic towards intact animals.

Carboxylic Ionophores. Carboxylic ionophore-mediated transport is detailed on the left of Figure 2. The form assumed within the membrane at the start of the transport cycle is an electrically neutral zwitterion, $M^+ \cdot I^-$; anionic free I^- is presumably too polar to be stable at that locus. When this species diffuses to the membrane interface, it is subject to solvation; the cation can be hydrated and removed from the complex. The resultant highly polar I^- is obliged to remain at the interface until a new charge partner, represented by $N^+ \cdot H_2O$, arrives. Once in position, N^+

exchanges its solvation H_2O for the oxygen liganding system of I^- forming lipid compatible $N^+.I^-$ which then diffuses across the membrane. There the process is reversed and N^+ is exchanged for M^+. The ionophore then reenters the membrane as M^+I^- thereby completing the catalytic cycle. The net reaction is the movement of N^+ across the membrane in exchange for M^+ without an accompanying net charge translocation. This is presumably an essential requirement for tolerance of appreciable concentrations of ionophores by animals, i.e. carboxylic ionophores are relatively nontoxic compared to neutral ionophores. In other words, the ability of carboxylic ionophores to alter physiological processes in a pharmacologically useful manner stems from their capability to alter transmembrane ion gradients without directly short circuiting the transmembrane potentials of electrically active cells.

The formation and dissociation of ionophore-cation complexes is equivalent to the displacement of the primary cation solvation sphere by the ionophore liganding atoms. The solvated liganding groups approach the solvated cation until they attain apposition. They then interact via an associative interchange mechanism analogous to an S_N2 mechanism (12). Formation of the transition state involves extension of the cation to both the entering ligand and the departing cation solvation sphere. In the process, the less rigorously defined solvation sphere of the ligand is also discharged. The ionophore then engulfs the cation, its liganding groups progressively displacing the molecules of the cation solvation shell in a concerted fashion. In the case of the carboxylic ionophores, the initial stage prior to the formation of the transition complex is a simple ion pair.

Although they vary widely in structure and conformation, the carboxylic ionophores feature a variety of heteroatoms constituting a liganding system which operates by means of induced dipoles. The magnitude of the dipoles increases progressively by induction as approached by the cation and ultimately produces a solvation system stronger than that of the bulk phase solvent. Whereas the individual solvation molecules, within the primary solvation sphere of a cation, exchange independently with the bulk solvent, the ligands of an ionophore, held together by a common backbone, must behave in a cooperative manner. Intramolecular hydrogen bonding and substituents which favor cyclic conformations (e.g. spirane systems) promote the stability of complexes. Consequently, the various cation affinity and selectivity patterns which characterize each ionophore arise from the precise spacial depolyment of liganding heteroatoms as determined by molecular conformation (13,14).

Conformational Studies of a Representative Carboxylic Ionophore, Salinomycin

Salinomycin, a representative carboxylic ionophore (Figure 3) (15), is a particularly suitable model for studying the dynamic

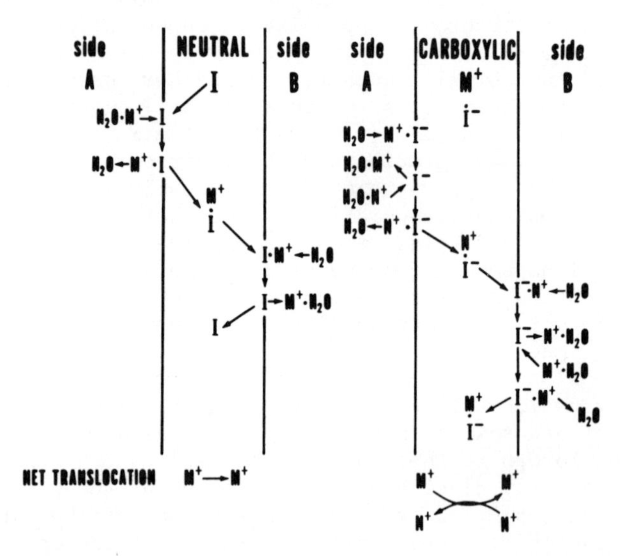

Figure 2. Different modes of ionophore-mediated transmembrane transport. Neutral ionophore-mediated transport is depicted on the left and carboxylic ionophore-mediated transport, on the right. The individual transport steps are detailed in the text.

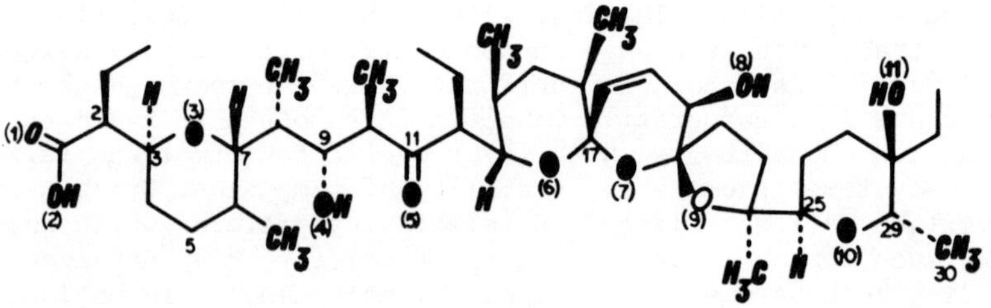

Figure 3. Structure of salinomycin (15)

conformational aspects of complexation. The circular dichroism (CD) arising from the $n \rightarrow \pi^*$ transition of the C-11 carbonyl is sensitive to molecular environment and serves as a probe to report the chirality in its vicinity. CD enables us to evaluate the conformational perturbations produced by altering the polar and protic properties of the solvent system. Systematic perturbation of the solution conformation of salinomycin by an appropriate choice of solvents reveals that ion affinity and selectivity are variable, conformationally determined, properties.

Representative CD spectra of protonated salinomycin, its K^+ complex and its uncomplexed anion are presented in Figure 4. No significant shift of the negative 290 nm peak occurs with solvent change or liganding state; Beer's law is obeyed from 10^{-4} to 10^{-6} M. The function most suitable for relating CD spectra to the conformation of a molecule is the rotational strength (R_o^T) of the observed electronic transition (16). Since the Gaussian approximation appears to hold for the salinomycin CD curves, R_o^T was calculated from [θ] and wavelength by a standard equation (17).

Figure 5 illustrates the effect of solvent changes on the R_o^T of the ionophore free acid and its anion. Kosower's Z values proved empirically an effective function for ranking solvents according to their integrated polar and protic properties (18). The $|R_o^T|$ of the free acid decreases linearly with a small positive slope as the Z values rise. In contrast, the $|R_o^T|$ of the uncomplexed anion, the species participating in complexation, drops sharply between Z values of 80 and 83, varying little above and below these values. Thus, the conformation of the anion tends toward one of two metastable states depending upon solvent Z value.

The role of the solvent in determining equilibrium solution conformation can best be understood in terms of functional group stabilization. In polar protic media the equilibrium conformation of the uncomplexed anionic ionophore is determined by the solvation of the carboxylate anion and the polar liganding groups. Thus, two distinct solvent effects are operative, solvation of the polar liganding groups resulting in conformational stabilization due to decreased dipole-dipole repulsion and maximization of the solvation energy of the anion. The protonated ionophore responds only to the solvation of polar liganding groups. Thus, Figure 5 provides insight into the relative importance of each of these factors in determining equilibrium solution conformation. The perturbation of conformation due to solvation of polar liganding groups alone, as in the protonated ionophore, causes only a slight change in conformation, i.e. a small change in $|R_o^T|$, over a large range of Z values. However, ionization of the protonated form of the ionophore profoundly changes its response to solvents. At Z values > 83, the carboxylate is stabilized by its protic, polar environment. The resulting solvation sphere influences the conformation strongly as evidenced by the very low $|R_o^T|$ values (Figure 3). As the Z values fall, and the solvent becomes less

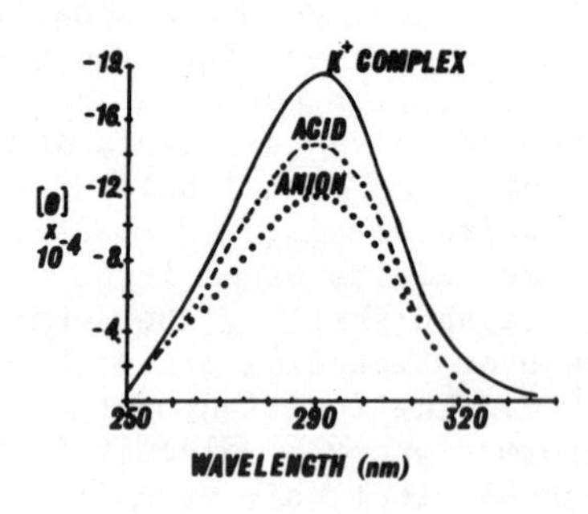

Figure 4. CD spectra of the carboxylic acid free anion and K^+ complex forms of salinomycin. The free anionic form was generated by the addition of excess tri-n-butylamine and the K^+ complex by the addition of excess KSCN.

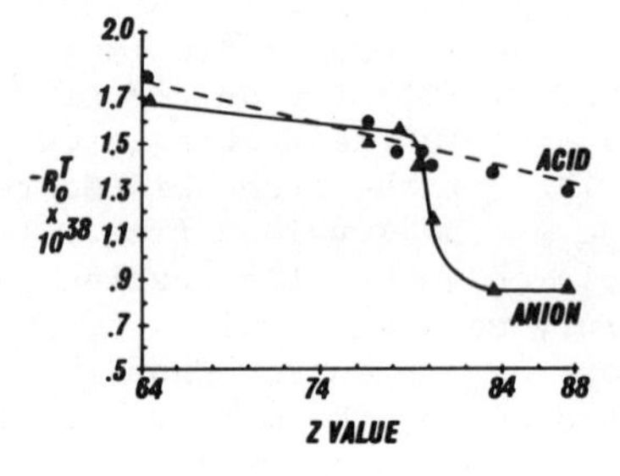

Figure 5. Rotational strengths of the carboxylic acid and free anion forms of salinomycin as a function of solvent Z values

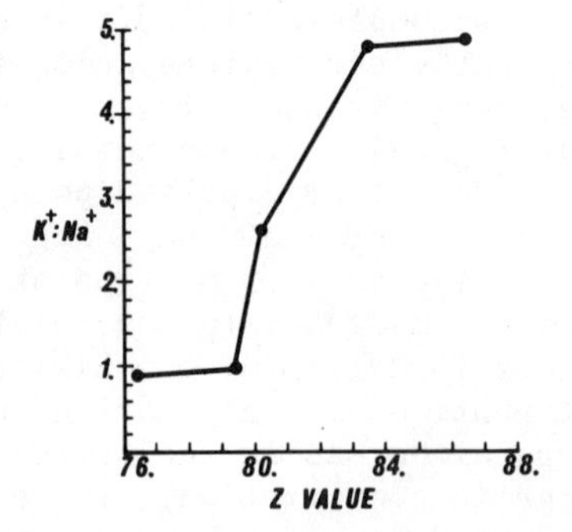

Figure 6. K^+:Na^+ selectivity ($1/K_{D_{Na^+}}$: $1/K_{D_{K^+}}$) of salinomycin as a function of solvent Z value

able to stabilize the charge, stabilization is achieved by a tight head-to-tail (O_1-$O_{17}H$) hydrogen bond. The formation of this bond results in a compression of the liganding cavity, the limit of which is determined by dipole-dipole repulsion. Application of the Octant Rule (16) to computer models of the anion corroborates that tightening of the head-to-tail bond should be accompanied by a concomitant increase in $|R_o^T|$.

Figure 4 indicates that CD can be employed to determine complexation K_D's (see Table I). The ratio of the Na^+:K^+ K_D's, i.e. K^+:Na^+ selectivity, also shows a sharp shift between Z values of 80 and 83 (cf. Figure 6). Thus, the ability of the complexing form of the ionophore to discriminate between ions depends strongly upon environmental influences on conformation. Changes in inter-ligand distances and ligand orientations effected by changes in ionophore conformation manifest themselves by a determinative alteration of the free energy of complexation.

CD was utilized to obtain the solvent dependency of the conformation of the cation-ionophore complex as well as K_D's. Saturation isotherms were plotted from linear computer fits of $1/[\text{cation}]$ versus $1/\Delta R_o^T$; the slopes yielded K_D's while extrapolation of R_o^T to infinite cation concentration provided the R_o^T's of the cation-saturated ionophore. It is important to note that the cation itself is a significant vincinal moiety, which by virtue of its charge, polarizability and location with respect to the chromophore of concern, can modify the rotational strength of the chromophore.

Comparison of the $|R_o^T|$ values for the Na^+ and K^+ complexes of salinomycin in Table I with the $|R_o^T|$ values for salinomycin anion in Figure 5 shows an increase in the magnitude of $|R_o^T|$ upon complexation in all solvents. This corresponds to a change in conformation upon complexation, i.e. reorientation of the ionophore about the cation. Application of the Octant Rule to computer generated models of salinomycin indicates that this reorientation is a constriction of the liganding oxygens which surround the cation. The extent of this constriction correlates with the stability of the complex indicated by its K_D (cf. Table I).

X-ray crystallographic studies confirm that all cationic complexes of carboxylic ionophores have their liganding atoms oriented toward a central cavity. The extent to which this conformation would be altered in the absence of a bound cation due to the mutual electrostatic repulsion of the dipolar oxygen atoms would, in turn, be modulated by the mobility of the backbone supporting the ligands.

We conclude that the dynamics of molecular conformation associated with salinomycin complexation in all likelihood extend at least to the other naturally occurring carboxylic ionophores. The influence of ionophore environment, e.g. solvent, on ionophore conformation is particularly significant when considering the environmental continuum encountered by an ionophore when transversing a biological membrane.

Table I Effect of Solvent Z Value on K_D and R_o^T of Na^+ and K^+ Complexes of Salinomycin

SOLVENT	Z	$K_D Na^+$	$K_D K^+$	$\|R_o^T\| Na^+ x10^{38}$	$\|R_o^T\| K^+ x10^{38}$
50% DIOXANE/H_2O	87.6	$1.84x10^{-3}$	$3.87x10^{-4}$	1.17	1.21
MeOH	83.6	$4.89x10^{-4}$	$1.03x10^{-4}$	1.47	1.75
80% DIOXANE/H_2O	80.2	$3.12x10^{-5}$	$1.17x10^{-5}$	1.77	1.80
EtOH*	79.6	$5.69x10^{-5}$	$5.52x10^{-5}$	1.69	1.73
90% DIOXANE/H_2O*	76.7	$5.45x10^{-5}$	$5.48x10^{-5}$	1.71	1.70

* *A priori* we would expect a progressive drop in K_D's as the solvent Z values decrease since the energies required to desolvate the cations (12) and ionophore (38) prior to complexation decrease progressively. The rise in apparent K_D values in solvents of low Z values can be accounted for by progressive increases in ion pairing which reduce the actual cation concentration, i.e. cation activity, available for complexation. Preliminary corrections for ion pairing by means of Bjerrum's equation, however, do not significantly alter the cation selectivity patterns reported here.

The extension to ionophore selectivity of a hypothesis based on analogy with the rigid matrices of ion selective glasses (19) is inconsistent with the dynamic conformational aspect of ion selectivity developed in the present paper. Furthermore, the conformational options of ionophores are not necessarily a graded function of environmental polarity but may display sudden shifts between metastable states over narrow polarity ranges. Electrostatic interactions between ions and induced dipoles undoubtedly play a determinative role in cation complexation by ionophores, but the ability of the ionophore to alter its conformation cannot be ignored as it is in the assumption of isosterism (19).

Pharmacological Properties of Carboxylic Ionophores

Pharmacological Effects. Although both neutral and carboxylic ionophores have been extensively employed as tools for in vitro studies of biological systems for the reasons detailed previously, only the carboxylic ionophores are sufficiently tolerated by intact animals to produce well defined pharmacological responses. We initially examined the cardiovascular effects of lasalocid because of its ability to transport the key biological control agents, Ca^{2+} and catecholamines (20,21). However, we later discovered that carboxylic ionophores selective for alkali ions were even more potent in evoking the same responses (22).

Figure 7 illustrates the two distinct primary cardiovascular effects produced by monensin. At low concentrations, 50 μg/kg, it produces a direct dilitation, i.e. relaxation of the smooth muscle of the coronary arteries, manifested by a multifold increase in coronary blood flow. At this level or below, no other effects occur. If the dose is increased to 0.2 mg/kg, an inotropic response follows the initial coronary dilitation. This response, an increase in cardiac contractility, can be monitored as the maximum rate of rise of pressure in the left ventricle, LV dP/dt max. Other parameters parallel the inotropic effect. Following an initial drop caused by dilitation of the systemic arteries, mean blood pressure rises as does pulse pressure, the interval between lowest (diastolic) and highest (systolic) transient pressures; the rate of blood pumped by the heart (cardiac output) also rises.

The two distinct effects are thus an increase in coronary flow, which rapidly follows injection of the ionophore, followed by an inotropic response, which only appears at higher doses.

The resolution by dosage of the two ionophore responses is clearly apparent in the dose-response plot of Figure 8. Coronary flow rises progressively until it plateaus at 10-50 μg/kg monensin. Higher doses cause a secondary increase in flow reflecting the rise in atrial pressure which drives blood through the coronaries. Only 2.5 μg/kg (i.e. 2.5 ppb) are suffucient to double the basal flow rate. It is possible to detect the increased flow of 1 μg/kg (1 ppb) with statistical confidence.

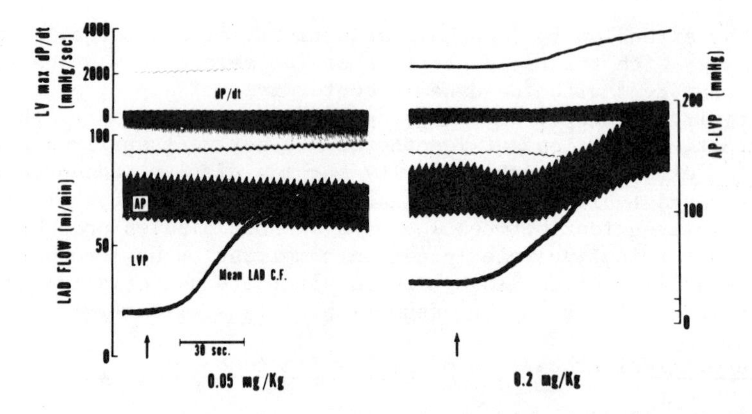

Figure 7. Cardiovascular response of a typical anesthetized dog to monensin. A low dose (0.05 mg/kg) was first introduced iv (dissolved in ethanol), and after an interval of an hour to permit the animal to return to basal conditions, a higher dose (0.2 mg/kg) was administered. The lowest tracing (mean LAD C.F.) is the time-averaged flow measured by a magnetic flow probe encircling the left anterior descending coronary artery. The AP trace gives the diastolic–systolic pressure range recorded from a catheter in the aorta. LV dP/dt *max, the index of cardiac contractility, was obtained from a manometer-tipped catheter inserted in the left ventricle. The measured pressure was converted to its derivative to record* dP/dt *directly.*

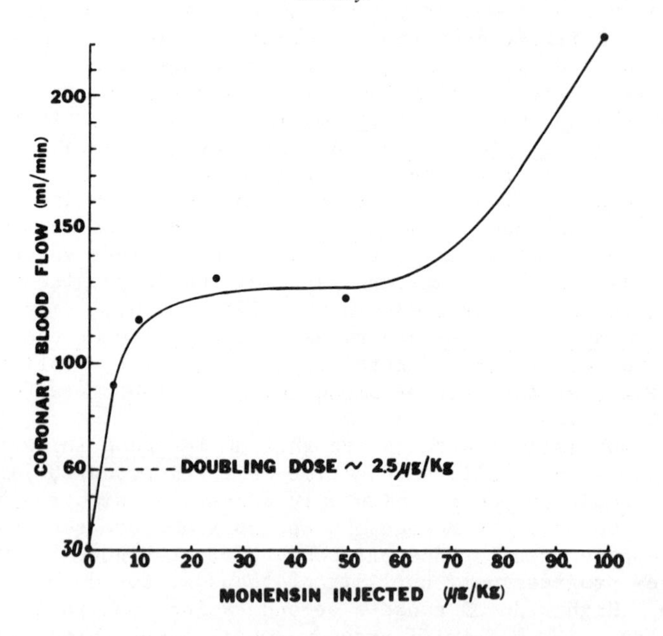

Figure 8. Dose–response curve of coronary flow vs. monensin in the dog. Data replotted from Ref. 37, as a function of dose at a fixed time interval of 5 min after injection.

Mechanism of the Pharmacological Effects. Table II compares the in vitro ion carrying capacity of a series of ionophores with their inotropic potency. Appreciable rates of Ca^{2+} or catecholamine (norepinephrine) transport are observed only for lasalocid, the ionophore of the group with the poorest inotropic potency. Extremely wide ranges of Ca^{2+} and norepinephrine transport capacity are seen with no correlation with inotropic potency. The Ca^{2+}-selective A-23187 gives only a sporatic inotropic response with the intact dog. The correlation between inotropic potency and Na^+ transport capacity is less negative and is within the realm of likely differences between the properties of the experimental solvent barrier system and those of actual biological membranes. When the activities of ionophores are compared on the basis of the quantity required to release a standard amount of K^+ from erythrocytes, chiefly in exchange for Na^+, the correlation with inotropic potency is even better.

Cells in general contain high K^+:Na^+ levels and are bathed in electrolytes containing low K^+:Na^+ ratios. Inducing ionophore-mediated exchange-diffusion transport thermodynamically favors loss of intracellular K^+ for a roughly equivalent amount of Na^+. Since the relative increase in cellular Na^+ induced by ionophores is considerably greater than the relative loss of K^+, we infer that the gain in intracellular Na^+, reflected by the more conveniently measured release of K^+, is more significant than the loss of K^+ per se. An additional factor is that different biological membranes, e.g. erythrocytes and mitochondria, respond differently to ionophores (23). All things taken into consideration, the data of Table II are reasonably supportive of a mechanism of action of ionophores involving initiation of an increase in intracellular Na^+.

Many of the effects of ionophores appear to involve an increase in intracellular Ca^{2+}. Increased contractility implies an increased availability of intracellular Ca^{2+} to trigger the interaction of actin and myosin. At higher concentrations, monensin progressively induces contraction of the resting heart (contracture) indicating that Ca^{2+} activity becomes too elevated to allow normal relaxation (24).

Increased intracellular Ca^{2+} activity also activates secretory cells (25). Inhibition studies indicate that the inotropic effect of monensin is mediated in part by the release of catecholamines from the adrenals and/or the heart itself (22). Monensin also discharges catecholamines from disaggregated bovine chromaffin cells in culture (26,27), and induces the release of acetylcholine at the neuromuscular junction (28). Thus, the secretion stimulatory activity of monensin also supports the concept that increased intracellular Na^+ activity produces a rise in intracellular Ca^{2+} activity sufficient to stimulate Ca^{2+}-activable cells.

Two hypotheses for the conversion of a primary increase in intracellular Na^+ activity to a subsequent increase in intracellu-

Table II Comparison of Inotropic Potency of Ionophores with *in vitro* Transport Properties

Ionophore	Inotropic Potency	Ca^{2+} Transport	Norepinephrine Transport	Na^{+} Transport	Erythrocyte K^{+} Release
Lasalocid	(1.0)	(1.0)	(1.0)	(1.0)	(1.0)
Lysocellin	1.5	-	-	-	4.1
Septamycin	2.0	-	-	-	-
Nigericin	2.8	.000009	0.001	1.4	16.4
Dianemycin	4.3	.00015	.1	27	18.0
Monensin	6.1	.000009	.003	31	7.2
X-206	7.7	.000025	.002	2	10.9
Salinomycin	12.1	-	-	-	10.0
A-204	13.1	.000025	.01	20	41
A-23187	±	.37	low	.002	-

Inotropic potencies were compared as the inverse of the ionophore dose required to double LV max dP/dt. Ca^{2+}, norepinephrine and Na^{+} transport rates were obtained in the vertically stacked three phase system described in ref. (39). Erythrocyte K^{+} release potency was measured as the inverse of the concentration required to release 10 mM K^{+} from washed human erythrocytes suspended in mock plasma containing 5 mM KCl, 145 mM NaCl and 10 mM TRIS chloride, pH 7.4.

lar Ca^{2+} are plausible. One would be an exchange-diffusion carrier in the plasma membrane permitting the large Ca^{2+} activity gradient (α 10^{-3} M extracellular, α 10^{-7} M interior) to permit entry of Ca^{2+} into the cell in exchange for Na^+. (On thermodynamic grounds one would expect the exchange ratio to be 3-4 Na^+ expelled for each Ca^{2+} taken up). Thus, making more intracellular Na^+ available for exchange, or in thermodynamic terms reducing the gradient against which Na^+ must move (α 10^{-2} M intracellular, α 10^{-1} M extracellular), would favor the entry of Ca^{2+}. A critical evaluation of this hypothesis has appeared in a recent review (29). An alternate mechanism would be the release of intracellularly bound Ca^{2+} by displacement by Na^+. This is feasible since the gross chemical Ca^{2+} intracellular concentration is ca. 10^{-3} M while it requires only 10^{-6} - 10^{-5} M Ca^{2+} activity to activate contraction or secretion. There might well exist purposeful Ca^{2+}-Na^+ ion-exchange sites within cells so designed that only a small relative Na^+ activity change in the mM range would trigger a large relative Ca^{2+} activity change in the μM range which would be sufficient to activate Ca^{2+}-dependent intracellular processes.

Impact of Ionophores on Man and Animals

Carboxylic Ionophores and Efficiency of Feed Conversion by Livestock. A strong note of relevance to studies of the chemical and pharmacological properties of carboxylic ionophores derives from the large scale use of monensin as a livestock feed additive. The rationale is that carboxylic ionophores control endemic coccidiosis in the poultry gut (30) and promote a more favorable fermentation of cellulose in the bovine rumen (31). In either case, the net result is the economically important increased efficiency of conversion of feed into meat.

Pharmacokinetics of Ionophore Absorption. We have developed a sensitive chemical assay for carboxylic ionophores (which will be published elsewhere) based on their ability to form lipid soluble complexes with cations. We can detect as little as 1 part per billion (ppb) monensin in 2 ml of blood plasma or tissue. For a comparison yardstick, current feeding regimens call for ca. 30 parts per million (ppm) in cattle feed (32) and as much as 100 ppm in poultry feed (33).

Typically, a cow ingests about 0.3 g (∿ 1 ppm) monensin/day. As previously observed in Figure 7, as little as 1 ppb (based on body weight) produces a detectable physiological effect on the dog.

In order to establish the pharmacokinetic relationships between orally ingested and intravenously injected monensin, we carried out preliminary studies of monensin blood levels in the dog. In Figure 9 we see that injected monensin clears from the plasma with a $t_{1/2}$ of ∿ 2.5 minutes which we presume is too rapid for the operation of normal elimination mechanisms. Hence, it is

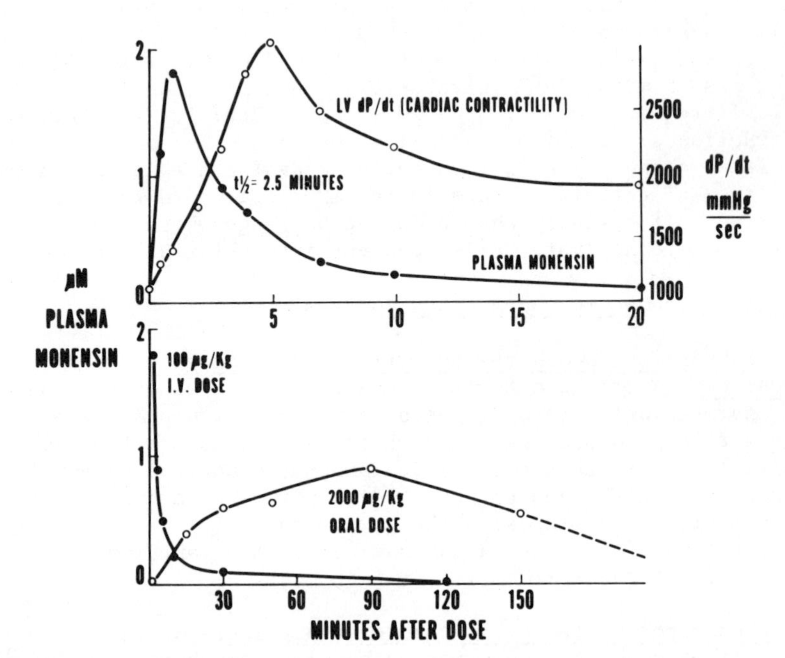

Figure 9. Pharmacokinetics of monensin in the dog. In the upper trace, 100 µg/kg monensin was injected into a barbiturate-anesthetized dog with a manometer-tipped catheter in the left ventricle to measure dP/dt. *Blood samples were taken at various periods and 2 mL samples of plasma obtained by centrifugation for ionophore assay. Note that the monensin cleared the blood rapidly and that the cardiac responses persisted. Subsequent assays revealed the monensin entered the dog tissues, particularly the lungs. The lower trace compares the pharmacokinetics of the injected dose with those obtained from a nonanesthetized dog that received the monensin orally (2 mg/kg) as a concentrate applied to a small quantity of feed. The plasma levels obtained by administration of an oral dose approached those obtained by injection, indicating that the major portion of the oral dose passed through the plasma and into the tissues before being eliminated.*

reasonable to assume that the ionophore leaving the plasma is taken up by the tissues. This would not at all be unexpected considering the high lipid:water partition coefficient of ionophores. It is supported by the delayed and persistent elevation of the ionophore-sensitive cardiac function parameter, LV dP/dt. Preliminary trials of a variation of our assay adapted for whole tissues indicate that in the rabbit the major portion of monensin appears in the tissues within 10 minutes following i.v. injection, at concentrations roughly paralleling the degree of blood perfusion: lung > heart > kidney > liver, muscle, fat.

The lower graph of Figure 8 compares the time course of appearance in the plasma of injected and orally administered monensin doses in the dog. The oral dose appears in the blood more slowly but produces more sustained ionophore blood levels. The time concentration integral gives an index of the quantity of the drug which passes through the plasma; rate of entry and clearance from the blood affect only the shape of the curve, not the net integral. The integral can be calibrated by comparison with the integral of a known dose administered directly into the blood. Although different animals and different dose levels were used, the ratio of the i.v.:oral dose integrals are approximately proportional to the 1:20 ratios of the net doses administered. This signifies that a major portion, if not all of the orally ingested monensin dose, passes through the blood stream of the dog before being eliminated. In the rabbit, a herbivore, one might predict absorption of oral doses would be slower. We can detect orally administered monensin doses in rabbit plasma, but only after a couple of hours following ingestion. We have not yet completed the more prolonged plasma level-time profiles in this species.

The Need for Increased Surveillance of the Exposure of Man to Ionophores. From the lipid solubility of monensin and other ionophores, we would predict they should have no trouble equilibrating across biological membrane systems including the gut. This is certainly the case for the two diverse species observed, the dog, a carnivore, and the rabbit, a herbivore. Accordingly, we infer that there is ample opportunity for monensin and other carboxylic ionophores administered orally to livestock to distribute systemically and exert a pharmacological effect on the recipient animal. Furthermore, the resultant physiological effects may be part of the mechanism by which ionophores produce their improved feed conversion efficiency.

There are further inferences which directly affect man. If the ionophores do pervade the tissues, it is possible that man may become exposed to pharmacologically competent and potentially detrimental levels of ionophores through his meat supply.

Based on limited pharmacokinetic and toxicological data, the F.D.A. has set upper permissible levels of 0.05 ppm in meat for human consumption (34). The isotope residue studies of Herberg et al. report that under current feeding procedures cattle liver may accumulate over ten times this level of monensin as a combination of parent compounds and metabolites of unknown pharmacological effects (35). This data was obtained 12 hours after administration of tagged monensin. One might surmise that residues would be appreciably higher for an animal butchered a shorter period of time following its last exposure to monensin. This is particularly significant in that literature supplied to farmers advises that no withdrawal period is necessary.

Currently available methods for assaying monensin involve cumbersome extraction procedures, thin layer chromatography and detection by means of bioautographs with microorganisms whose sensitivity to ionophores and their metabolites (36) may or may not parallel mammalian sensitivity. The simple chemical assay method we have developed can provide a more rational basis for assigning permissible residue levels, for routinely monitoring products arriving at the market, and ascertaining whether stipulated ionophore withdrawal periods are being complied with.

Additional complications yet to be evaluated derive from the notably poor biodegradability of monensin. Reports indicate that cattle fecally eliminate 75% of ingested monensin without degradation. Furthermore, 60-70% of the monensin survives 10 weeks incubation at 37° (34). Current manuring practices render it prudent to determine whether crops or garden produce take up significant quantities of carboxylic ionophores or whether the obviously large soil burdens of such compounds find their way into water supplies.

We have long been interested in the possibility that the cardiovascular effects of carboxylic ionophores could be harnessed to provide new drugs for the treatment of disease states such as heart failure and shock. There may, however, be subpopulations of man for whom ionophores may be particularly toxic. For example, a toxic interaction between monensin and digitalis on the dog heart has been reported (37). Our oral absorption data do indicate that if a useful human therapeutic application can be established, ionophores could be administered as drugs orally.

Summary

We have described how the unique physical properties of ionophore molecules lead to better understanding of their unique biological effects. Ionophores have been applied as tools for biological research, as commercially important livestock feed additives for increasing the efficiency of meat production, and investigated as potentially useful drugs in man. Expertise derived from studies of the molecular properties of ionophores has been utilized to design a simple assay procedure which gives promise for providing more rational safeguards for man in the widespread use of ionophores in food production. Lastly, in view of the burgeoning increases in the scale of commercial ionophore usage, it appears urgent that we increase our understanding in depth of the physiological and metabolic effects of ionophores and their pharmacological and toxicological ramifications.

Acknowledgements

We wish to acknowledge the assistance of Ms. Georgina Del Valle and Mr. Frank Lattanzio in the development of the ionophore assay and Drs. L. Allen and M. Kolber in helping program the computer studies. We are indebted to Eli Lilly for samples of monensin and A.H. Robbins and Kaken Chemical Co. (Japan) for salinomycin. These studies were supported in part by NIH grant HL-23932 and a grant from the Florida Affiliate of the American Heart Association.

Literature Cited

1. Pressman, B.C.; Harris, E.J.; Jagger, W.S.; Johnson, J. Proc. Natl. Acad. Sci. U.S.A., 1969, 58, 1949-1956.
2. Fukuyama, T.; Akasaka, K.; Karanewsky, D.S.; Wang, C.-L.J.; Schmid, G.; Kishi, Y. J. Am. Chem. Soc., 1979, 101, 262-263.
3. Pressman, B.C. Ann. Rev. Biochem., 1976, 45, 501-530.
4. Ovchinnikov, Yu.A.; Ivanov, V.T.; Shkrob, A.M. "Membrane-Active Complexones"; Elsevier:New York, 1975; Vol. 12.
5. Westley, J.W. "Kirk-Othmer Encyclopedia of Chemistry and Technology"; Wiley:New York, 1978; pp. 47-64.
6. Shemyakin, M.M., Ovchinnikov, V.T., Ivanov, V.K., Antanov, A.M., Shkrob, A.M., Mikholeva, I.I., Enstratov, A.V.; Malenkov, G.G. Biochem. Biophys. Res. Commun., 1967, 29, 834-841.
7. Hanada, M.; Nanata, Y.; Hayashi, T.; Ando, K.J. Antibiotics, 1974, 27, 555-557.
8. Pedersen, C.J. J. Am. Chem. Soc.,1967, 89, 7017.
9. Liotta, C.L.; Harris, H.P. J. Am. Chem. Soc., 1973, 95, 225.
10. Pinkerton, M.; Steinrauf, L.K. J. Mol. Biol., 1970, 49, 533-546.
11. Pressman, B.C. Fed. Proc., 1968, 27, 1283-1289.
12. Burgess, J. "Metal Ions in Solution"; Wiley:New York, 1978; pp. 318-326.
13. Urry, D.W. "Enzymes of Biological Membranes"; Plenum Pub. Corp.:New York, 1976; Vol. I; ed. Martinosi, A., pp. 31-69.
14. Urry, D.W. J. Am. Chem. Soc., 1974, 94, 77-81.
15. Kinashi, H., Ōtake, N., Yonehara, H. Acta Chrystallographica, 1977, B31, part 10, 2411-2415.
16. Djerassi, C. "Optical Rotatory Dispersion: Applications to Organic Chemistry"; McGraw-Hill:New York, 1960; pp. 41-48.
17. Moffitt, W.; Woodward, R.B.; Moscowitz, A.; Klyne, W.; Djerassi, C. J. Am. Chem. Soc., 1961, 83, 4013-4018.
18. Kossower, E.M. J. Am. Chem. Soc., 1958, 80, 3253-3260.

19. Eisenman, G.; Ciani, S.; Szabo, G. J. Membrane Biol., 1969, 1, 294-345.
20. Pressman, B.C. Fed. Proc., 1973, 32, 1698-1705.
21. deGuzman, N.T.; Pressman, B.C. Circulation, 1974, 69, 1072-1077.
22. Pressman, B.C.; deGuzman, N.T. Ann. N.Y. Acad. Sci., 1975, 264, 373-386.
23. Haynes, D.H.; Wiens, T.; Pressman, B.C. J. Membrane Biol., 1974, 18, 23-38.
24. Shlafer, M.; Somani, P.; Pressman, B.C.; Palmer, R.F. J. Mol. Cell. Cardiol., 1978, 10, 333-346.
25. Douglas, W.W. "Secretory Mechanisms of Exocrine Glands"; Munksgaard:Copenhagen, 1974; p. 116.
26. Hochman, J.; Perlman, R.L. Biochem. Biophys. Acta, 1976, 421, 168-175.
27. Rubin, R.W.; Corcoran, J.; Pressman, B.C. J. Cell Biol., 1979, 83, 434a.
28. Kita, H., Van der Kloot, W. Nature, 1974, 250, 658-659.
29. van Breemen, C.; Aaronson, P.; Loutzenhiser, R. Pharmacol. Rev., 1979, 30, 167-208.
30. Shunnard, R.F.; Callender, M.E. "Antimicrobial Agents in Chemotherapy"; 1966; pp. 369-377.
31. Richardson, L.F.; Raun, A.P.; Potter, E.L.; Cooley, C.O.; Rathmacher, R.P. J. Animal Sci., 1976, 43, 657-664.
32. Perry, T.W.; Beeson, W.M.; Mohler, M.T. J. Animal Sci., 1976, 42, 761-765.
33. Chappel, L.R.; Babcock, W.E. Poultry Sci., 1979, 58, 304-307.
34. Feinman, S.E.; Matheson, J.C. "Draft Environmental Impact Statement: Subtherapeutic Antibacterial Agents in Animal Feeds"; available from Hearing Clerk, Food and Drug Administration, Room 4-65, 5600 Fishers Lane, Rockville, Maryland 20857, 1978; pp. A100-A108.
35. Herberg, R.; Manthey, J.; Richardson, L.; Cooley, C.; Donoho, A. J. Agric. Food Chem., 1978, 26, 1087-1089.
36. Donoho, A.; Manthey, J.; Occolowitz, J.; Zornes, L. J. Agric. Food Chem., 1978, 26, 1090-1095.
37. Saini, R.K.; Hester, R.K.; Somani, P.; Pressman, B.C. J. Cardiovasc. Pharmacol., 1979, 1, 123-138.
38. Pressman, B.C.; Haynes, D.H. "The Molecular Basis of Membrane Function"; Prentice-Hall:New York, 1969; ed. Tosteson, D.C., pp. 221-246.
39. Pressman, B.C.; deGuzman, N.T. Ann. N.Y. Acad. Sci., 1974, 227, 380-391.

RECEIVED July 17, 1980.

2

Possible Functions and Medical Significance of the Abstruse Trace Metals

FORREST H. NIELSEN

United States Department of Agriculture, Science and Education Administration, Human Nutrition Laboratory, Grand Forks, ND 58202

Since 1970, a number of reports have suggested that several metals present in minute quantities in animal tissues are essential nutrients. The trace metals include cadmium, lead, nickel, tin and vanadium. Findings suggesting that cadmium, lead and tin are essential have come from one laboratory (1,2,3) and have not been confirmed in another laboratory. Minor growth depression in suboptimally growing rats was the main criterion for demonstrating the essentiality of cadmium, lead and tin. That criterion is of questionable physiological significance. The evidence is more substantial for the essentiality of nickel and vanadium. Also, apparent progress has been made in determining essential functions for those elements. Thus, in this chapter the possible medical significance and essential functions of nickel and vanadium are emphasized.

Nickel

Essentiality. Nickel is an essential nutrient for animals and probably for humans. Signs of nickel deprivation have been described for five animal species - chick, rat, minipig, goat and sheep. Briefly, the signs of deficiency include the following:

I (4) reported that the signs of nickel deprivation in chicks included depressed levels of liver phospholipids, oxidative ability of the liver in the presence of α-glycerophosphate, yellow lipochrome pigments in the shank skin, hematocrits and ultrastructural abnormalities in the liver.

I (5) found the signs of nickel deprivation in the rat included elevated perinatal mortality, unthriftiness characterized by a rough coat and/or uneven hair development in the young, pale livers, elevated rate of α-glycerophosphate oxidation by liver homogenates, and ultrastructural changes in the liver. Nickel deprivation also apparently depressed growth and hematocrits, but these signs were not consistently significant, especially in adult rats. In a series of studies, summarized recently, Schnegg and Kirchgessner (6) developed a set of nickel deprivation signs for the rat that appear divergent to those of Nielsen et al. (5). Schnegg and Kirchgessner found that, at age 30 days, rats exhibited significantly depressed growth, hematocrits, hemoglobin levels, erythrocyte counts, levels of urea, ATP and glucose in serum, levels of triglycerides, glucose and glycogen in liver, levels of iron, copper and zinc in liver, kidney and spleen, and activities of several liver and kidney enzymes. They also found that the signs of nickel deprivation were less severe in older rats and in rats fed 100 μg instead of 50 μg of iron/g of diet. Schnegg and Kirchgessner suggested that some of the signs resulted from impaired iron absorption induced by nickel deprivation.

Anke et al. (7,8) found that nickel-deprived minipigs and goats exhibited depressed growth, delayed estrus, elevated perinatal mortality, unthriftiness characterized by a rough coat and scaly and crusty skin, depressed levels of calcium in the skeleton and of zinc in liver, hair, rib and brain. Spears et al. (9,10) found that nickel-deprived lambs showed depressed growth, total serum proteins, erythrocyte counts, and total lipids and cholesterol in liver, and copper in liver. Iron contents were elevated in liver, spleen, lung and brain.

The discussed findings show that nickel meets the requirements for essentiality as defined by Mertz (11). That definition states that an element is essential if its deficiency reproducibly results in impairment of a function from optimal to suboptimal.

Biological Function. The evidence showing that nickel is essential does not clearly define its metabolic function. However, recent findings show that nickel may function as a cofactor or structural component in specific metalloenzymes or metalloproteins, or as a bioligand cofactor facilitating the intestinal absorption of the Fe(III) ion.

Himmelhoch et al. (12) were first to report findings suggesting that nickel has a role as a structural component of a metalloprotein. They fractionated human serum by column chromatography and found a metalloprotein that contained nickel, but nondetectable levels of Ca, Mg, Sr, Ba, Fe, Zn and Mn. Nomoto et al. (13) used a technique basically the same as that of Himmelhoch et al. to demonstrate the presence of a nickel-containing macroglobulin, which they named "nickeloplasmin", in rabbit serum. Subsequently, Sunderman et al. (14) isolated nickeloplasmin from human serum. Originally, Sunderman et al. (14) stated that the nickeloplasmin of humans and rabbits was an α_2-macroglobulin. Later, however, immunologic studies by Nomoto et al. (15) indicated that rabbit serum nickeloplasmin reacts as an α_1-macroglobulin that is apparently homologous to human α_2-macroglobulin. They cautioned, however, that the apparent relationship between rabbit α_1-macroglobulin and human α_2-macroglobulin was complicated when Saunders et al. (16) found that five components of human α_2-macroglobulin can be distinguished on the basis of electrophoretic and enzyme-binding properties. Other characteristics of nickeloplasmin were an estimated molecular weight of 7.0 x 10^5, nickel content of 0.90 g atoms/mole, positive reaction to periodic acid Schiff stain for glycoproteins, and esterolytic activity (on the basis of its capacity to hydrolyze tritiated tosyl-arginine methyl ester at a pH of 7.5 in tris-HCl buffer) (14,15). Decsy and Sunderman (17) found that the nickel in nickeloplasmin was not readily exchangeable with ^{63}Ni(II) *in vivo* or *in vitro*. It was necessary to administer a relatively large dose of ^{63}Ni(II) to obtain rapid labelling of serum nickeloplasmin. Decsy and Sunderman (17) offered two possible explanations for their findings. One was that nickel occurs in a different valence state, such as Ni(III), when bound to nickeloplasmin, and thus, labelling of nickeloplasmin was limited by the *in vivo* oxidation of ^{63}Ni(II) to the requisite valence. The other possibility was that nickeloplasmin preferentially binds nickel as an organic complex that is not synthesized readily by the rabbit *in vivo*. The findings of Decsy and Sunderman (17) suggested that nickeloplasmin was a ternary complex of serum α_1-macroglobulin with a Ni-constituent of serum. Sunderman (18) noted that Haupt et al. (19) isolated from human serum a 9.55-α_1-glycoprotein that strongly bound Ni(II) and thus suggested that nickeloplasmin might represent a complex of the 9.55-α_1-glycoprotein with serum α_1-macroglobulin. To date, there is no clear indication as to the physiological significance or function of nickeloplasmin.

The hypothesis that nickel in animals may function as an enzyme cofactor has been stimulated by the discovery that urease from several plants and microorganisms is a nickel metalloenzyme (20-25). Dixon et al. (20) found that highly purified urease (E.C.3.5.1.5) from jack beans (*Canavalia ensiformis*) contained stoichiometric amounts of nickel, 2.0 ± 0.3 g atom of nickel per 105,000 g of enzyme. The active-site nickel ion was tightly bound, being similar to the zinc ion in yeast alcohol dehydrogenase (E.C.1.1.1.1) and manganous ion in chicken liver pyruvate carboxylase (E.C.6.4.1.1). Jack bean urease was stable and fully active in the presence of 0.5 mM EDTA at neutral pH. The nickel ion was removed only upon exhaustive dialysis in the presence of chelating agents (21), and then it was not possible to restore nickel with reconstitution of enzymatic activity. Jack bean urease has relatively low reactivity of the active-site sulfhydryl group (26). According to Dixon et al. (21), this could be explained by coordination of the active-site nickel with the unreactive cysteine.

The biological role of urease apparently is the conversion

of urea to inorganic ammonia that can be used by plants (24,25). Dixon *et al.* (21) suggested the following mechanism for that conversion: The amide nitrogen of urea coordinates with the enzyme-bound nickel. Nucleophilic attack or general base catalysis by a suitable active-site group would then lead to an active-site, nickel-ammonia complex.

Thus, a specific biological role is known for nickel in plants. No such specific role has been defined for animals. Nickel can activate many enzymes *in vitro* (Table I), but its role as a specific cofactor for any enzyme has not been shown in animals.

The specific manner in which nickel acts in animals is unknown, but recent findings suggest that it has a role in the passive absorption of the Fe(III) ion. I found in rats that the form of dietary iron might explain the apparent differences in data for growth and hematocrits between my early studies (5) and the studies of Schnegg and Kirchgessner (6). In my early studies (5), I supplied 50 μg of iron/g of diet as iron sponge dissolved in HCl (determined to be ferric chloride), whereas Schnegg and Kirchgessner (6) supplied 50 μg of iron/g of diet as the sulfate. Schnegg and Kirchgessner indicated, by personal communication, that they had used ferrous sulfate, but I (27) could not obtain growth and hematocrit findings similar to theirs unless iron was supplied as ferric sulfate. When I studied the relationship between nickel and iron further in factorially designed experiments, nickel and iron interacted to affect hematocrit and hemoglobin, but apparently only when dietary iron was mostly in a relatively unavailable form, such as ferric sulfate.

In three experiments, female weanling rats were fed a basal diet containing about 10 ng of nickel and 2.3 μg of iron/g and supplemented with graded levels of nickel and iron. Iron was supplemented to the diet at 0, 25, 50 and 100 μg/g in all experiments. Iron was supplied as $Fe_2(SO_4)_3 \cdot nH_2O$ in Experiments 1 and 3, and as a mixture of 40% $FeSO_4 \cdot nH_2O$ and 60% $Fe_2(SO_4)_3 \cdot nH_2O$ in Experiment 2. An extra level, 12.5 μg/g, was added in Experiment 3. In all experiments, nickel was supplemented to the diet at 0, 5 and 50 μg/g. After 9-10 weeks, especially when the dietary iron supplement was only ferric sulfate, the interaction between iron and nickel affected several parameters examined. Data for hematocrit and hemoglobin appear in Tables II and III. In Experiments 1 and 3, when dietary ferric sulfate was low, hematocrit and hemoglobin were lower in nickel-deprived than -supplemented rats, especially when iron/g of diet was 25 μg. Experiment 1 nickel-deprived rats had an average hematocrit of 36.3% and hemoglobin level of 10.09 g/100 ml, whereas rats fed nickel at 5 and 50 μg/g of diet had hematocrits of 40.8% and 42.0% and hemoglobin levels of 11.77 and 12.09 g/100 ml, respectively. In Experiment 3, nickel-deprived rats had an average hematocrit of 26.8% and

Table I

Enzymes "activated" by nickel[a]

Enzyme	E.C. No.	Source
Acetyl coenzyme A synthetase	6.2.1.1	Bovine heart mitochondria
Amino acid decarboxylase	-	E. coli, C. welchii
Amylase	3.2.1.1	Human saliva
Arginase	3.5.3.1.	Bovine liver Canavalia ensiformis Jack bean Yeast
Ascorbic acid oxidase	1.10.3.3	Broad bean leaf
ATPase (14s, 30s dynein)	3.6.1.-	Tetrahymena cilia
3,4 Benzpyrene hydroxylase	1.14.14.2	Lung
Carboxypeptidase	3.4.12.-	
Citritase	4.1.3.6	
Deoxyribonuclease I	3.1.4.5	Bovine pancreas
Desoxyribonuclease	3.1.4.-	Bovine thymus
Enolase	4.2.1.11	
Esterase	-	Porcine liver
Hexokinase	2.7.1.1	Yeast
Histidine decarboxylase	4.1.1.22	Lactobacillus 30a
Oxalacetic carboxylase	-	Parsley root
Pepsin	3.4.23.1	Porcine mucosa
Phosphodeoxyribomutase	2.7.5.6	E. coli
Phosphoglucomutase	2.7.5.5	Rabbit muscle
Phospholipase A	3.1.1.-	Crotalus atrox venom
Phosphorylase phosphatase	3.1.3.17	Bovine adrenal cortex
Protease	3.4.--	Human senile lens
Pyridoxal phosphokinase	2.7.1.35	
Pyruvate kinase	2.7.1.40	Rabbit muscle
Pyruvic acid oxidase	1.2.3.4	Proteus vulgaris
Ribonuclease	3.1.4.-	Bovine pancreas
Ribulose diphosphate carboxylase	4.1.1.39	Spinach
Thiaminokinase	2.7.6.2	Rat liver
Trypsin	3.4.21.4	
Tyrosinase	1.14.18.1	Mouse melanoma Potato
Urease	3.5.1.5	Jack bean Lemna paucicostata Rumen bacterial Soybean

[a]Compiled by Nielsen (33).

Table II

Effects on rats of nickel, iron, and their interaction on hematocrits

Treatment[a]		Hematocrit		
Ni	Fe	Experiment 1	Experiment 2	Experiment 3
µg/g	µg/g		%	
0	0	14.3	18.1	14.2
0	12.5	-	-	22.1
0	25	36.3	42.2	26.8
0	50	41.9	42.4	38.3
0	100	42.1	41.8	39.1
5	0	15.2	19.3	17.0
5	12.5	-	-	25.8
5	25	40.8	41.6	32.1
5	50	42.3	41.9	38.9
5	100	42.0	41.5	39.3
50	0	20.1	20.4	16.0
50	12.5	-	-	23.2
50	25	42.0	41.1	33.8
50	50	40.5	42.2	39.0
50	100	41.3	42.2	40.2

Analysis of Variance - P Values

	Experiment 1	Experiment 2	Experiment 3
Nickel effect	0.002	.01	.0001
Iron effect	0.0001	.0001	.0001
Nickel x iron	0.0001	NS	.006
Error mean square (df)	2.8(70)	1.6(58)	4.9(75)
Scheffé values[b]			
Treatment means - 6 s-test 6	3.3	6.4	4.1
Iron effect means - 18 s-test 18	1.1	2.1	1.4
Nickel effect means - 24 s-test 24	0.8	-	1.2
Nickel effect means - 30 s-test 30	-	1.2	-

[a]Levels of supplements in diet: Ni (nickel chloride) and Fe (ferric sulfate) in Experiments 1 and 3; Fe was a mixture of 40% ferrous and 60% ferric sulfate in Experiment 2.

[b]The Scheffé test (28) is a method for performing multiple comparisons between group means. Means differing by more than the value given are significantly different ($P < 0.05$). As it assumes all possible comparisons are performed, it is regarded as a conservative test.

Table III

Effects on rats of nickel, iron, and their interaction on hemoglobin levels

Treatment[a]		Hemoglobin Level		
Ni	Fe	Experiment 1	Experiment 2	Experiment 3
μg/g	μg/g		g/100 ml	
0	0	2.65	3.73	2.55
0	12.5	-	-	4.90
0	25	10.09	13.27	6.19
0	50	12.65	13.42	10.85
0	100	13.05	13.23	11.61
5	0	3.03	4.08	3.36
5	12.5	-	-	5.81
5	25	11.77	13.15	8.31
5	50	12.98	13.26	10.92
5	100	13.07	13.13	11.41
50	0	3.88	4.46	2.95
50	12.5	-	-	5.01
50	25	12.09	13.01	8.92
50	50	12.62	13.29	11.13
50	100	12.80	13.26	11.65
Analysis of Variance - P Values				
Nickel effect		.003	.003	.0002
Iron effect		.0001	.0001	.0001
Nickel x iron		.0001	NS	.0002
Error mean square (df)		0.31(71)	0.16(58)	0.56(75)
Scheffé values[b]				
Treatment means - 6 s-test 6		1.04	2.14	1.38
Iron effect means - 18 s-test 18		0.34	0.71	0.47
Nickel effect means - 24 s-test 24		0.25	-	0.41
Nickel effect means - 30 s-test 30		-	0.42	-

[a]Levels of supplements in diet: Ni (nickel chloride) and Fe (ferric sulfate) in Experiments 1 and 3; Fe was a mixture of 40% ferrous and 60% ferric sulfate in Experiment 2.

[b]The Scheffé test (28) is a method for performing multiple comparisons between group means. Means differing by more than the value given are significantly different ($P < 0.05$). As it assumes all possible comparisons are performed, it is regarded as a conservative test.

hemoglobin level of 6.19 g/100 ml; whereas rats fed nickel at 5 and 50 µg/g of diet had hematocrits of 32.1% and 33.8% and hemoglobin levels of 8.31 and 8.92 g/100 ml, respectively. The difference between nickel-deprived and supplemented rats in Experiments 1 and 3 were significant by the Scheffé test (28). Dietary nickel apparently did not affect hematocrit or hemoglobin when the diet contained 100 µg of iron/g. Nickel and iron did not interact to affect hematocrit and hemoglobin when iron was supplied as ferric-ferrous sulfate.

The form of dietary iron also influenced the effect of nickel on hematocrit and hemoglobin. When ferric sulfate was fed (Experiments 1 and 3), both parameters were significantly lower in nickel-deprived than -supplemented rats. In Experiment 2 the effect of nickel was much less marked than in Experiments 1 and 3. In Experiment 2, the greatest difference was in rats fed no supplemental iron.

There were some differences between Experiments 1 and 3, especially when the diet contained 25 or 50 µg of iron/g. In Experiment 3, hematocrit and hemoglobin levels were significantly depressed in all groups fed 25 µg iron/g of diet, although the depression was less severe in nickel-supplemented than -deprived rats. In Experiment 1, with 25 µg of iron/g of diet, hematocrit and hemoglobin were depressed only in nickel-deprived rats; values were near normal in rats fed 5 or 50 µg of nickel/g of diet. In Experiment 3, the hematocrit and hemoglobin data indicated that rats fed 50 µg of iron/g of diet as ferric sulfate were still slightly iron-deficient. In Experiment 1, hematocrit and hemoglobin apparently were normal in rats fed 50 µg of iron/g of diet. Possibly, the iron supplement was most highly contaminated with the ferrous form in Experiment 1. The iron supplement was ascertained to be 92% in the ferric form in Experiment 3, but was not tested in Experiment 1.

The observations that the form of dietary iron apparently affected the response of rats to nickel deprivation and nickel and iron interacted suggest that nickel affects iron absorption. The apparent dependence of that interaction upon the relatively insoluble ferric salt suggests that nickel has a role in the absorption of the Fe(III) ion. Fe(III) salts are extremely insoluble in neutral/alkaline biofluids (29). Thus, for absorption by the duodenum, the Fe(III) must be complexed, or converted to the more soluble Fe(II) form. According to May *et al.* (29), only ligands, such as porphyrin-like molecules, that form high-spin complexes and thereby increase the electrode potential stabilize Fe(II) over Fe(III). Most other bioligands lower the electrode potential and thus enhance the stability of the Fe(III) state. Therefore, the preferred chelated state of iron *in vivo* is probably Fe(III) and the reduction to Fe(II) occurs spontaneously only in the presence of high local concentrations of a reducing metabolite, or under the influence of special enzyme mechanisms. Nickel might interact with iron through one of those mechanisms but probably does not. The finding that 50 μg of nickel/g of diet was not much better than 5 μg in improving hematocrits and hemoglobin levels in nickel-deprived rats fed low levels of iron as ferric sulfate is apparently inconsistent with the possibility that nickel acts as, or part of, a reducing agent converting Fe(III) to Fe(II). The idea that nickel might act in a special enzyme mechanism that converts Fe(III) to Fe(II) is attractive, but no such mechanism is known.

The most attractive possibility is that nickel promotes the absorption of Fe(III) *per se* by enhancing its complexation to a lipophilic molecule. Evidence shows that both active and passive transport mechanisms have roles in iron absorption. Active transport to the serosal surface is relatively specific for the divalent cation (30), which indicates that the Fe(III) ion is absorbed by passive transport. Passive transport is diffusion-controlled and only permits the transit of lipophilic molecules. Substantial evidence shows lipophilic Fe(III) complexes traverse biomembranes in the same manner as lipophilic complexes of other metals. Nickel could affect the metabolism of the lipophilic Fe(III) complexes in at least two ways. Nickel might either act in an enzymatic reaction that forms a lipophilic iron transport molecule or simply preserve a transport ligand, such as citrate, by complexing with it until replaced by the Fe(III) ion.

The hypothesis that nickel has a role in the passive diffusion of Fe(III) is supported by my data for hematocrit and hemoglobin discussed previously. Dowdle *et al.* (31) suggested that the active transport mechanism for iron would become important if passive diffusion were restricted. Thus, at the lower levels of iron supplementation as a ferric-ferrous mixture, there was some ferrous ions available for active transport, and nickel deprivation did not significantly affect levels of hematocrit or hemoglobin. On the other hand, when only ferric iron was fed, the active transport mechanism could not operate, and in nickel deprivation, the passive diffusion of lipophilic Fe(III) complexes apparently was inhibited. As a result, levels of hematocrit and hemoglobin differed between nickel-deprived and -supplemented rats at low levels of iron supplementation. At high levels of supplementation, perhaps there was enough Fe(II) present in the diet to prevent any differences as the iron supplement was approximately 92% Fe(III).

Medical Significance. An initial impression is that nickel nutriture would not be of practical significance. I (4) reported that 50 μg of nickel/kg of diet satisfied the dietary nickel requirement of chicks, and Schnegg and Kirchgessner (6) reported a similar requirement for rats. If animal data were extrapolated to man, the dietary nickel requirement of humans would probably be in the range of 16-25 μg/1000 Cal (32). Limited studies indicate that the oral intake of nickel by humans ranges between 170 and 700 μg per day (33) which would be ample to meet the hypothesized nickel requirement.

However, the finding that nickel may be important in the absorption and metabolism of iron might help define situations in which nickel would have medical significance. I am defining medical significance as the unintentional production of a nutritional disorder in humans. Possibly for individuals who consume unavailable, or deficient amounts of, iron, or have an elevated need for iron, nickel nutriture might be of concern. For example, many women consume inadequate iron. Nickel allergy is a common disorder. About 10% of tested individuals reacted positively to the nickel-patch test and incidence was highest among women (34). Because recent reports indicate that dietary nickel may be important in hand-eczema caused by nickel (35,36), one treatment for nickel allergy is reduction of dietary nickel. Extrapolation from animal findings suggests that care should be exercised with such treatment to assure that proper nickel and iron nutriture is maintained to avoid adverse consequences.

Vanadium

Essentiality. Evidence for the nutritional essentiality of vanadium is not conclusive. Strasia (37) found that rats fed less than 100 ng of vanadium/g of diet exhibited slower growth, higher plasma and bone iron, and higher hematocrits than controls fed 0.5 μg of vanadium/g of diet. However, Williams (38) was unable to duplicate the findings of Strasia (37), even in the same laboratory under similar conditions. Schwarz and Milne (39) reported that a vanadium supplement of 25 to 50 μg/100 g of a semi-purified diet gave a positive growth response in rats. On the other hand, Hopkins and Mohr (40) reported that the only effect of vanadium deprivation on rats was an apparent impaired reproductive performance (decreased fertility and increased perinatal mortality) that became apparent only in the fourth generation.

Studies with chicks also gave inconsistent signs of deficiency. Hopkins and Mohr (41,42) found that vanadium-deprived chicks exhibited significantly depressed wing and tail feather development, depressed plasma cholesterol at age 28 days, elevated plasma cholesterol at age 49 days, and, in a subsequent study (40), elevated plasma triglycerides at age 28 days. I reported that vanadium-deprivation depressed growth, elevated hematocrits and plasma cholesterol, and adversely affected bone development (43).

I became concerned about the inconsistency of the effect of vanadium deprivation on chicks and rats, and attempted to establish a definite set of signs of vanadium deprivation for these species. In 16 experiments, in which chicks were fed several diets of different composition, vanadium deprivation adversely affected growth, feathering, hematocrits, plasma cholesterol, bone development, and the levels of lipid, phospholipid and cholesterol in liver. In several experiments with rats, vanadium deprivation adversely affected perinatal survival, growth, physical appearance, hematocrits, plasma cholesterol, and lipids and phospholipids in liver. Unfortunately, no sign of vanadium deprivation in chicks, or rats, was found consistently throughout all experiments.

Apparently inconsistency of vanadium deprivation signs is related to the fact that vanadium metabolism is sensitive to changes in the composition of the diet (44). Perhaps diet composition affects the form of dietary vanadium. Vanadium has a rich and varied chemistry, especially in the (IV) and (V) state. The form of vanadium, usually an oxyanion (i.e. VO_3^-, VO_2^+), depends upon its concentration in, and pH of, the medium (45). Perhaps, one form is more readily available for absorption, or active in metabolism, than another. Thus, a diet that is relatively low in vanadium might be nutritionally either deficient or adequate depending on the form of the vanadium.

Nonetheless, because the evidence is inconsistent, further studies are necessary to definitely establish vanadium as an essential nutrient. It might be necessary to find a specific physiological role for vanadium in order to establish its essentiality.

Biological Function. The most recent findings that suggest vanadium does have a physiological role, have come not from nutritional, but from *in vitro* studies with (Na, K) ATPase and ATP phosphohydrolase (E.C.3.6.1.3). Although Rifkin (46) was first to report that vanadium potently inhibits (Na, K)-ATPase, Cantley *et al.* (47) were first to find that pentavalent orthovanadate was a naturally occurring inhibitor of that enzyme. Vanadate was shown to inhibit (Na, K) ATPase from kidney (46,47,48), brain (48), heart (48,49), red blood cells (50,51), shark rectal gland and eel electroplax (49). ATP-phosphohydrolase from various dynein fractions, commonly known as dynein ATPase, also was potently inhibited by vanadate (52,53,54). Josephson and Cantley (55) found that vanadate did not potently inhibit other ATPase systems, such as Ca-ATPase, mitochondrial coupling factor F, and actomyosin. Cande and Wolniak (54) found that vanadate did not potently inhibit glycerinated myofibril contraction or myosin ATPase activity. Those findings suggest that vanadate would be an ideal specific inhibitor of (Na, K)-ATPase or dynein ATPase.

Magnesium and potassium facilitate vanadate inhibition of (Na, K)-ATPase activity and they both appear to bind synergistically with vanadate (56). ATP depressed vanadate inhibition of enzyme activity (48). On the other hand, Gibbons

et al. (53) found that dynein ATPase inhibition by vanadate did not depend upon the magnesium concentration or on the presence or absence of potassium. Furthermore, ATP had no obvious affect on vanadate inhibition of dynein ATPase.

Cantley et al. (50) found that vanadate binds to one high-affinity and one low affinity site per (Na, K)-ATPase enzyme molecule. The low-affinity site was apparently responsible for inhibition of (Na, K)-ATPase activity and was the high-affinity ATP site where sodium-dependent protein phosphorylation occurs. Cantley et al. (56) proposed that the unusually high affinity of vanadate for (Na, K)-ATPase was due to its ability to form a trigonal bipyramidal structure analogous to the transition state for phosphate hydrolysis.

Cantley et al. (50) found that vanadate was transported to the red blood cell where it inhibited the sodium pump by binding to (Na, K)-ATPase from the cytoplasmic side (the site of ATP hydrolysis). They suggested that the vanadium in mammalian tissue acts as a regulatory mechanism for the sodium pump that maintains a high intracellular K^+ to Na^+ ratio by coupling with ATP hydrolysis.

In addition to acting as an inhibitor of dynein and (Na, K)-ATPase, vanadium is also a potent inhibitor of RNase (57) and alkaline and acid phosphatases (58,59). This suggests that vanadium generally tends to inhibit enzymes of phosphate metabolism. However, according to Gibbons et al. (53), the mechanism of inhibition is not the same in each enzyme. The inhibition of RNase and alkaline phosphatase is greater by oxyvanadium (IV) than by vanadium (V).

Thus, the findings to date suggest that vanadium has a biological function in controlling one or more enzymatic reactions concerned with phosphate metabolism. However, further in vivo studies are necessary before a conclusive statement can be made.

Medical Significance. The medical significance of vanadium is unclear because knowledge is incomplete of the conditions necessary to produce vanadium deficiency, dietary components that affect vanadium metabolism, and its biological function. It is difficult to suggest a vanadium requirement for animal species, including humans. However, at least four independent laboratories have found that diets with less than 25 ng of vanadium/g adversely affect rats and chicks under certain conditions. If animal data could be extrapolated to humans, then a 70 kg man consuming 1 kg of diet per day (dry basis) would have a daily requirement of about 25 μg of vanadium under certain dietary conditions.

Recent studies have shown that the vanadium content of most foods is very low (60,61,62,63,64), generally not more than a nanogram/g. Myron et al. (63) reported that nine institutional diets supplied 12.4-30.1 μg of vanadium daily,

and intake averaged 20 μg. Byrne and Kosta (64) stated that the dietary intake of vanadium is in the order of a few tens of micrograms and may vary widely. This suggests that vanadium intake is not always optimal in humans.

In addition to nutritional deficiency, nutritional vanadium toxicity may have medical significance. The findings discussed previously suggest that because vanadium is a potent inhibitor of several enzymes, any undue elevation in tissue vanadium content might adversely affect biochemical systems that depend upon normal phosphate metabolism. Even relatively small amounts of dietary vanadium could be toxic in some situations. For example, Hunt (65) found that the addition of 500 μg of chromium as the acetate/g of diet made 5 μg of vanadium/g of diet toxic to chicks. Those chicks exhibited depressed growth and hematocrits, elevated plasma cholesterol, kidney (Na, K) ATPase, and liver/body weight ratio. Morphology of their proximal tibiae was drastically altered; the growth plate was abnormally thick and the zone of calcified cartilage abnormally thin. Metaphyseal bone was nonexistent. Transmission electron microscopic examinations revealed a disorganized growth plate and the presence of an abnormal, electron-dense matrix component around the chondrocytes in the proliferative zone. Five μg of vanadium/g of diet without chromium supplementation had no obvious effect on chicks.

Cadmium, Lead, and Tin

Essentiality. At present, the evidence suggesting that cadmium, lead and tin are essential does not fulfill the requirements for essentiality as defined by Mertz (11). Although dietary supplements of cadmium, lead, or tin slightly improved the growth of suboptimally growing rats, these supplements did not result in optimal growth (1,2,3). Thus, it cannot be stated unequivocally that cadmium, lead, or tin deficiency reproducibly results in an impairment of a function from optimal to suboptimal.

Apparently, the suboptimal growth in all rats in the cadmium, lead and tin studies was due to riboflavin deficiency (66). Unfortunately, the death of the principal investigator of cadmium, lead and tin essentiality (Klaus Schwarz) prevented further studies which would have answered the question whether deficiencies of those elements would depress growth in rats which were not riboflavin-deficient. This question may remain unanswered for some time because, to my knowledge, studies concerned with the essentiality of cadmium, lead, and tin are not currently pursued in another laboratory.

The reports which suggest the essentiality of cadmium, lead and tin can also be criticized in the following manner:

1. The basal diets were not adequately described, thus preventing the confirmation of the growth findings in another

laboratory.

2. The statistical methods used for the analysis of the growth data were questionable. It was not obvious why covariance analysis was used for the analysis of this type of data. Perhaps the preferable analysis of variance would have not given significant findings. Furthermore, some of the significant findings apparently were obtained through the method of combining experiments, thus increasing the statistical term n (no. of animals) (2,3). Combining experiments before statistical treatment of the data is inappropriate.

3. The small growth difference between "deficient-controls" and supplemented rats (about 5 to 7 grams after 25 to 30 days on experiment) may be of questionable physiological meaning. Perhaps this growth response was due to the supplemental metals partially preventing the breakdown of some essential nutrient such as riboflavin, or substituting for some trace element lacking in the diet.

4. The addition of suggested essential metals to the diet was of no apparent benefit to deficient-control animals in subsequent studies. For example, in the tin studies, the deficient-controls gained about 1.3 to 1.9 g/day; tin-supplemented rats, 1.7 to 2.2 g/day. However, even with the addition of tin, and some other elements subsequently found possibly essential, such as fluorine and silicon, the deficient-controls in the lead study still gained only 1.5 to 2.1 g/day; lead-supplemented rats, 1.6 to 2.2 g/day. Deficient-control and cadmium-supplemented rats also exhibited similar daily weight gains. No explanation was given for the finding that deficient-controls weighed the same in each of the tin, lead and cadmium studies, even though one would expect the deficient-controls would show better growth rates in latter studies because their diets contained more essential elements.

Because of the previously discussed questions and criticisms, I conclude that cadmium, lead and tin should not be included in the list of essential trace metals at the present time.

Biological Function and Medical Significance. Until more conclusive evidence is found suggesting cadmium, lead and tin are essential, the description of any possible biological function seems inappropriate. The toxicologic aspects of cadmium, lead and tin are of medical significance. However, a proper discussion of the toxicology of those elements is beyond the scope of this presentation and is adequately done elsewhere (67,68,69).

Summary

The evidence to date has established nickel as an essential nutrient for several animal species. The essentiality of vanadium has not been conclusively proven. Some findings suggest that nickel has a biological function as a cofactor or structural component in specific metalloenzymes or metalloproteins, or as a bioligand cofactor facilitating the intestinal absorption of the Fe(III) ion. Vanadium may function as a regulator of some specific enzymes involved with phosphate metabolism. Thus, nickel and vanadium might be of medical significance.

Abstract

Since 1970, a number of reports have suggested that several metals, including nickel, vanadium, cadmium, lead, and tin, present in minute quantities in animal tissues, are essential nutrients. Findings that have indicated the essentiality of cadmium, lead, and tin are limited, unconfirmed and of questionable physiological significance. The evidence is more substantial for the essentiality of nickel and vanadium. Also, apparent progress has been made in determining essential functions for those elements. Nickel has been shown to be an integral part of the macroglobulin nickeloplasmin isolated from human and rabbit serum, and of the enzyme urease isolated from various plants and microorganisms. Vanadium may be a regulator of (Na, K) ATPase because physiological amounts of vanadate potently inhibit that enzyme *in vitro*. In addition, important biological interactions between nickel and iron, and vanadium and chromium have been described. Thus, nickel and vanadium may also be of medical significance through their interaction with other trace metals.

Literature Cited

1. Schwarz, K.; Spallholz, J. Growth effects of small cadmium supplements in rats maintained under trace-element controlled conditions. *Fed. Proc.*, 1976, **35**, 255.

2. Schwarz, K. New essential trace elements (Sn, V, F, Si): Progress report and outlook. In: "Trace Element Metabolism in Animals-2", eds: W.G. Hoekstra, J.W. Suttie, H.E. Ganther, and W. Mertz. University Park Press, Baltimore, MD, 1974, pp. 355-380.

3. Schwarz, K.; Milne, D.B.; Vinyard, E. Growth effects of tin compounds in rats maintained in a trace element-controlled environment. *Biochem. Biophys. Res. Commun.*, 1970, **40**, 22-29.

4. Nielsen, F.H.; Myron, D.R.; Givand, S.H.; Ollerich, D.A. Nickel deficiency and nickel-rhodium interaction in chicks. *J. Nutr.*, 1975, **105**, 1607-1619.

5. Nielsen, F.H.; Myron, D.R.; Givand, S.H.; Zimmerman, T.J.; Ollerich, D.A. Nickel deficiency in rats. J. Nutr., 1975, 105, 1620-1630.

6. Schnegg, A.; Kirchgessner, M. Ni deficiency and its effects on metabolism. In: "Trace Element Metabolism in Man and Animals-3", ed: M. Kirchgessner. Tech. Univ. Munchen, Freising-Weihenstephan, West Germany, 1978, pp. 236-243.

7. Anke, M.; Grün, M.; Dittrich, G.; Groppel, B.; Hennig, A. Low nickel rations for growth and reproduction in pigs. In: "Trace Element Metabolism in Animals-2", eds: W.G. Hoekstra, J.W. Suttie, H.E. Ganther, and W. Mertz. University Park Press, Baltimore, MD, 1974, pp. 715-718.

8. Anke, M.; Hennig, A.; Grün, M.; Partschefeld, M.; Groppel, B.; Lüdke, H. Nickel-ein essentielles Spurenelement. Arch. Tierernährung, 1977, 27, 25-38.

9. Spears, J.W.; Hatfield, E.E.; Forbes, R.M.; Koenig, S.E. Studies on the role of nickel in the ruminant. J. Nutr., 1978, 108, 313-320.

10. Spears, J.W.; Hatfield, E.E.; Fahey, G.C., Jr. Nickel depletion in the growing ovine. Nutr. Repts. Internat., 1978, 18, 621-629.

11. Mertz, W. Some aspects of nutritional trace element research. Fed. Proc., 1970, 29, 1482-1488.

12. Himmelhoch, S.R.; Sober, H.A.; Vallee, B.L.; Peterson, E.A.; Fuwa, K. Spectrographic and chromatographic resolution of metalloproteins in human serum. Biochemistry, 1966, 5, 2523-2530.

13. Nomoto, S.; McNeely, M.D.; Sunderman, F.W., Jr. Isolation of a nickel α_2-macroglobulin from rabbit serum. Biochemistry, 1971, 10, 1647-1651.

14. Sunderman, F.W., Jr.; Decsy, M.I.; McNeely, M.D. Nickel metabolism in health and disease. Ann. N.Y. Acad. Sci., 1972, 199, 300-312.

15. Nomoto, S.; Decsy, M.I.; Murphy, J.R.; Sunderman, F.W., Jr. Isolation of ^{63}Ni-labeled nickeloplasmin from rabbit serum. Biochem. Med., 1973, 8, 171-181.

16. Saunders, R.; Dyce, B.J.; Vanner, W.E.; Haverback, B.J. The separation of alpha-2 macroglobulin into five components with differing electrophoretic and enzyme-binding properties. J. Clin. Invest., 1971, 50, 2376-2383.

17. Decsy, M.I.; Sunderman, F.W., Jr. Binding of ^{63}Ni to rabbit serum α_1-macroglobulin in vivo and in vitro. Bioinorg. Chem., 1974, 3, 95-105.

18. Sunderman, F.W., Jr. A review of the metabolism and toxicology of nickel. Ann. Clin. Lab. Sci., 1977, 7, 377-398.

19. Haupt, H.; Heimburger, N.; Kranz, T.; Baudner, S. Human serum proteins with a high affinity for carboxymethyl cellulose. III. Physical-chemical and immunological characterization of a metal-binding 9.55-α_1-glycoprotein (CM-Protein III). Z. Physiol. Chem., 1972, 353, 1841-1849.

20. Dixon, N.E.; Gazzola, C.; Blakeley, R.L.; Zerner, B. Jack bean urease (E.C.3.5.1.5) is a metalloenzyme. A simple biological role for nickel? J. Amer. Chem. Soc., 1975, 97, 4131-4133.

21. Dixon, N.E.; Gazzola, C.; Blakeley, R.L.; Zerner, B. Metal ions in enzymes using ammonia or amides. Science, 1976, 191, 1144-1150.

22. Fishbein, W.N.; Smith, M.J.; Nagarajan, K.; Scurzi, W. The first natural nickel metalloenzyme: urease. Fed. Proc., 1976, 35, 1680.

23. Spears, J.W.; Smith, C.J.; Hatfield, E.E. Rumen bacterial urease requirement for nickel. J. Dairy Sci., 1977, 60, 1073-1076.

24. Polacco, J.C. Nitrogen metabolism in soybean tissue culture. II. Urea utilization and urease synthesis require Ni^{2+}. Plant Physiol., 1977, 59, 827-830.

25. Gordon, W.R.; Schwemner, S.S.; Hillman, W.S. Nickel and the metabolism of urea by Lemna paucicostata Hegelm. 6746. Planta, 1978, 140, 265-268.

26. Gorin, G.; Chin, C.-C. Urease. IV. Its reaction with N-ethylmaleimide and with silver ion. Biochim. Biophys. Acta, 1965, 99, 418-426.

27. Nielsen, F.H.; Zimmerman, T.J.; Collings, M.E.; Myron, D.R. Nickel deprivation in rats: Nickel-iron interactions. J. Nutr., 1979, 109, 1623-1632.

28. Scheffé, H. "The Analysis of Variance", John Wiley & Sons, Inc., New York, NY, 1959, pp. 68-72.

29. May, P.M.; Williams, D.R.; Linder, P.W. Biological significance of low molecular weight iron (III) complexes. In: "Metal Ions in Biological Systems, Volume 7: Iron in Model and Natural Compounds", ed: H. Sigel. Marcel Dekker, New York, NY, 1978, pp. 29-76.

30. Manis, J.G.; Schachter, D. Active transport of iron by intestine: Features of the two-step mechanism. Amer. J. Physiol., 1962, 203, 73-80.

31. Dowdle, E.B.; Schachter, D.; Schenker, H. Active transport of Fe^{59} by everted segments of rat duodenum. Amer. J. Physiol., 1960, 198, 609-613.

32. Nielsen, F.H. "Newer" trace elements in human nutrition. Food Tech., 1974, 28, 38-44.

33. Nielsen, F.H. Nutrient and growth regulators: Chemistry and physiology - nickel. In: "CRC Handbook Series in Nutrition and Food", ed: M. Recheigl, Jr., CRC Press, West Palm Beach, FL, accepted for publication.

34. National Academy of Sciences. "Nickel. Report of the Subcommittee on Nickel", NAS Committee on Medical and Biologic Effects of Environmental Pollutants, National Academy of Sciences, Washington, D.C., 1975, pp. 124-143.

35. Christensen, O.B.; Möller, H. External and internal exposure to the antigen in the hand eczema of nickel allergy. Contact Dermatitis, 1975, 1, 136-141.

36. Spruit, D.; Bongaarts, P.J.M. Nickel content of plasma, urine and hair in contact dermatitis. Dermatologica, 1977, 154, 291-300.

37. Strasia, C.A. "Vanadium: Essentiality and Toxicity in the Laboratory Rat". Ph.D. Thesis, Purdue University, University Microfilms, Ann Arbor, MI, 1971.

38. Williams, D.L. "Biological Value of Vanadium for Rats, Chickens, and Sheep". Ph.D. Thesis, Purdue University, University Microfilms, Ann Arbor, MI, 1973.

39. Schwarz, K.; Milne, D.B. Growth effects of vanadium in the rat. Science, 1971, 174, 426-428.

40. Hopkins, L.L., Jr.; Mohr, H.E. Vanadium as an essential nutrient. Fed. Proc., 1974, 33, 1773-1775.

41. Hopkins, L.L., Jr.; Mohr, H.E. The biological essentiality of vanadium. In: "Newer Trace Elements in Nutrition", eds: W. Mertz and W.E. Cornatzer. Marcel Dekker, Inc., New York, NY, 1971, pp. 195-213.

42. Hopkins, L.L., Jr.; Mohr, H.E. Effect of vanadium deficiency on plasma cholesterol of chicks. Fed. Proc., 1971, 30, 462.

43. Nielsen, F.H.; Ollerich, D.A. Studies on a vanadium deficiency in chicks. Fed. Proc., 1973, 32, 929.

44. Nielsen, F.H. Evidence for the essentiality of arsenic, nickel, and vanadium and their possible nutritional significance. In: "Advances in Nutritional Research", ed: H.H. Draper, Plenum Publishing Corp., New York, NY, 1979, in press.

45. Pope, M.T.; Dale, B.W. Isopoly-vanadates, -niobates and -tantalates. Quart. Rev. (London), 1968, 22, 527-548.

46. Rifkin, R. In vitro inhibition of Na-K and Mg_2 ATPase by mono-, di and trivalent cations. Proc. Soc. Exp. Biol. Med., 1965, 120, 802-804.

47. Cantley, L.C., Jr.; Josephson, L.; Warner, R.; Yanagisawa, M.; Lechene, C.; Guidotti, G. Vanadate is a potent (Na, K)-ATPase inhibitor found in ATP derived from muscle. J. Biol. Chem., 1977, 252, 7421-7423.

48. Nechay, B.R.; Saunders, J.P. Inhibition by vanadium of sodium and potassium dependent adenosinetriphosphatase derived from animal and human tissues. J. Environ. Path. Toxicol., 1978, 2, 247-262.

49. Quist, E.E.; Hokin, L.E. The presence of two (Na^+ + K^+)-ATPase inhibitors in equine muscle ATP: Vanadate and a dithioerythritol-dependent inhibitor. Biochim. Biophys. Acta, 1978, 511, 202-212.

50. Cantley, L.C., Jr.; Resh, M.D.; Guidotti, G. Vanadate inhibits the red cell (Na^+, K^+) ATPase from the cytoplasmic side. Nature, 1978, 272, 552-554.

51. Beaugé, L.A.; Glynn, I.M. Commercial ATP containing traces of vanadate alters the response of (Na^+ + K^+) ATPase to external potassium. Nature, 1978, 272, 551-552.

52. Kobayashi, T.; Martensen, T.; Nath, J.; Flavin, M. Inhibition of dynein ATPase by vanadate, and its possible use as a probe for the role of dynein in cytoplasmic motility. Biochem. Biophys. Res. Comm., 1978, 81, 1313-1318.

53. Gibbons, I.R.; Cosson, M.P.; Evans, J.A.; Gibbons, B.H.; Houck, B.; Martinson, K.H.; Sale, W.S.; Tang, W.-J.Y. Potent inhibition of dynein adenosinetriphosphatase and of the motility of cilia and sperm flagella by vanadate. Proc. Natl. Acad. Sci. USA, 1978, 75, 2220-2224.

54. Cande, W.Z.; Wolniak, S.M. Chromosome movement in lysed mitotic cells is inhibited by vanadate. J. Cell Biol., 1978, 79, 573-580.

55. Josephson, L.; Cantley, L.C., Jr. Isolation of a potent (Na-K) ATPase inhibitor from striated muscle. Biochemistry, 1977, 16, 4572-4578.

56. Cantley, L.C., Jr.; Cantley, L.G.; Josephson, L. A characterization of vanadate interactions with the (Na, K)-ATPase. Mechanistic and regulatory implications. J. Biol. Chem., 1978, 253, 7361-7368.

57. Lindquist, R.N.; Lyon, J.L.; Lienhard, G.E. Possible transition-state analogs for ribonuclease, the complexes of uridine with oxyvanadium (IV) ion and vanadium (V) ion. J. Amer. Chem. Soc., 1973, 95, 8762-8768.

58. Lopez, V.; Stevens, T.; Lindquist, R.N. Vanadium ion inhibition of alkaline phosphatase-catalyzed phosphate ester hydrolysis. Arch. Biochem. Biophys., 1976, 175, 31-38.

59. Van Etten, R.L.; Waymack, P.P.; Rehkop, D.M. Transition metal ion inhibition of enzyme-catalyzed phosphate ester displacement reactions. J. Amer. Chem. Soc., 1964, 96, 6782-6785.

60. Söremark, R. Vanadium in some biological specimens. J. Nutr., 1967, 92, 183-190.

61. Welch, R.M.; Cary, E.E. Concentration of chromium, nickel, and vanadium in plant materials. J. Agric. Food Chem., 1975, 23, 479-482.

62. Myron, D.R.; Givand, S.H.; Nielsen, F.H. Vanadium content of selected foods as determined by flameless atomic absorption spectroscopy. J. Agric. Food Chem., 1977, 25, 297-300.

63. Myron, D.R.; Zimmerman, T.J.; Shuler, T.R.; Klevay, L.M.; Lee, D.E.; Nielsen, F.H. Intake of nickel and vanadium by humans. A survey of selected diets. Amer. J. Clin. Nutr., 1978, 31, 527-531.

64. Byrne, A.R.; Kosta, L. Vanadium in foods and in human body fluids and tissues. Sci. Total Environ., 1978, 10, 17-30.

65. Hunt, C.D. "The Effect of Dietary Vanadium on ^{48}V Metabolism and Proximal Tibial Growth Plate Morphology in the Chick". Ph.D. Thesis, University of North Dakota, 1979.

66. Moran, J.K.; Schwarz, K. Light sensitivity of riboflavin in amino acid diets. Fed. Proc., 1978, 37, 671.

67. Fulkerson, W.; Goeller, H.E. "Cadmium the Dissipated Element", Oak Ridge National Laboratory Report ORNL-NSF-EP-21, 1973.

68. Goyer, R.A.; Mushak, P. Lead toxicity laboratory aspects. In: "Toxicology of Trace Elements, Advances in Modern Toxicology, Vol. 2", eds: R.A. Goyer and M.A. Mehlman, John Wiley & Sons, New York, NY, 1977, pp. 41-77.

69. Baines, J.M.; Stoner, H.B. The toxicology of tin compounds. Pharmacol. Rev., 1959, 11, 211.

RECEIVED May 21, 1980.

INVOLVEMENT OF METAL COMPLEXES IN DISEASE PROCESSES

3

Metal Carcinogenesis in Tissue Culture Systems

MAX COSTA
Division of Toxicology, University of Texas Medical School at Houston, P.O. Box 20708, Houston, TX 77025

MARCIA K. JONES and ORRIN LINDBERG
Department of Pharmacology and Toxicology, College of Medicine, Texas A&M University, College Station, TX 77843

Metals present a human carcinogenic hazard primarily by inhalation, and therefore, induce primary neoplasms of the respiratory tract. A primary consideration in metal carcinogenesis is the particle size of the material, since it must be relatively small (< 2 microns) to penetrate into the alveolar spaces and induce lung cancer. A second important consideration is the identity of the specific metal compound because not all compounds of the same metal have similar carcinogenic activities. A case in point is the observation that crystalline Ni_3S_2 is a very potent carcinogen in experimental animals. When Ni_3S_2 was administered by various routes to experimental animals, a large proportion of these animals developed tumors. In contrast, similar animal exposure to amorphous NiS resulted in no neoplasia (1,2). Additional important considerations include: 1) water or lipid solubility properties of the metal carcinogen, 2) dosage and predisposition of the individual toward the development of cancer. The latter point is particularly important in light of some of the preliminary results presented in the experimental section of this paper showing that pretreatment of cells with benzopyrene potentiates the carcinogenesis of Ni_3S_2. Benzopyrene (found in cigarette smoke) and other xenobiotics to which we are differentially exposed to in our environment may greatly alter the course of metal induced carcinogenesis.

In this chapter we wish to address ourselves specifically to tissue culture systems for assaying metal carcinogenic activity. The rationale for this is the same as that for developing simple, rapid, reliable, and inexpensive *in vitro* cancer tests for organic compounds. It is impossible in terms of time and cost to evaluate hazards of all the potentially carcinogenic metal compounds using the conventional animal carcinogenesis testing systems. However, a single *in vitro* test system is probably insufficient in assigning carcinogenic activity to a particular compound.

0-8412-0588-4/80/47-140-045$07.25/0

Additionally, in vitro tests should not elimate animal testing but may help in tests with animals. The proper use of several in vitro test systems will greatly aid our understanding of which metal compounds are hazardous. These screening tests may also allow us to identify carcinogens in air samples from the industrial environment.

Metals Have Caused Cancer in Humans. Numerous recent reviews have been compiled which describe the various epidemiological studies implicating arsenic, cadmium, chromium, and nickel as causes of human cancer (3-11). These studies have analyzed the industrial worker's exposure to these metals, or have examined selected populations having excessive exposure to these metals in their drinking water, the atmosphere adjacent to their habitation, etc. The reader is referred to these reviewers for a more comprehensive discussion of the available data pertaining to metal carcinogenesis (3-11). It should, however, be pointed out that nickel and its compounds have been the most extensively documented class of metal carcinogens in humans. There are several other metals, including beryllium, which have also been linked to human cancer, but the most important metal carcinogens have been listed above (11).

Metals Induce Cancer When Administered to Experimental Animals. Again, the reader is referred to many recent comprehensive reviews which discuss the evidence that certain metals cause cancer in experimental animals (9,10,11). The most credible data implicates cadmium, chromium, cobalt, and nickel as carcinogens in experimental animals. It should be noted that while arsenic has been shown to be responsible for the induction of human cancer, attempts to induce cancer in experimental animals with arsenic and its compounds have not been successful. In contrast, while cobalt induced cancer in experimental animals, numerous epidemiological studies have failed to show a correlation between excessive human exposure to cobalt and the induction of human neoplasia. One of the most studied metal carcinogens are the nickel compounds, of which crystalline Ni_3S_2 appears to be the most potent (1,2,12). Ni_3S_2 has been shown to induce cancer at the site of administration, resulting in muscle tumors if given by intramuscular injection (1), renal tumors if administered by intrarenal injection (2), testicular tumors if injected into the testes (2), or lung tumors if administered by inhalation (12).

In vitro Analysis of Metal Carcinogenesis. Methods of exploring metal carcinogenesis will be grouped into three separate sections. The first group, biochemical studies, includes those studies that deal with the basic interactions of isolated enzyme systems extracted from living cells. The second group considers the effect of metals on bacterial systems, and the third reviews present techniques using tissue culture of cells in assessing metal carcinogenesis.

Biochemical Studies

Most of the studies that have examined the interactions of metals in biochemical systems have attempted to prove the hypothesis that carcinogenic metals induce alterations in the base pairing and fidelity of mRNA transcription *in vitro*. These studies have suggested that a principal mechanism of metal carcinogenesis involves distortions in this transcriptional processes. One such *in vitro* study examined the interaction between complementary and uncomplementary strands of synthetic polynucleotides which were mixed together and allowed to anneal (13). When synthetic strands of poly C·U (polycytosine monophosphate·polyuridine monophosphate) or poly C·A (polycytosine monophosphate·polyadenosine monophosphate) are mixed with the purine polynucleotide poly I (polyinosine monophosphate) in the absence of any divalent metal ion, the strands hybridized because the cytosine residues formed hydrogen bonds with their complementary base, the hypoxanthine residues of the inosine (13). If magnesium (2+) ions or nickel (2+) ions are added to this system, the degree of hybridization was increased, suggesting the occurrence of uncomplementary base pairing. Normally, the uracil and adenine nucleotides should not pair with the poly I, but in the presence of magnesium or nickel, the mispairing occurred (13). Cadmium has a similar affect on the hybridization of poly C·A or poly C·U with poly I (14). Cadmium at concentrations of 1 mM provoked mispairing between hypoxanthine and uracil, but not between hypoxathine and adenine nucleotides (14). In addition to increasing the number of mispairings between uncomplementary nucleotides, carcinogenic metals alter the structure of the double stranded DNA helix. In the presence of magnesium, calcium, zinc, or manganese the secondary structure of poly A (polyadenosine monophosphate) in solution consists of both single and double helixes, but in the presence of nickel and cobalt, poly A appears only as a single stranded helix (15). Copper and cadmium result in the formation of randomly coiled structures of poly A (15). Thus, some potentially carcinogenic metals appear to alter the secondary structure of polynucleotides, as well as promoting the mispairing of uncomplementary nucleotides. Collectively, these studies suggest that the interaction of potentially carcinogenic or mutagenic metal ions with solutions of polynucleotides causes uncomplementary base pairings and distortion of the secondary or tertiary structure of DNA. These effects *in vitro* can be correlated with the *in vivo* situation, since alterations of this nature could result in a significant increase in DNA mutation frequencies. The mispairing of uncomplementary nucleotides would, for the most part, cause point mutations when the DNA is replicated. The end result of this type of mutation, assuming the mistake was not repaired, would be the substitution of an incorrect amino acid in the primary structure of a cellular constituent or enzyme. If a number of unrepaired point mutations occur, it is conceivable that a sufficient number of alterations in cellular proteins would result in cell transformation. This transformed cell could then proliferate abnormally, and with proper promotion, develop into a neoplasm.

Additional experiments have suggested that carcinogenic or mutagenic metals may decrease fidelity of DNA replication or transcription. The fidelity of DNA transcription is assayed as follows: a synthetic template is mixed with the DNA polymerase enzyme isolated from a bacteria or a virus. Other reagents required for the transcription of DNA are added, including two radiolabeled triphosphate nucleotides; one complementary to the strand being replicated and labeled with [^{3}H], and the other uncomplementary and labeled with [^{14}C]. Using this system investigators have found that the presence of 1-10 mM beryllium (2+) resulted in an increased misincorporation of the uncomplementary cytosine triphosphate into nucleotides copied from poly A·T (16). These results could have been attributed in part to exonuclease contamination of the DNA polymerase preparation used. The contaminating exonuclease would incorporate cytosine by repair rather than by replication. However, Sirover and Loeb (17) found that beryllium decreased the fidelity of DNA polymerase with a similar system using avian myoblast virus DNA polymerase which lacks the exonuclease repair enzyme. Their results suggest that beryllium salts actually reduce the fidelity of transcription. The uncomplementary cytosine was incorporated once for every 1,100 complementary dTTP nucleotides polymerized into the daughter strand using a poly·A oligo dT template and natural conditions. When beryllium was added to this mixture the error frequency increased form 1 per 1,100 to 1 per 75 (17). In these studies beryllium was found to bind with the DNA polymerase enzyme; thus, this interaction probably caused the decrease in the fidelity of transcription. Additional studies showed that cobalt, manganese, and nickel could substitute for magnesium as cofactors of DNA polymerase enzyme (18). However, even in the presence of magnesium, additions of cobalt and manganese significantly impaired the fidelity of transcription. Various carcinogenic and mutagenic metals such as Ag, Be, Cd, Cr, Cu, Mn, Ni, and Pb decreased the fidelity of DNA transcription, resulting in the enhanced incorporation of the uncomplementary nucleotide (19). In contrast, soluble salts of Al, Ba, Ca, Fe, K, Mg, Na, Rb, Sr, and Zn did not affect the fidelity of DNA synthesis even at concentrations high enough to inhibit the synthesis of DNA (19). These biochemical studies highlight two important concepts: 1) they illustrate that the mechanism of metal carcinogenesis for a wide variety of carcinogenic or mutagenic metals may involve a decrease in the fidelity of DNA replication; 2) these studies have suggested that this biochemical transcription system can be used as a screen for the detection of potentially carcinogenic or mutagenic environmental metal carcinogens. The problem with this system as a screen is the requirement that the metals be in solution to rapidly interact in vitro with the DNA polymerase during the allotted assay time of 30-60 minutes. Many carcinogenic molecules are not readily water soluble. In addition, if they undergo dissolution, they may be chemically changed from the original molecule. For example, when crystalline nickel subsulfide is allowed to solubilize in water, it forms nickel oxide and nickel hydroxide. It can be more rapidly dissolved in dilute acid solution, for example HCl, but the adveous product is $NiCl_2$, not the original nickel compound.

Carcinogenic or mutagenic metal salts have varying effects on the initiation of new RNA chains. Salts of Co (2+), Cd (2+), Cu (2+), Mn (2+), and Pb (2+) enhance the rate of initiation of new RNA chains at concentrations that inhibited overall RNA synthesis rates (20). However, non-carcinogenic metals such as K^+, Li^+, Mg^{++}, Na^+, and Zn^{++} inhibited the initiation of new RNA chains at concentrations which depressed overall DNA synthesis (20). Since the metals which have been shown to be carcinogenic or mutagenic are able to activate points of RNA initiation at concentrations which inhibit total RNA synthesis, while non-carcinogenic metals are not able to activate new RNA chain initiation at concentrations which are inhibitory to RNA synthesis it is presumed that this technique may be utilized to measure the carcinogenic or mutagenic activity of specific metal salts (20). These observations also point to other possible mechanisms by which metals cause mutations and/or cancer in cells. The carcinogenic metals appear to initiate new RNA chains and therefore may lead to the expression of new genes which under normal conditions would not be expressed. Moreover, the carcinogenic metals may cause the excessive initiation and expression of selected gene products.

Effects of Carcinogenic Metals in Bacterial Systems

Microbial studies have demonstrated a link between carcinogenesis and mutagenesis and will be reviewed briefly with reference to the individual metals tested.

The metal compound most studied in bacterial mutagenesis systems is chromium. Hexavalent chromium is a potent mutagen in several bacterial systems. Venitt and Levy (21) tested the mutagenicity of three chromate compounds: Na_2CrO_4, K_2CrO_4, and $CaCrO_4$ using various bacterial strains. All three chromium compounds in concentrations from 0.05 to 0.2 mM per plate yielded positive mutation frequencies approximately 3 times greater than the control level (21). the hexavalent chromium ions bind to G-C base pairs and result in transitions from G-C to A-T when the DNA is replicated (21). Other studies have also confirmed that hexavalent chromium is mutagenic in a variety of bacterial systems.

In addition to hexavalent chromium ions, other metal ions have induced mutations in bacteria. These metals include As (3+), Cd (2+), Hg (2+), Mo (6+), Se (4+), Te (4+), Te (6+), and

V(4+) (22, 23, 24, 25). These results have also shown that the metal ions listed above were mutagenic in a wide variety of bacterial strains (22,23,24,25). However, certain carcinogenic metal ions were not mutagenic in bacterial systems. The most important of these carcinogens was the nickel ion, which did not induce mutations in a variety of the bacterial systems tested (25). The bacterial mutagenesis assay is useful for detecting potential organic carcinogens, but appears to have less value in assessing the carcinogenic activity of metal carcinogens. A major problem with bacterial systems is that the microsomal activation system is not present and must be added to activate the procarcinogen to the ultimate carcinogen. This introduces considerable variations in the assay. The influence of microsomal enzyme systems on metal mutagenesis or carcinogenesis has not been well studied and requires further investigation.

Studies in Cell Culture Systems

Introduction. In recent years the use of tissue culture has become one of the most important techniques in the determination of the toxic and carcinogenic activity of xenobiotics. Morphological transformation and cytological changes are induced in tissue culture following exposure to carcinogenic metals and their compounds. Tissue culture systems have certain advantages over microbial assays and *in vivo* systems because: 1) the cells are similar to those found *in vivo*, but unlike the *in vivo* situation a homogenous population is treated, 2) neoplastic transformation is induced in tissue culture, while microbial systems detect mutations, 3) control of variables in tissue culture systems is much simpler than with *in vivo* systems. Tissue culture systems are especially suited for assaying potentially carcinogenic metals since metal carcinogens, unlike the organics, are thought to be primary carcinogens. Metals are thought to be able to interact directly with target tissues, requiring metabolic activation for cancer induction, whereas organic carcinogens generally require activation by microsomal enzymes. Exposure of tissue culture cells to many organic carcinogens will induce transformation only if the activating enzymes are present. This is a drawback to the use of tissue culture methods for certain types of carcinogenesis research with organic chemicals. Activation by microsomal enzymes, with inorganic metal carcinogens, may not play a primary role in the carcinogenesis process if current hypotheses are correct. However, the preliminary data presented in this paper suggests that microsomal enzymes may, in fact, alter the course of metal carcinogenesis in tissue culture.

The effects of metals on tissue culture systems will be divided into 3 sections. The first section involves toxic effects of the same metals on cell cultures; the second section will discuss the effects of metals on transcriptional, translational, and mutagenic processes; and the third section will discuss the in vitro transformation of cells in culture by exposure to metal carcinogens.

Toxic Effects

Metals have been shown to have a variety of toxic effects on cells in culture. In one study (26), exposure of rat embryo muscle cells to nickel subsulfide resulted in a depression of cell division and induced abnormal mitotic spindles, distorted bipolar spindles, lagging chromosomes, and unequal cytoplasmic division (26). Mitotic arrest occurred in telephase, and the post-telephase period was consistent with a mechanism of action involving a disturbance in function of the mitotic spindle mechanism (26). Chromosomal breakage was observed in leukocyte cultures after a 2 day exposure to sodium arsenate (Na_2HAsO_4) at concentrations from 0.1-10 µg/ml (27,28). The abberations incurred by exposures of leukocytes in vitro to arsenic were similar to those observed in vivo with patients who were treated chronically with arsenic compounds (27,28.29). The clinical picture included a variety of abnormal mitotic and chromosomal configurations in the leukocytes of humans treated with arsenic (27,28,29). In a similar study As, Sb, and Te salts, but not Be, Cd, Co, Fe, Hg, Ni, Se, and V salts caused chromosomal abberations in human leukocytes (30). Exposure of human leukocytes to cadmium sulfide induced numerous alterations within the chromosomal structure (31). Further experimentation showed that the exposure of tissue culture fibroblasts to $CdSO_4$ caused chromosomal breakage and aberrations (32). Other studies have demonstrated that treatment of eukaryotic cells with As, Cd, Cr, Ni, Sb, and Te alters chromosomal structures (33,34,35). These results suggest that the metal compounds (particularly arsenic, cadmium, chromium, and nickel) altered the normal mitotic processes. An additional study has shown that exposure of Chinese Hamster Ovary (CHO) cells to nickel subsulfide, a potent carcinogen, induced elongation of these cells (36). Following exposure of Chinese Hamster Ovary cells to this nickel carcinogen, the cell shape changes from a rounded form to an elongated fibroblastic-like cell structure (36). The changes in the morphology of CHO cells resembled those that were caused by exposure of these cells to agents that elevated cellular cAMP levels (36).

Mutagenic Effects

Arsenic competes with phosphate for incorporation into precursors of DNA or RNA such as dATP, rATP, dGTP, rGTP, etc. (28,37,38,39). If arsenic is incorporated into molecules such as ATP, then it may also be added to serine and threonine residues of cellular proteins, since protein kinase uses the terminal phosphate of ATP in the phosphorylation of a variety of cellular proteins.

Hexavalent chromium was shown to inhibit DNA synthesis in hamster fibroblasts at concentrations of 0.1 mM (40,41). In these studies it was found that hexavalent chromium was more efficient than trivalent chromium in the inhibition of DNA synthesis, and that the effects on tissue culture cells of hexavalent chromium were more pronounced in a salts/glucose solution rather than a complete media, since the latter had fetal bovine serum and other undefined factors that promoted the reduction of hexavalent chromium to trivalent chromium (40). Different studies have confirmed that other carcinogenic metals, such as soluble salts of nickel, inhibited DNA synthesis in tissue culture cells. These are in agreement with the studies conducted *in vivo* demonstrating that certain carcinogenic metals also inhibit hepatic DNA synthesis during liver regeneration.

Transformation of Cells in Tissue Culture by Carcinogenic Metals and Their Compounds

Mammalian cell cultures have been used as the basis of several systems in detecting the potential carcinogenic activity of chemicals. Basically, two general approaches have been utilized: continuous cell lines and primary cell cultures. Cell lines have the advantage of ease of use, in that cultures do not have to be obtained fresh from animals prior to each test, but may be maintained for months to years by proper subculturing techniques. They have the disadvantage of possessing one or more "transformed" characteristics (e.g., immortality). In some cases cell lines may also lack certain enzyme systems required for metabolic activation of chemicals. Some of the cell lines used for transformation assays include the murine (BALB/3T3) A31 system (42), and the baby kidney-21 (BHK-21) systems (43).

Primary cell cultures have advantages over cell lines in that the cells are not initially immortal, and usually have none of the transformed characteristics which may be seen in some cell lines (44,45). Additionally, the embryonic cells which are most widely used for transformation tests generally maintain enzyme systems characteristic of the original host for at least a few subcultures. Primary cell cultures may have the disadvantage, however, of requiring harvesting and preparation of cells periodically, or for each test. This problem can be minimized by preparation of large numbers of cells and cryopreservation of known sensitive samples.

A useful *in vitro* transformation assay which may be particularly applicable for metals, is the the enhancement of hamster embryo cell transformation by simian adenovirus. Casto, et al (46) and DiPaolo et al (47) reported that all metal salts (of a series of 38 tested) with known carcinogenic activity, increased the frequency of simian adenovirus SA-7 induced transformations.

Metals were divided into three groups: high (positive at 0.05 mM), moderate (positive at 0.05 to 0.6 mM), and low (positive only at concentration greater than 0.9 mM) activity.

Activity	Salts of
High	Antimony, Arsenic, Cadmium, Chromium, Platinum
Moderate	Berrylium, Cobalt, Copper, Lead, Mangagnes, Mercury, Nickel, Silver, Thallium, Zinc
Low	Iron

In every cell culture based transformation assay, several characteristics are monitored as indicators of potential oncogenic transformation. These characteristics, for transformed and non-transformed cells, are summarized in the following statements. Normal cells: 1) grow in an orderly fashion with little cell criss-crossing, 2) are incapable of forming 3 dimensional colonies in soft agar media, and 3) do not form tumors when administered to athymic "nude" mice. The ultimate test of neoplastic transformation is the ability of cells to form tumors in "nude" mice. A recent study has shown that the various criteria which have been applied to transformation, such as disordered growth, plant lectin agglutination, and growth in soft agar do not necessarily indicate neoplastic transformation; that is, the ability of cells to form tumors when administered to experimental animals (48).

A number of investigators have treated established cell lines with carcinogenic metals and found that morphological transformation was induced in these cultures. Fradkin et al (49) found that treatment of cell lines with hexavalent chromium resulted in disordered growth of baby hamster kidney (BHK-12) cells exposed to 0.25 or 0.5 μg/l of $CaCrO_4$. The characteristic change in the growth pattern of these cells was a loss of order in colonies and extensive cell piling which was not present in untreated cultures (49). After the cells were treated with $CaCrO_4$, they acquired the ability to grow in a semisolid media. Normal cells were not able to proliferate in this media. Treatment of cultures of mouse fetal cells with $CaCrO_4$ resulted in morphological alterations (35). Cells were exposed to $CrCl_3$ or $K_2Cr_2O_7$, and exposures to both hexavalent and trivalent chromium compounds resulted in colonies of cells that piled up in a randomly oriented fashion surrounded by various giant cells (35). The authors of this study questioned whether or not the morphological alterations induced by these chromium compounds represented any neoplastic changes (35).

In a series of studies by Costa et al (12,50-57), secondary cultures of Syrian hamster fetal cells were exposed to either crystalline nickel subsulfide or amorphous nickel sulfide. Following exposure to the nickel compounds the cultures were washed free of the metals and the cells seeded to form colonies. Treatment with the potent carcinogen, nickel subsulfide, induced a concentration dependent incidence of morphological transformation, while similar treatment with amorphous nickel sulfide, the non-carcinogen, did not result in any of these changes. The nickel subsulfide transformed colonies were cloned, derived into immortal cell lines, and tested for their ability to grow in soft agar and to form tumors in athymic "nude" mice. All morphologically transformed cell lines tested were able to proliferate in soft agar and formed tumors in "nude" mice. Similar cloning of normal cultures was unsuccessful, and none of the normal mass cultures derived from hamster embryos, and tested for colony formation in soft agar or formation of tumors in "nude" mice were positive.

Materials and Methods

Test Compounds. Nickel subsulfide (crystalline αNi_3S_2, particle size < 5 μm) was provided by Dr. Edward Kostiner, University of Connecticut, and its purity and crystal structure were verified by emission spectroscopy and X-ray diffractometry as previously described (2,58). Amorphous nickel monosulfide (NiS) was precipitated by addition of ammonium sulfide to a solution of $NiCl_2$ that was prepared from carbonyl-derived Ni dust and ultrapure HCl. The amorphous NiS was devoid of crystal structure, based upon X-ray diffractometry. The αNi_3S_2 and NiS powders were sterilized by washing in acetone immediately prior to suspension in tissue culture medium.

Morphological Transformation Assay Using Syrian Hamster Fetal cells. Syrian hamster embryo cells were isolated as previously described (50,51,52). Tertiary passage cultures were prepared by plating about 1 X 10^6 cells into 100 mm diameter plates. The cells were allowed to attach to the monolayer for one or two days and selected cultures were then pretreated with 3 μg/ml of benzopyrene for 24 h. Pretreatment of cells with 2 μg/ml of benzopyrene was shown to increase the microsomal protein content by 2 fold within 24 h in these cells (data not shown). Cultures were then treated with the appropriate metal compounds three times for a period of two days for each treatment. Following this procedure cells were removed from the monolayer by trypsinization (0.25% trypsin in Puck's saline A), and 5,000 and 10,000 cells were replated to form colonies in 100 mm diameter plates containing 10 ml of Dulbecco's medium supplemented with 10% fetal bovine serum (Hy-clone, Sterile Systems, Inc.). Foci assays were conducted by plating approximately 50,000 treated or untreated cells into 100 mm diameter tissue culture plates. The media was replenished about one time each week during the subsequent two week incubation period. Cultures seeded to form colonies were incubated 12-14 days. Cells that were seeded for the foci assay were incubated for 23 days. At the end of the appropriate incubation period the monolayer of cells was washed two times with a phosphate buffered normal saline solution, fixed with 95% ethanol, and stained with a 0.5% (w/v) crystal violet-ethanol mixture. Where appropriate, the total number of surviving colonies were counted in each plate. Each colony or foci was evaluated for morphological transformation using a light microscope by an observer who was not aware of the treatment conditions.

Mutagenesis Assay Using Chinese Hamster Ovary Cells. Chinese Hamster Ovary cells were grown in monolayer culture within a humidified atmosphere composed of 5% CO_2 and 95% air using McCoy's 5a medium supplemented with 10% fetal bovine serum (Hy-clone, Sterile Systems, Inc.). Cultures of Chinese Hamster Ovary cells were exposed to the appropriate metal compounds for 24 h and then allowed to undergo approximately twelve divisions (doubling time 14-16 h) prior to being placed in the selection media. At the end of this division period approximately 7 X 10^6 cells (monolayer cultures of each had formed) were incubated in the selection medium consisting of complete McCoy's media supplemented with 3 µg/ml 8-azaguanine (8-AG) and 6 µg/ml 6-thioguanine (6-TG). Cultures were incubated in the selection media for approximately 3 weeks. The selection medium was replaced with fresh media containing the 8-AG and 6-TG every three or four days during the selection period. At the end of this incubation, cells were fixed with 95% ethanol and stained with a 0.5% (w/v) crystal violet-ethanol mixture. The number of mutant colonies in each plate was then counted. By definition a colony consisted of 50 or more cells in a cluster. As expected, nearly all of the cells in the plates died off, leaving only a few colonies in plates treated with mutagenic agents. These procedures are similar to other mutation assays reported in the literature (59).

Toxicity tests were conducted in Chinese Hamster Ovary cells to determine the concentration of either Ni_3S_2 or NiS which affected cell plating efficiency following a 24 h exposure period. Proliferating cultures of Chinese Hamster Ovary cells were exposed to the nickel compounds for 24 h, and then the cells were trypsinized from the monolayer. Cell numbers were determined with a hemocytometer and 400 cells were plated to form colonies in 100 mm diameter tissue culture plates. The cells were incubated for about 9 days, fixed, stained, and the total number of surviving colonies in each plate were counted and expressed as a function of the total number of cells plated.

Uptake Studies

Log phase cultures were prepared in Leighton Tubes (tubes containing a plastic microscopic slide which provides a surface for cell growth). Cells were exposed to the metal compunds and following the exposure period were washed two times with normal saline, fixed with 95% ethanol and stained with a crystal violet solution (0.5% crystal violet in ethanol). The cells were observed with a light microscope for the presence of intracellular nickel compounds.

For electron microscopy studies, cells were prefixed in 3% gluteraldehyde solution buffered with 0.05 M Phosphate Buffer pH 7.4 containing 0.05 M sucrose. Cells were rinsed in 0.05 M Phosphate Buffer pH 7.4 containing 0.5 M sucrose five times for 10 min and then dehydrated using sequential ethanol and acetone washes. The cells were embedded in an epoxy resin mixture and then sectioned with a microtone. The sections were post stained with Uranyl Acetate and Lead Citrate and then examined with an electron microscope.

Results

Morphological Transformation of Syrian Hamster Fetal Cells. Figure 1 shows the orderly growth pattern of cells in a normal untreated colony of Syrian hamster fetal cells. In contrast, Figure 2 shows the changes in the growth pattern of normal cells resulting from treatment with a metal carcinogen such as crystalline Ni_3S_2. Note that the transformed cells grew in a disorderly pattern, with cells invading each other's boundaries. This net-like, disordered growth pattern is the morphological alteration characteristic of neoplastic transformation. Our laboratory has cloned a number of colonies having disordered growth patterns similar to those shown in Figure 2 (50,51,52). These clones were shown to produce tumors in athymic "nude" mice following subcutaneous injection and to form 3 dimensional colonies in soft agar medium (50,51,52). Cells having a normal growth pattern as shown in Figure 1 were administered to "nude" mice and plated to form colonies in soft agar medium. However, these cells did not produce tumors in "nude" mice, or 3 dimensional colonies in soft agar medium.

Concentration Dependent Morphological Transformation by Ni_3S_2 Table 1 shows the results of a typical experiment where cultures of Syrian hamster embryo cells were exposed to several concentrations of Ni_3S_2 or NiS. Control cultures, which were untreated or treated with amorphous NiS, had no significant incidence of morphological transformation. However, the incidence of Ni_3S_2 induced morphological transformation was dependent upon the concentration of Ni_3S_2 which the cells were exposed.

Influence of Benzopyrene Pretreatment on Metal Induced Transformation. The induction of cancer by metal carcinogens is thought to be independent of microsomal enzyme activation. These conclusions are not based upon experimental evidence as such, but on an idea that has developed from at least two observations: 1) carcinogenic metals and their com-

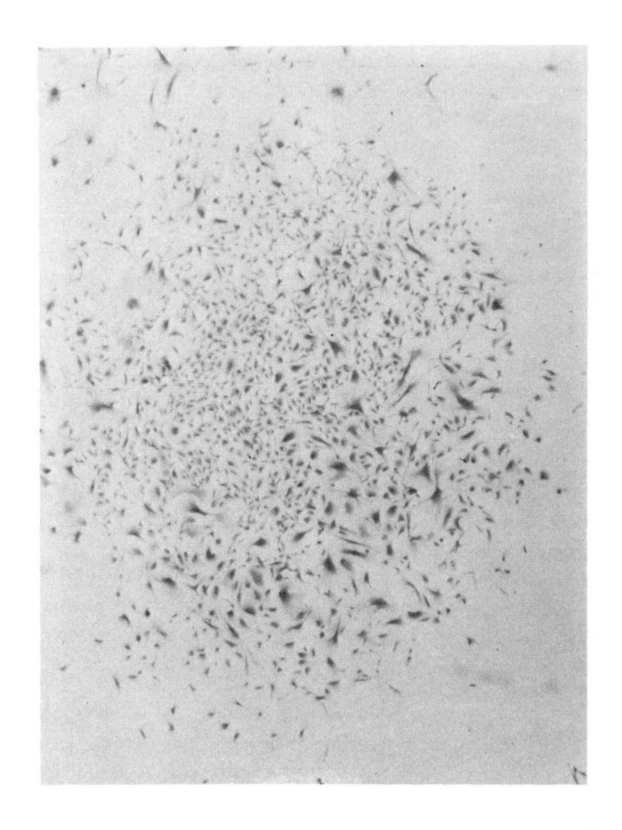

Figure 1. Photograph showing the ordered growth pattern of a typical "normal" colony

Figure 2. Photograph showing the disordered growth pattern of a Ni_3S_2 transformed colony

TABLE I

EFFECT OF CRYATALLINE Ni_3S_2 AND AMORPHOUS NiS ON THE TRANSFORMATION OF SYRIAN HAMSTER FETAL CELLS IN TISSUE CULTURE

Test Compound	Concentration μg/ml	No. of Plates[a]	Proportion of Transformed Colonies[b]
Controls	0	36	2/2045 [0.1%]
αNi_3S_2	0.1	12	25/599[c] [4.2%]
	1.0	12	40/584[c] [6.8%]
	5.0	12	26/271[c] [8.9%]
	10.0	6	4/20 [20%]
NiS	0.1	6	1/203 [0.5%]
	1.0	6	0/261 [0.0%]
	5.0	6	2/291 [1.0%]

[a]5,000 cells were plated into 35 mm tissue culture plates to form colonies.

[b]Ratio of the number of transformed colonies to the number of total colonies on all culture plates (with percentage in brackets [])

[c]$p < 0.005$ versus corresponding ratio for control plates [x^2 test].

pounds are relatively simple molecules which are not extensively metabolized in vivo, and 2) the induction of cancer by a carcinogenic metal occurs at the exposure site even in tissues, such as muscle lacking the capability to activate and metabolize drugs, as well as other foreign compounds.

In Table 2 we present some preliminary data suggesting that induction of microsomal enzymes, in particular cytochrome P 448 associated hydroxylases, enhanced the Ni_3S_2 induced transformation of Syrian hamster embryo cells. Cultures treated with Ni_3S_2 or benzopyrene had incidences of transformation in the colony assay ranging from 3.2-3.6% of the surviving colonies when 10,000 treated cells were challenged to form colonies. However, cultures pretreated with benzopyrene and then exposed to Ni_3S_2 had considerably higher proportion of morphologically transformed colonies. In this experiment, cultures were pretreated with benzopyrene using a single 24 h exposure interval while the same cultures were exposed to the metal compounds for 2 days using 3 separate exposures. The enhancement of Ni_3S_2 transformation in cultures pretreated with benzopyrene was greater than the incidence of transformation induced by each agent alone. These results suggested that benzopyrene induces a process which enhanced the carcinogenesis of Ni_3S_2. It is possible that this inducible process represents the activation of aryl hydrocarbon hydroxylases, and their subsequent interaction with the Ni_3S_2 enhances its carcinogenesis.

Toxicity of Ni_3S_2 and Benzopyrene in Syrian Hamster Embryo Cells. Table 3 shows the effect on the plating efficiency of Syrian hamster embryo cells exposure to NiS or Ni_3S_2. Note that exposure of cells to 2 μg/ml of Ni_3S_2 reduced cell plating efficiency by one-half the values obtained in untreated cultures. Results from other experiments which we have conducted suggests that a single 24 h treatment of cells with 3 μg/ml of benzopyrene only reduces cell plating efficiency by approximately 30%. It appears from the data shown that pretreatment with benzopyrene enhanced the Ni_3S_2 induced toxicity to Syrian hamster cells greater than the toxicity displayed for each compound individually. However, the Ni_3S_2 carcinogenic enhancement by benzopyrene was greater than the toxic enhancement.

Mutagenesis of Chinese Hamster Ovary Cells by Ni_3S_2. Preliminary experiments (Table 4) suggest that Ni_3S_2, a potent carcinogen has weak to no mutagenic activity in Chinese Hamster Ovary cells. The incidence of colonies resistant to the toxicity of 6-thioguanine and 8-azoguanine was not significantly increased in cultures that had been pretreated with Ni_3S_2 for 24 h. When Chinese Hamster Ovary cells were treated with NiS (a non-carcinogen), a lesser degree of resistance to the selecting agents was found. Chinese Hamster Ovary cells were exposed for 24 h to various levels of the NiS or Ni_3S_2 to determine the

TABLE II
EFFECT OF PRETREATMENT WITH BENZOPYRENE UPON THE TRANSFORMATION OF SYRIAN HAMSTER EMBRYO CELLS BY NICKEL COMPOUNDS

Treatment Conditions	Proportion of Transformed Colonies		No. of Transformed Foci per Plate
	5,000 Cells Plated	10,000 Cells Plated	
Ni_3S_2 (2 μg/ml) and Benzopyrene (3 μg/ml pretreatment for 24 h)	7/34 = 21% (N=3)	20/92 = 22% (N=4)	5.3 (N=4)
Ni_3S_2 (2 μg/ml)	28/305 = 9.2% (N=3)	16.494 = 3.2% (N=4)	1.0 (N=3)
NiS (2 μg/ml)	1/72 = 1.4% (N=2)	1/308 = 0.3% (N=2)	0 (N=3)
Benzopyrene (3 μg/ml)	------------	20/559 = 3.6% (N=4)	2.0 (N=4)
Control (untreated)	0/171 = 0% (N=2)	0/336 = 0% (N=2)	0 (N=2)

TABLE III

EFFECT OF BENZOPYRENE PRETREATMENT ON PLATING EFFICIENCY IN CULTURES SUBSEQUENTLY TREATED WITH Ni_3S_2

Treatment Conditions	Plating Efficiency (%)	
	5,000 Cells	10,000 Cells
Control (no treatment)	1.60%	1.71%
Ni_3S_2 (2 μg/ml)	0.89%	0.71%
NiS (2 μg/ml)	1.54%	0.72%
Ni_3S_2 (2 μg/ml) and benzopyrene (3 μg/ml pretreatment for 24 h)	0.23%	0.21%

Third passage log phase cultures of Syrian hamster fetal cells were treated as described in the table. Cultures treated with benzopyrene and Ni_3S_2 were exposed to benzopyrene for 24 h prior to treatment with the metal. Cultures were treated with the metal compounds three times using a two-day exposure for each treatment. Cells were then removed from the plate by trypsinization and the number of cells present in each plate was determined with a hemocytometer. Five thousand or ten thousand cells were replated to form colonies into 100 mm diameter plates and the number of surviving colonies in each plate was counted. Each number shown in the table is the mean of four tissue culture plates.

TABLE IV

MUTAGENESIS OF CHINESE HAMSTER OVARY CELLS BY SPECIFIC NICKEL COMPOUNDS

24h treatment Condition	Concentration (μg/ml)	Number of 6-TG and 8-AG Resistant Colonies per Plate x ± sd
No treatment	0	0.2 ± 0.4 (N=5)
Amorphous NiS	0.5	0.5 ± 0.58 (N=4)
	1.0	0.5 ± 0.58 (N=4)
	1.5	1.3 ± 1.53 (N=4)
Crystalline Ni_3S_2	0.5	1.0 ± 0.8 (N=4)
	1.0	2.3 ± 0.5 (N=4)
	1.5	2.5 ± 1.0 (N=4)

Proliferating cultures of Chinese Hamster Ovary cells were exposed for 24 h to the compounds shown in the table. The cells were allowed to undergo twelve divisions in complete growth medium and then about 7 X 10^6 cells attached to the monolayer were placed in selection medium (McCoy's 5a medium supplemented with 10% fetal bovine serum, 8 μg/ml 8-azaguanine (8-AG) and 6 μg/ml 6-thioguanine (6-TG). Cultures were incubated in selection medium for about three (3) weeks. The selection medium was replenished with fresh medium about two (2) times each week. At the end of the selection period the cultures were washed two (2) times with normal saline, fixed with 95% ethanol and stained with an ethanol-crystal violet solution (0.5% w/v crystal violet in 95% ethanol). A colony was defined as a cluster of 50 or more cells.

toxicity of these compounds. These results are shown in Table 5 and show that concentrations of Ni_3S_2 ranging from the LD_{50} to LD_{30} were used in attempting to obtain mutations.

Uptake of Ni_3S_2 Into Cells

Figures 3 and 4 show light microscope photographs of CHO and Syrian hamster embryo cells respectively which have phagocytized Ni_3S_2 particles. Table 6 shows that Syrian hamster embryo cells actively take up Ni_3S_2 particles and undergo morphological transformation following exposure to this compound. Similar exposure to amorphous NiS results in no significant transformation and little uptake of Ni S particles. Table 7 shows that the approximate half life of Ni_3S_2 particles in cells is 40 h. The particles may be altered to a form not visible with the light microscope or result in cell lysis.

Discussion

The development of an *in vitro* metal carcinogenesis test system which is reliable, rapid, and inexpensive has been the subject of several recent reports (47,48,50-57). These results were discussed in earlier sections of this chapter. The purposes of the present report is to review some of the work conducted on the effects of metals using *in vitro* assays, and to relate these findings specifically to current views on the mechanisms of metal carcinogenesis. The preliminary data presented in this chapter suggests some new points of view relating to the carcinogenesis of metals using *in vitro* systems. The experiments suggest that the benzopyrene pretreatment enhances the toxicity and carcinogenicity of Ni_3S_2. The enhancement of transformation by combined treatment of cells with Ni_3S_2 and benzopyrene is greater than the summation of that induced by individual compounds. Further work is required to clarify the mechanisms involved in the enhancement. These additional studies may involve a study of the effect of microsomal enzymes on metal carcinogenesis.

The preliminary benzopyrene experiments described in this paper have important implications in assessing carcinogenic hazards associated with human exposure to metal carcinogens. If smokers (benzopyrene is found in cigarette smoke as well as other inducers of microsomal enzymes) are exposed to metal carcinogens, the relative risks of contracting neoplasms of the respiratory systems are greater in these individuals than in those who do not smoke.

The additional preliminary data concerned the possible mutagenic activity of Ni_3S_2. Current theories concerning possible modes of cancer induction favor a cellular mutation or a series of mutagenic events for the initiation of neoplastic transformations. The data reported here indicates that Ni_3S_2 displays little mutagenic activity. However, further experiments are

TABLE V

TOXICITY OF CRYSTALLINE Ni_3S_2 AND AMORPHOUS NiS IN CHINESE HAMSTER OVARY CELLS

24 h Treatment Conditions	Exposure Concentration (μg/ml)	No. of Surviving Colonies per Plate	Plating Efficiency (%) No. Surviving Colonies / per 400 cells plated
Control (no treatment)	0	373	93
Crystalline Ni_3S_2	30.0	5	1
	10.0	68	17
	3.0	88	22
	1.0	136	34
	0.3	144	36
	0.1	240	60
Amorphous NiS	30.0	120	30
	10.0	164	41
	3.0	176	44
	0.3	236	59

Proliferating monolayer cultures of Chinese Hamster Ovary cells were exposed for 24 h to the concentration of the compounds shown in the table. The monolayer of cells were dislodged by treatment with trypsin, counted, and then 400 cells were replated to form colonies. The cultures were allowed to form colonies for about 2 weeks with media changes two times each week. The plates were then fixed, stained, and the number of surviving colonies per plate were counted.

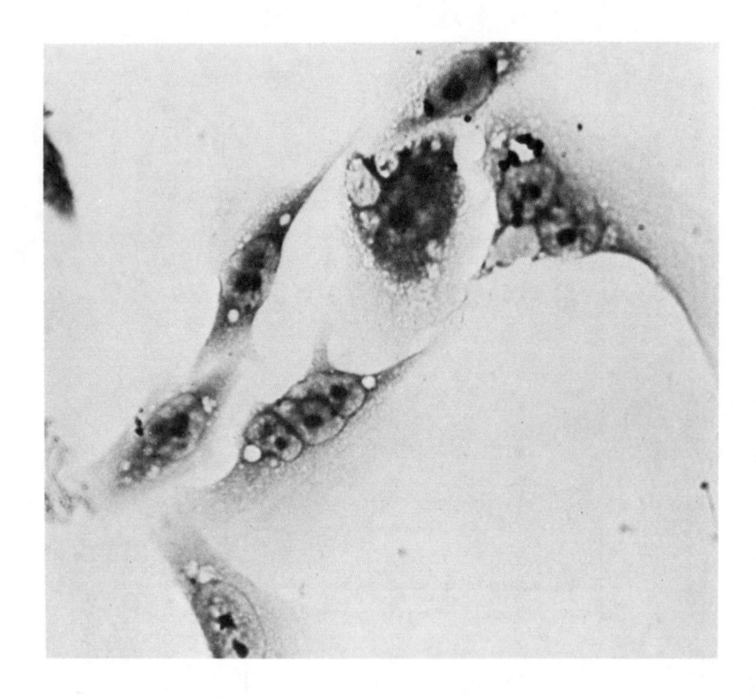

Figure 3. Light microscope photograph of Chinese hamster ovary cells that have phagocytized Ni_3S_2 particles. Note the vacuoles and dark Ni_3S_2 particles in some of the vacuoles.

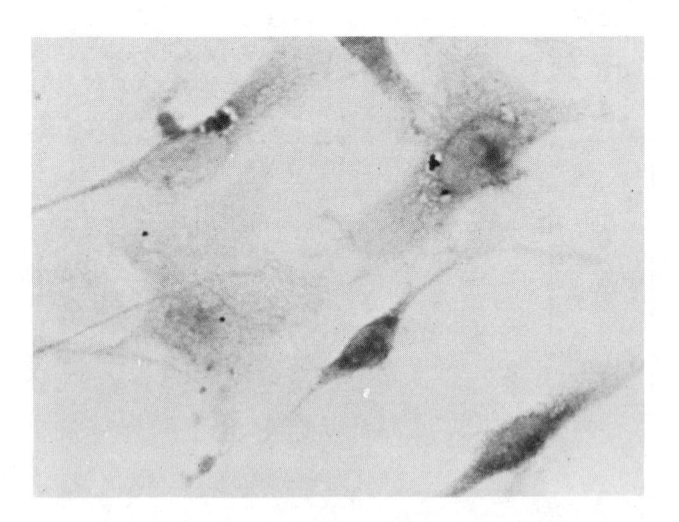

Figure 4. Light microscope photograph of Syrian hamster embryo cells that have phagocytized Ni_3S_2 particles

TABLE VI
PHAGOCYTOSIS AND MORPHOLOGICAL TRANSFORMATION
BY Ni_3S_2 AND NiS IN SYRIAN HAMSTER EMBRYO CELLS

Chemicals	Concentration (μg/ml)	Transformed Colonies / Total Surviving Colonies	Phagocytosis (Percentage of cells having nickel particles)
Amorphous NiS	1	0/222 (0%)	0.10%
	5	0/189 (0%)	0.40
	10	0/166 (0%)	0.80
Crystalline Ni_3S_2	1	6/214 (2.8%)	6.70
	5	12/138 (8.7%)	23.60
	10	11/93 (11.8%)	42.90

Secondary cultures of Syrian hamster embryo cells were exposed to the particulate nickel compounds shown in the table for three separate exposure for a period of 48h (transformation assay) or 24h (uptake assay). To assess morphological transformation the free metal compounds were removed from contact with normal saline. The cells were then dislodged from the monolayer by trypsinization, and replated (1,000-5,000 cells) into 100 mm tissue culture plates to form colonies. Following 12 days of incubation the colonies were fixed, stained, and evaluated for morphological transformation. The number of transformed colonies was expressed as a function of the total number of surviving colonies. Each transformation ratio represents the mean of 6 separate plates. For the uptake studies, log-phase monolayer cultures grown on plastic microscopic slides were exposed to the metal compounds. Following the exposure period the cells were washed two times with normal saline, fixed with 95% ethanol, and stained with a methanol-crystal violet solution. One thousand cells were examined with a light microscope (see Figure 2 and 3) in each slide for the presence of nickel particles.

TABLE VII
HALF-LIFE OF Ni_3S_2 PARTICLES
IN SYRIAN HAMSTER EMBRYO CELLS

Cells with intracellular nickel particles	Time after removal of free metal
67.3	0 h
66.4	4 h
50.8	8 h
43.8	24 h
27.6	48 h
6.7	96 h
3.0	120 h

Log phase Hamster embryo cells were exposed to 20 μg/ml of Ni_3S_2 for 24 h. Following this exposure the media containing Ni_3S_2 was removed and the cells were washed two time with normal saline. The cells were then placed in fresh complete media and at various time intervals were fixed, stained and 1,000 cells examined with the light microscope for intracellular nickel particles. Each number shown in the table is the mean of 2 slides where a total of 2,000 cells were examined for both slides. The cells having nickel particles are expressed as a percentage of those examined.

required to evaluate the mutagenic activity of Ni_3S_2, one of the most potent metal carcinogens. Other studies (40,41) have suggested that several carcinogenic metals are mutagenic in bacterial systems, but nickel compounds have shown no mutagenic activity in bacterial systems. Very few studies have evaluated the mutagenic activity of carcinogenic metals in mammalian cell culture systems. Most of these studies have demonstrated effects on chromosomal structure and function (26-35). It is important to note that in all of our studies we have used amorphous NiS (a non-carcinogen) as a negative control in addition to an untreated control. Amorphous NiS was neither carcinogenic nor mutagenic in our two test systems.

In using the Syrian hamster embryo system for metal carcinogenesis testing it was important to demonstrate that neoplastic changes were associated with morphological transformation (52). Additionally, when testing various metal samples for carcinogenic activity, it is important to use positive (Ni_3S_2) and negative (NiS) controls in every experiment to evaluate the consistency of each assay. If Ni_3S_2 does not induce morphological transformation in a concentration dependent manner, then the validity of the entire assay should be suspect. Similarly, if untreated cultures and NiS treated cultures have a high incidence of morphological transformation the results of the assay are not valid.

A possible explanation for why Ni_3S_2 is a potent carcinogen while amorphous NiS lacks activity was presented in the uptake studies. From other experiments it appears that the carcinogenic activity of particulate metal compounds is proportional to their cellular uptake. In future studies we hope to concentrate on this phagocytosis as a possible mechanism of metal induced carcinogenesis.

Abstract. We have reviewed work conducted in our laboratory and other laboratories investigating the carcinogenic and mutagenic effects of metals and their compounds upon *in vitro* systems. Preliminary data is also presented which shows the following: 1) Pretreatment of Syrian hamster embryo cells with benzopyrene, an inducer of aryl hydrocarbon hydroxylase, potentiates the morphological transformation of Syrian hamster embryo cells induced by Ni_3S_2. The incidence of Ni_3S_2 transformation in cultures pretreated with benzopyrene was in some instances 10 fold greater than those transformations caused by similar exposure to either Ni_3S_2 or benzopyrene alone. 2) Ni_3S_2 treatment of Chinese Hamster Ovary cells caused the appearance of 2-3 6-thioguanine and 8-azoguanine resistant colonies (per plate, 7×10^6 cells plated) while untreated Chinese Hamster Ovary cells averaged 0.2 resistant colonies per plate for a similar number of cells at risk. Therefore, Ni_3S_2 displays very weak or no mutagenic activity in the mutagenesis system tested, and 3) carcinogenic activity of particulate metal compounds such as Ni_3S_2 is proportional to their cellular uptake. Cells actively phagocytized particulate Ni_3S_2 but did not take up amorphous NiS particles to a significant degree. The latter observation may help understand why specific metal compounds are carcinogenic.

ACKNOWLEDGMENT

This work was supported by grant #ES02254 from the National Institute of Environmental Health Sciences.

LITERATURE CITED

1. Sunderman, F.W., Jr.; and Maenza, R.M. Res. Commun. Chem. Pathol. Pharmacol., 1976, 14, 319.

2. Sunderman, F.W., Jr.; Maenza, R.M.; Allpass, P.R.; Mitchell, J.M.; Damjanov, I.; and Goldblatt, P.J. "Inorganic and Nutritional Aspects of Cancer", Plenum Publ. Corp., New York, 1978; 57-67.

3. Doll, R.; Mathews, J.D.; and Morgan, L.G. Br. J. Industr. Med., 1977, 34, 102.

4. Kreyberg, L. Br. J. Industr. Med., 1978, 35, 109.

5. Lessard, R.; Reed, D.; Maheux, B.; and Lambert, J. J. Occup. Med., 1978, 20, 815.

6. National Institute for Occupational Safety and Health. "Criteria for a Recommended Standard: Occupational Exposure to Nickel", U.S. Department of Health, Education, and Welfare, Washington, D.C., 1977; 1-282.

7. Barton, R.T. J. Otolaryngol., 1977, 6, 412.

8. International Agency for Research on Cancer. "Evaluation of Carcinogenic Risk of Chemicals to Man: Nickel Compounds", World Health Organization, Geneva, 1976; 11, 75-112.

9. Sunderman, F.W., Jr. Prev. Med., 1976, 5, 279.

10. Sunderman, F.W., Jr. Fed. Proc., 1978, 37, 40.

11. Sunderman, F.W., Jr. Biol. Trace Element Res., (in press).

12. Ottolenghi, A.D.; Haseman, J.K.; Payne, W.W.; Salk, H.L.; and MacFarland, H.M. J. Natl. Cancer Inst., 1977, 54, 1165.

13. Eichorn, G.L.; Richardson, C.; and Pitha, J. "162nd National Meeting", Amer. Chem. Soc. Abstr. #17, Biol. Chem. Div., Washington, D.C., 1971.

14. Murray, M.J.; and Flessel, C.P. Biochim. Biophys. Acta, 1976, 425, 256.

15. Shin, Y.A; Heim, J.M.; and Eichorn, G.L. Bio-inorg. Chem., 1972, 1, 149.

16. Luke, M.Z.; Hamilton, L.; and Hollocher, T.C. Biochem. Biophys. Res. Commun., 1975, 62, 497.

17. Sirover, M.A.; and Loeb, L.R. Proc. Amer. Assn. Cancer Res., 1976, 17, 113 (Abstr. #451).

18. Sirover, M.A.; and Loeb, L.A. J. Biol. Chem., 1977, 252, 3605.

19. Sirover, M.A.; and Loeb, LA. Science, 1976, 194, 1434.

20. Hoffman, D.J.; and Niyogi, S.K. Science, 1977, 198, 513.

21. Venitt, S.; and Levy, L.S. Nature, 1974, 250, 493.

22. Nishioka, H. Mutation Res., 1975, 31, 185.

23. Yagi, T.; and Nishioka, H. Doshisha Daigaku Rikogaku Kenkyu Hokoku, 1977, 18, 63.

24. Rossman, T.G; Meyn, M.S.; and Troll, W. Environ. Health Perspect., 1977, 19, 229.

25. Green, H.H.L.; Muriel, W.J.; and Bridges, B.A. Mutation Res., 1976, 38, 33.

26. Swierenga, S.H.H.; and Basrur, P.K. Lab. Invest., 1968, 19, 663.

27. Petres, J.; and Hundeiker, M. Arch. Klin. Exptl. Dermatol., 1968, 231, 336.

28. Petres, J.; Baron, D.; and Hagedorn, M. Environ. Health Perspect., 1977, 19, 223.

29. Nordenson, I.; Beckman, G.; Beckman, L.; and Nordstrom, S. Hereditas, 1978, 88, 47.

30. Paton, G.A.; and Allison, A.C. Mutation Res., 1972, 16, 332.

31. Shiraishi, Y.; Kurahashi, H.; and Yosida, T.H. Proc. Japan Acad., 1972, 48, 133.

32. Rohr, G; and Bauchinger, M. Mutation Res., 1976, 40, 125.

33. Felton, T.L. Diss. Abstr. Int. B., 1978, 38, 4635.

34. Tsudo, H.; and Kato, K. Gann, 1976, 67, 469.

35. Raffetto, G.; Parodi, S.; Parodi, C.; de Farrari, M.; Troiano, R.; and Brambilla, G. Tumori, 1977, 63, 503.

36. Costa, M. Toxicol. Appl. Pharmacol., 1978, 44, 555.

37. Jung, E.G.; and Trachsel, B. Arch. Klin. Exp. Dermatol., 1970, 237, 819.

38. Jung, E.G.; Trachsel, B.; and Immich, H. Germ. Med. Mth., 1969, 14, 614.

39. Rosen, P. J. Theor, Biol., 1971, 32, 425.

40. Levis, A.G.; and Buttignol, M. Brit. J. Cancer, 1977, 35, 496.

41. Levis, A.G.; Buttignol, M.; Bianchi, V.; and Sponza, G. Cancer Res., 1978, 38, 110.

42. Quarles, J.M.; Tennant, R.W. Cancer Res., 1975, 35, 2637.

43. Stoker, M.; Macpherson, I. Nature, 1964, 203, 1355.

44. DiPaolo, J.A.; Nelson, R.L.; Donovan, P.S. et al. Arch. Pathol., 1973, 95, 380.

45. Quarles, J.M.; Sega, M.W.; Schenley, C.K.; and Tennant, R.W. Natl. Cancer Inst. Mono., 1979, 51, 257.

46. Casto, B.C.; Meyers, J.; and DiPaolo, J.A. Cancer Res., 1979, 39, 193.

47. DiPaolo, J.A.; Nelson, R.L.; and Casto, B.C. Brit, J. Cancer, 1978, 38, 452.

48. Stanbridge, E.J.; and Wilkinson, J. Proc. Natl. Acad. Sci. U.S.A., 1978, 75, 1466.

49. Fradkin, A.; Janoff, A.; Lane, B.P.; and Kuschner, M. Cancer Res., 1975, 35, 1058.

50. Costa, M. "Ultratrace Metal Analysis in Biological Sciences and Environment", Adv. Chem. Ser., Amer. Chem. Soc., Washington, D.C., 1979; 172, 73.

51. Costa, M. "Molecular Basis of Environmental Toxicity", Amer. Soc. Biol. Chemists and Amer. Soc. Environ. Sci., Ann Arbor Science Publ., Inc., Ann Arbor (in press).

52. Costa, M.; Nye, J.S.; Sunderman, F.W., Jr.; Allpass, P.R.; and Gondos, B. Cancer Res., 1979, 39, 3591.

53. Costa, M. "Molecular Basis of Environmental Toxicity", Amer. Soc. Biol. Chem. Div. Environ. Chem., 176th ACS meetings, Abstr. #92, 1978.

54. Costa, M. "Inorganic Chemistry in Biology and Medicine", Amer. Chem. Soc., Washington, D.C., 1979.

55. Costa, M. Amer. Chem. Soc., CTEM, 1977, 30-32.

56. Costa, M.; Nye, J.; and Sunderman, F.W., Jr. Fed. Proc., 1978, 37, 102.

57. Costa, M.; Nye, J.; and Sunderman, F.W., Jr. "Morphological Transformation of Syrian Hamster Fetal Cells Induced

by Nickel Compounds", IUPAC Kristiansand Conferences by Nickel Toxicology, 1978.

58. Damjanov, I.; Sunderman, F.W., Jr.; Mitchell, J.M.; and Allpass, P.R. Cancer Res., 1978, 38, 268.

59. Sharp, J.D.; Capecchi, M.E.; and Capecchi, M.R. Proc. Natl. Acad. Sci. U.S.A., 1973, 70, 3145.

RECEIVED April 24, 1980.

4

Metal Ion–Nucleic Acid Interactions

Aging and Alzheimer's Disease

G. L. EICHHORN, J. J. BUTZOW, P. CLARK, H. P. VON HAHN, G. RAO, J. M. HEIM, and E. TARIEN

Gerontology Research Center, National Institute on Aging, National Institutes of Health, Baltimore City Hospitals, Baltimore, MD 21224

D. R. CRAPPER and S. J. KARLIK[1]

University of Toronto, Ontario M5S 1A8, Canada

In this paper we discuss some studies on the interaction of aluminum with DNA that were carried out because of the apparent relationship of aluminum with Alzheimer's disease. We then consider how metal ions are involved in genetic information transfer, and may influence the aging process, and finally we discuss the use of metal ions in probing the aging process.

Aluminum, DNA and Alzheimer's Disease

Alzheimer's disease is one of the senile dementias; in fact, it is estimated that 70% of the people who have senile dementia have a form of Alzheimer's disease. The cause and treatment of Alzheimer's disease is therefore of utmost importance. Crapper and his collaborators at the University of Toronto have reported that autopsies of Alzheimer's patients reveal an accumulation of aluminum ions in localized areas of the brain (1). They also studied the effect of intracranially injecting experimental animals with aluminum, and they found that cats so treated accumulate aluminum in brain cells in concentrations similar to those found in Alzheimer's disease (2). These animals also exhibit structural alterations in brain cells that are similar but not identical to the alterations in Alzheimer's disease.

DeBoni and Crapper (3) have demonstrated that aluminum accumulates in the chromatin of cells. Fluorescent microscopy of cells in mitosis, stained with aluminum-staining morin dye, shows aluminum bound to chromatin. It is therefore of some potential relevance to Alzheimer's disease to investigate the interaction of aluminum and DNA.

Let us first consider what kinds of effects metal ions generally have on DNA. Metal ions bind primarily at two positions on DNA. They can bind to the bases, and in so doing they can destroy the hydrogen-bonded structure. Therefore, they destabilize

[1] Current Address: Gerontology Research Center, NIA, NIH, Balto. City Hospitals, Balto., MD 21224.

0-8412-0588-4/80/47-140-075$05.00/0

the DNA double helix. On the other hand, metal ions binding to phosphate stabilize the double helix. The reason for this stabilization is that the metal ions neutralize the negatively charged phosphate groups on the surface of the molecule; these would repel each other and cause the molecule to unwind (4). The two different effects that metal ions have on the stability of DNA are dramatically illustrated by the effects of magnesium and copper ions on the DNA "melting" curves, which show the transitions between double helical DNA, which has a relatively low absorbance, and single stranded DNA, which has a high absorbance (5). An absorbance-temperature plot therefore follows the unwinding of DNA; the midpoint in the transition is called the melting temperature (T_m). Mg^{2+}, which binds to phosphate, raises this T_m, while Cu^{2+}, which binds to the bases, lowers it. Mg^{2+} stabilizes the double helix, and Cu^{2+} destabilizes it. The effects of these two metals demonstrate that metals can stabilize DNA by binding to phosphate or destabilize it by binding to bases. The melting curves of DNA in the presence of these two metal ions, and metal ions generally, are relatively simple: they produce a monophasic transition.

Aluminum turned out to produce more complicated effects. This was perhaps to be expected, since Al has a complex chemistry; in aqueous solution it exists in a large variety of species (6). In addition to hydrated aluminum ion, Al^{3+} or $[Al(H_2O)_6]^{3+}$, there are $Al(OH)^{2+}$, $Al(OH)_2^+$, $Al(OH)_3$, $Al(OH)_4^-$, as well as $[Al_{13}O_4(OH)_{24}(H_2O)_{12}]^{7+}$. The relative amounts of these species varies with pH. DNA melting curves were obtained therefore at different pH values and at different aluminum concentrations. Some of the melting curves exhibit biphasic transitions; i.e., part of the DNA complex melts out in one temperature region and another part melts out in another region. Melting curves are presented as derivative curves, in which transitions become peaks (Fig. 1). Note the existence of a high melting aluminum-DNA complex even above 100°C, e.g. at pH 7.5 and 0.6 Al/DNA as well as a low melting aluminum-DNA complex, as at pH 5.0 and 0.4 Al/DNA. A third aluminum-DNA complex melts out in an intermediate temperature range, e.g. at pH 6.0 and 0.6 Al/DNA. Analysis of the data over a pH range from 5.0 to 7.5 and an Al/DNA concentration range of from 0 to 0.7 leads to the conclusion that all the melting areas are accounted for by these three complexes and uncomplexed DNA. We propose the structures shown in Figure 2 for the three Al-DNA complexes. We consider that the high melting complex I, stable at relatively high pH, contains hydroxylated Al, perhaps $Al(OH)^{2+}$ ion. The metal concentration dependence of the melting temperature is that produced by a divalent ion, and the binding of this ion to the phosphate would stabilize the DNA molecules. The low melting complex II, stable in the acidic region, presumably involves Al^{3+}, hydrated aluminum ions, binding to the bases of the DNA and thereby destabilizing the DNA double helix. The third complex, which occurs at high aluminum concen-

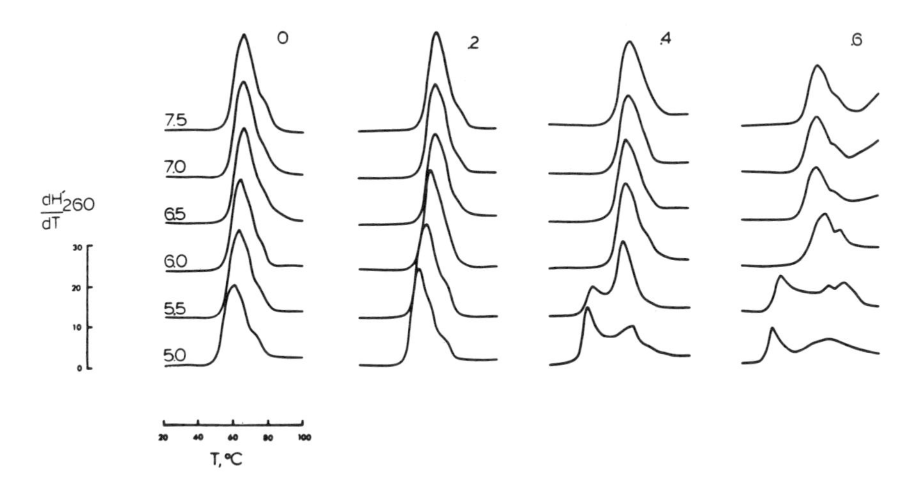

Figure 1. Derivative melting curves of solutions containing 6 × 10^{-5}M DNA (residue), 5 × 10^{-3}M $NONO_3$, and a mole ratio of Al/DNA residue indicated on top of the columns. pH is shown to the left of the curves for DNA without Al.

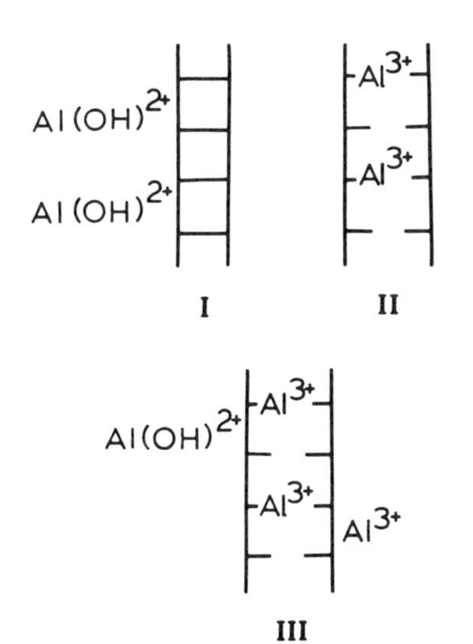

Figure 2. Proposed structures of Al–DNA complexes

trations, probably contains both phosphate binding $Al(OH)^{2+}$ and base binding Al^{3+}.

It has previously been shown that Cu^{2+} ions produce crosslinks between DNA strands (12, 13, 14, 15). Al^{3+} also produces such crosslinks, as demonstrated in the following way. When calf-thymus DNA is heat-denatured, and then cooled, the absorbance does not decrease to the level characteristic of double-helical DNA (Fig. 3A). The double helix is not regenerated because the bases in the denatured state are out of register. The slight decrease in absorbance on cooling is attributed to limited intrastrand hydrogen bonding, or hairpin formation. The low-melting Al-DNA complex, on the other hand, does not even form these hairpins on cooling - the absorbance remains constant (Fig. 3B). However, removal of Al by EDTA or by the introduction of a high electrolyte concentration brings the absorbance back to that of native DNA. The explanation of this reversibility of DNA denaturation is that the aluminum ions crosslink the nucleotides of the DNA strands during the unwinding of these strands; when the solution has been cooled the DNA strands are held together in such a way that it is now impossible to form hairpins, and when the aluminum is then removed with EDTA or with high salt, the double helix is reformed, because the crosslinking Al ions are able to maintain the complementary bases in register. Crosslinking of the DNA strands could of course account for deleterious biological effects, and it is tempting to speculate that defects in brain structure characteristic of Alzheimer's disease could be due to such structures. At this point there is no evidence that such structures exist in diseased brain.

Metal Ions, Genetic Information Transfer and Aging

The possible involvement of aluminum in Alzheimer's disease is of interest in aging because senile dementia is sometimes associated with aging. Metal ions may be involved in the aging process in more general ways, as we shall try to demonstrate.

It is generally accepted that aging is genetically determined. The dependence of longevity on species and sex, for example, cannot be readily explained in any other way. If aging is genetically determined, there must be changes in genetic information transfer, which involves the replication of DNA in the cell nucleus, transcription of the information contained in DNA onto messenger RNA, which moves from the nucleus to the cytoplasm, where its nucleotide sequence is translated into the amino acid sequence of proteins. Many laboratories have demonstrated that age changes do occur in genetic information transfer. Clark and Eichhorn have recently shown that there is an age difference in the accessibility of DNA from the chromatin of old and young rat liver cells to the action of micrococcal nuclease, which splits internucleotide bonds of DNA (11). As already indicated, this is only one of many examples of age changes in genetic information

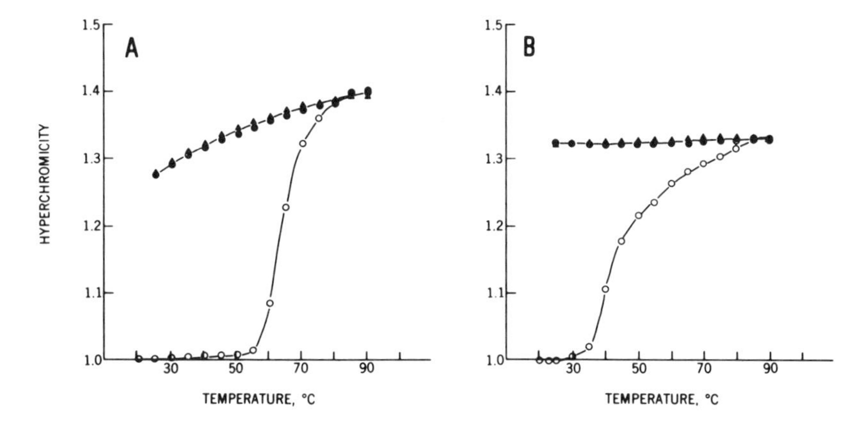

Figure 3. Melting (○) and cooling–reheating (● ▲) curves for 6×10^{-5}M DNA (residue) in 5×10^{-3}M N_2NO_3: (A) pH 6.3, without Al; (B) pH 5.1, 0.6 Al/DNA (residue)

transfer. None of these studies have led to an understanding of the basic cause of aging. It is useful to consider what, if anything could be done about aging, if the basic cause (or causes) of the aging process were ever discovered. Perhaps some form of genetic engineering could become feasible, but genetic engineering is associated with difficult moral problems. If there is an impact from the environment onto genetic information transfer, it could be easier to deal with such an environmental impact, and it would be morally less difficult (11).

Metal ions enter cells of living organisms from the environment. Some of these are essential metal ions and others are nonessential. Metal ions are involved in every step of genetic information transfer. They affect the structure of chromatin; it has been demonstrated by electron microscopy that the concentration of magnesium ions in cell nuclei determines the packing of the chromatin (12). Some studies carried out in our laboratory indicate that metal ions may be involved in age changes in the structure of chromatin (13). Cell nuclei were isolated from the liver of mature (12 mo.) and old (26 mo.) rats, and from the chromatin obtained from these nuclei, the histones were chromatographed on a Sephadex column. Four peaks were produced from mature rat liver chromatin (Fig. 4A); two of these peaks were substantially diminished in the chromatogram from the old rats (Fig. 4B). The nuclei in both instances had been isolated in the presence of magnesium. If the histones from mature rat liver chromatin were obtained from nuclei isolated in the absence of magnesium, or even in the presence of EDTA, the same peaks were diminished as in the case of the material from the old nuclei (Fig. 4C). Thus, the absence of metal ions in the isolation of the nuclei produces a similar affect as aging. It seems that metal ions are involved in the organization of the nuclear matter, and something in this organization changes with age.

As has been indicated above, metal ions are essential in every aspect of genetic information transfer. Nevertheless, metal ions can also cause deleterious effects in information transfer either if they are present in the wrong kind or in the wrong concentration. Let us consider an example of each of these possiblities; first, that in which metals are present in the wrong kind.

In RNA synthesis, the RNA polymerase enzyme must be capable of differentiating between a ribonucleotide and a deoxynucleotide; i.e., it must insert only those nucleotides that have an OH group in the 2'-ribosyl position. One of the following metal ions, Mg^{2+}, Co^{2+}, or Mn^{2+}, is required for the activity of RNA polymerase. Manganese is the most effective for the correct incorporation of the ribonucleotides into RNA. However, manganese is the only one of these three metal ions that causes substantial incorrect introduction of deoxynucleotides into RNA (14, 15). Thus, even though magnesium is less effective than manganese for the correct

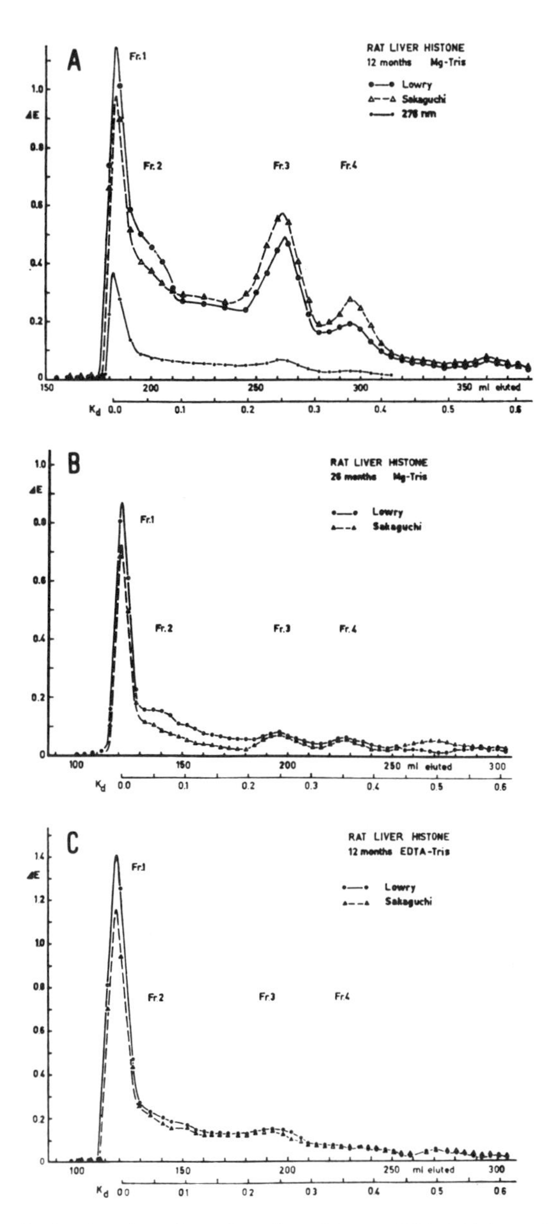

Figure 4. Sephadex fractionation of rat liver histone (13): (A) from 12-month old rat liver nuclei, isolated in the presence of Mg^{2+}; (B) from 26-month old rat liver nuclei, isolated in the presence of Mg^{2+}; (C) from 12-month old rat liver nuclei, isolated in the presence of Mg^{2+}

incorporation of ribonucleotides, magnesium or cobalt are to be preferred because these make fewer mistakes.

Now let us consider the case of essential metal ions that, nevertheless, produce deleterious effects in the wrong concentration. Magnesium ions are required for protein synthesis, yet Mg^{2+} ions in too high concentration lead to errors, as is illustrated by the studies of Szer and Ochoa (16) on the incorporation of phenylalanine and leucine in a ribosomal preparation using poly(U) as the messenger RNA. UUC as well as the UUU codon in poly(U) code for phenylalanine, so that incorporation of the latter represents correct translation. UUA and UUG code for leucine, so that leucine incorporation in this system is "incorrect." At low Mg^{2+} concentration only phenylalanine is in fact incorporated. Phenylalanine incorporation is maximal at 10mM Mg^{2+}; as the Mg^{2+} concentration is increased, however, leucine also becomes incorporated and its maximal incorporation is at 20mM Mg^{2+} (16).

A possible explanation for the correct translation at low Mg^{2+}, followed by error at higher concentration, arises from the stabilization of nucleic acid strand interaction by phosphate-binding metal ions. At low Mg^{2+} concentration, therefore, this interaction is relatively weak, allowing only the most stable hydrogen-bonding, which is the complementary base hydrogen-bonding. Since the recognition of the anticodon on transfer RNA molecules by the codon on messenger RNA molecules is through hydrogen bonding of complementary bases, only these bases will bond, and as a consequence only the correct amino acid will be incorporated. At high Mg^{2+} concentration, on the other hand, strand interaction is so strong that even relatively weak hydrogen bonds can form and result in the mispairing of bases that ultimately leads to errors in the incorporation of amino acids into proteins. We have demonstrated that low magnesium ion concentration does lead to specificity in base pairing, while high Mg^{2+} concentration leads to mispairing (11).

Metal ions can produce a large variety of other effects on nucleic acids that could be deleterious if they occur during genetic information transfer. Metal ions can bring about the degradation of RNA (17, 18, 19), changes in the specificity of enzymes that act on DNA (20), changes in the conformation of polynucleotides and nucleic acid - protein complexes (21).

It is also known that cellular metal ion concentrations change with age. An illustration of such age changes in human lens nuclei is given in Table I (22). We hypothesize that these changing concentrations of metal ions that are essential to genetic information transfer, yet can alter it, can affect information transfer and therefore contribute to the changes that are associated with the aging process.

It is of some interest that the lifespan of rotifers can be considerably lengthened if they are grown in the presence of chelating agents (Fig. 5, 23). There is no indication that this

Table I.

Elemental Analysis of Human Lens Nuclei, μg/g dry wt.

	Age		
	10-20	50-60	70-85
Co	11.7	90.3	61.3
Ir	< .001	.04	.09
La	< .002	.170	.270
Ni	< .005	7.1	10.0
Se	.17	.46	.81
Cu	17.4	2.0	.79
Fe	18.9	1.1	.25
K	11.5	8.1	8.2
Mg	82.4	80.0	34.3
Mn	11.7	15.5	9.4
Zn	17.6	1.5	.02

Taken from Swanson, A. A. and Truesdale, A. W., reference 22.

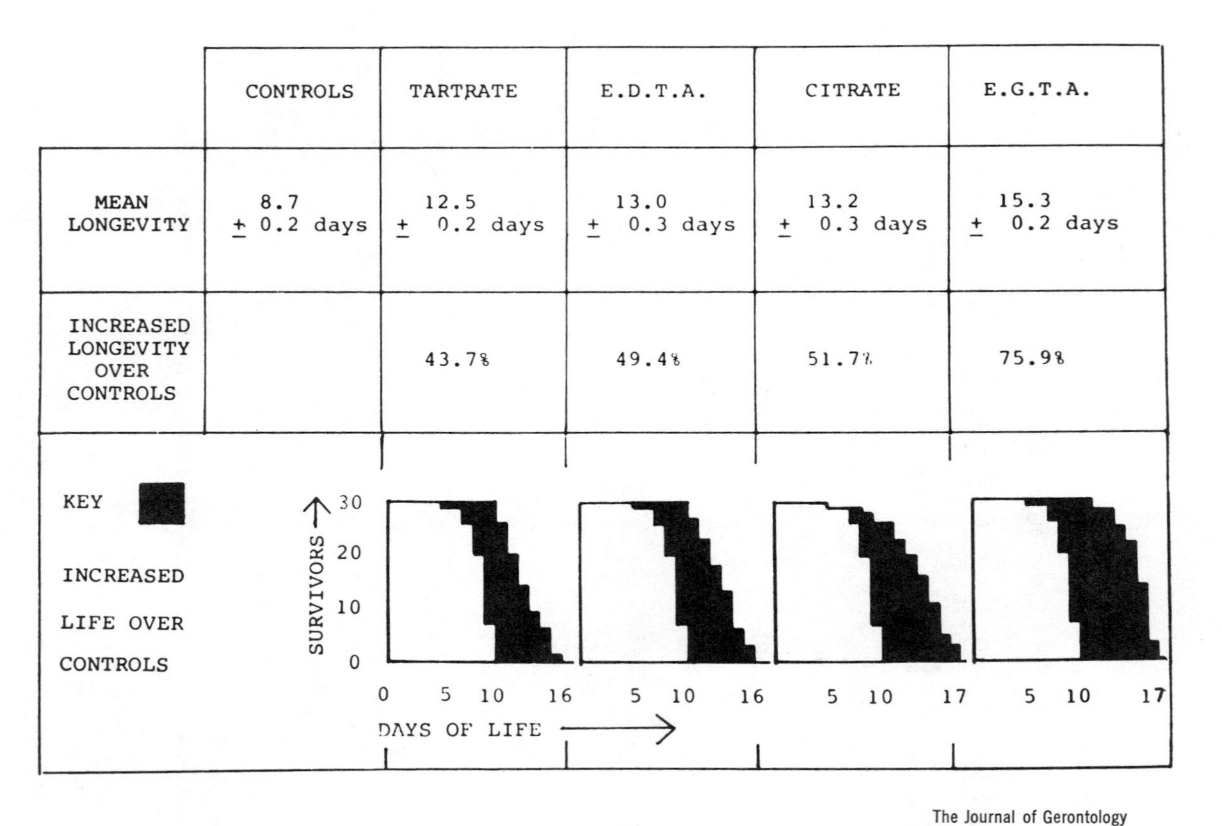

The Journal of Gerontology

Figure 5. Change in lifespan of rotifers treated with chelating agents (23). Control survival on left of each graph, chelate treatment survival on right.

phenomenon has any relation to genetic information transfer, of course, since metal ions have effects on many aspects of biochemistry.

Metal Ions as Probes of Age Changes

Finally, we discuss the use of metal ions in probing cellular age changes, whether or not these changes were induced by metal ions.

We have considered the amino acid misincorporation experiment of Szer and Ochoa (16), which was carried out with *E. coli* ribosomes. Mammalian ribosomes can also be stressed with metals to produce misincorporation; however, higher metal ion concentrations are required to produce the error (24). With 13mM magnesium the leucine/phenylalanine ratio is 63% with *E. coli* ribosomes, but only 25% with rabbit reticulocyte ribosomes. With *E. coli* ribosomes 25% error occurs already at 9mM Mg^{2+}; thus, when the ribosomes are stressed with high concentrations of magnesium, the *E. coli* ribosomes are more readily fooled than the reticulocyte ribosomes. Therefore, the metal ion concentration required to produce a certain level of error is an index of ribosomal fidelity. It occurred to us that age changes in ribosomes could perhaps be detected by challenging the ribosomes from young and old cells with magnesium ions. If the ribosomes deteriorated with age, one might expect that it would require a lower concentration of magnesium ions to cause misincorporation in old ribosomes than in young ribosomes. Figure 6 shows the effect of Mg^{2+} on poly(U) directed phenylalanine and leucine incorporation into protein with ribosomes from young and old rat liver. It is evident that phenylalanine incorporation is maximal at a lower magnesium ion concentration than leucine incorporation. Again the correct incorporation of phenylalanine occurs at a lower magnesium concentration than the incorrect incorporation of leucine. There is very little difference between the ability of the ribosomes of old and young cells to withstand the stress of high concentrations of magnesium ions which tend to prevent correct protein synthesis. Nevertheless, this experiment reveals the way in which metal ion effects can be used as probes in the study of aging.

Conclusion

Aluminum appears to be involved in Alzheimer's disease and concentrates in the nucleus; therefore, we have looked at aluminum binding to DNA, and we find that aluminum forms crosslinks with the DNA strands. Metal ions generally produce deleterious effects in genetic information transfer when present in the wrong kind or concentration. They are, however, essential in genetic information transfer. The concentration of metal ions in various cells changes with age, and we have hypothesized that perhaps these metal ions have an impact on genetic information transfer processes

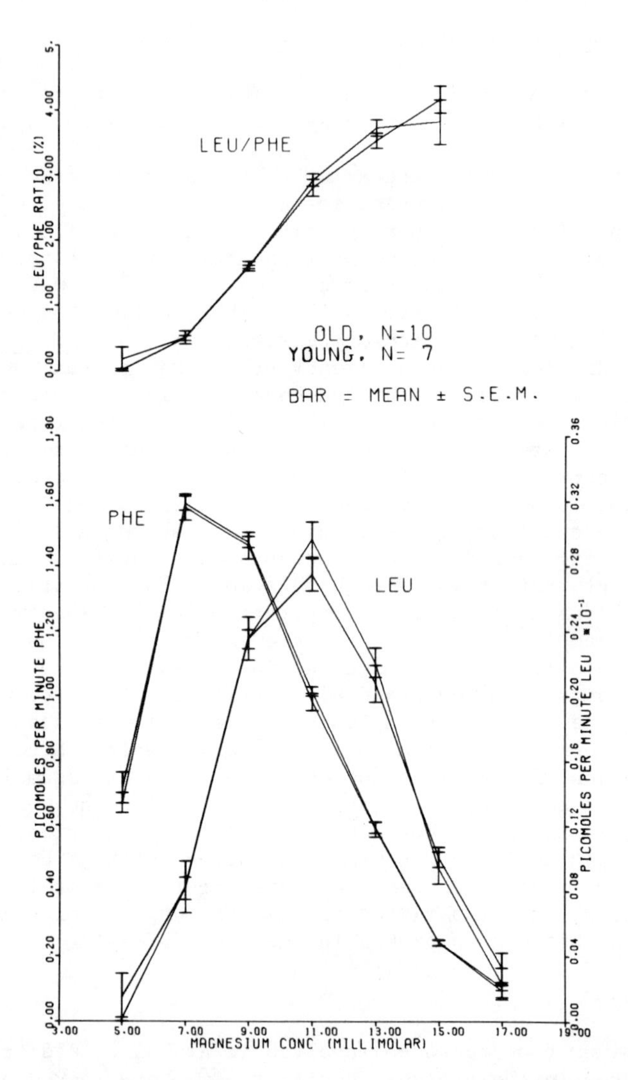

Figure 6. Poly(U) directed protein synthesis with ribosomes from 6–10 and 23–24 month-old rat livers, as a function of Mg^{2+} concentration: Phe, phenylalanine or "correct" incorporation; Leu, leucine or "incorrect" incorporation; Leu/Phe, "incorrect"/"correct" incorporation. Curves for young and old are virtually identical.

which are involved in aging. Finally, metal ions can be used in probing aging processes.

Abstract

The aging of organisms is accompanied by dramatic changes in the cellular concentration of metal ions. Much evidence indicates that aging is genetically determined. Since genetic information transfer processes require metal ions in suitable concentrations, and since excess metal ions produce errors in information transfer, it is reasonable to expect that the age changes in metal concentrations affect the aging process. Metal ions can induce the mispairing of nucleotide bases and the incorporation of the wrong amino acids into proteins; this phenomenon has been used to probe old and young rat liver ribosomes to determine whether any changes in fidelity of translation have taken place. Evidence has been obtained for an age change in chromatin structure that is associated with metal binding. There is good evidence that localized accumulations of aluminum in the brain accompany Alzheimer's disease, the most prevalent form of senile dementia (2). Aluminum concentrates in the cellular chromatin, and in a pH-dependent reaction can either stabilize or destabilize the DNA double helix. Stabilization appears to result from the binding of $Al(OH)^{2+}$ to DNA, and destabilization from the binding of Al^{3+}. The latter forms crosslinks between DNA strands; crosslink formation can be reversed by sequestering the aluminum with EDTA.

Literature Cited

1. Crapper, D. R., Krishnan, S. S. and Quittkat, S., Brain, 1976, 99, 67.
2. Crapper, D. R., Krishnan, S. S. and Dalton, A. V., Science, 1973, 180, 511.
3. DeBoni, U., Scott, J. W. and Crapper, D. R., Histochemistry, 1974, 40, 31.
4. Eichhorn, G. L., "Inorganic Biochemistry", Elsevier, Amsterdam, 1973; pp. 1212-1217.
5. Eichhorn, G. L., Nature, 1962, 194, 474.
6. Baes, Jr., C. F. and Mesmer, R. E., "The Hydrolysis of Cations", New York, Wiley, 1976; pp. 112-123.
7. Eichhorn, G. L. and Clark, P., Proc. Natl. Acad. Sci. USA, 1965, 53, 586.
8. Zimmer, C. and Venner, H., Studia Biophysica, 1967, 207.
9. Hiai, S., J. Mol. Biol., 1965, 11, 672.
10. Coates, J. H., Jordan, D. O. and Srivastava, V. K., Biochem. Biophys. Res. Commun., 1965, 20, 611.
11. Eichhorn, G. L., Mech. Ageing Dev., 1979, 9, 291.
12. Monneron, A. and Moulé, Y., Exp. Cell Res., 1968, 51, 531.
13. von Hahn, H. P., Miller, J. and Eichhorn, G. L., Gerontologia, 1969, 15, 293.

14. Hurwitz, J., Yarbrough, L. and Wickner, S., Biochem. Biophys. Res. Commun., 1972, 48, 628.
15. Rao, K. G., Su, F. Y. U., Sethi, S. and Eichhorn, G. L., Abst. 174th Nat. Meet. Am. Chem. Soc., Chicago, (1977).
16. Szer, W. and Ochoa, S., J. Mol. Biol., 1964, 8, 823.
17. Butzow, J. J. and Eichhorn, G. L., Biopolymers, 1965, 3, 95.
18. Butzow, J. J. and Eichhorn, G. L., Biochemistry, 1971, 10, 2019.
19. Butzow, J. J. and Eichhorn, G. L., Nature, 1975, 254, 358.
20. Clark, P. and Eichhorn, G. L., Biochemistry, 1974, 13, 5098.
21. Shin, Y. A. and Eichhorn, G. L., Biopolymers, 1977, 16, 225.
22. Swanson, A. A. and Truesdale, A. W., Biochem. Biophys. Res. Commun., 1971, 45, 1488.
23. Sincock, A. M., J. Gerontol., 1975, 30, 289.
24. Friedman, S. M., Berezney, R. and Weinstein, I. B., J. Biol. Chem., 1968, 243, 5044.

RECEIVED July 14, 1980.

RADIOPHARMACEUTICALS FOR DIAGNOSIS AND TREATMENT

5

Technetium-99*m* Radiopharmaceuticals

Uses, Problems, Potential Applications in Nuclear Medicine

LUIGI G. MARZILLI[1], ROBERT F. DANNALS, and H. DONALD BURNS

Department of Chemistry and Division of Nuclear Medicine,
The Johns Hopkins University, Baltimore, MD 21218

The object of this review is to introduce inorganic chemists to the subject of radiopharmaceuticals. Radiopharmaceuticals are compounds containing radioactive nuclides which are useful in the diagnosis or treatment of certain disease states. Since most research is presently focused on diagnosis, the chemical aspects of diagnostic radiopharmaceuticals will be treated here. In addition, radiopharmaceuticals can be used either *in vitro* or *in vivo* and only the latter type will be discussed.

After *in vivo* administration of a given radiopharmaceutical, the distribution can be determined through detection of its emitted radiation. This involves quantification and localization of the spatial distribution of the radioactivity introduced into the human body by specialized detection systems. The pharmacological, chemical and biological properties of the radiopharmaceutical influence its tissue or organ specificity and therefore its diagnostic usefulness.

Two principal objectives in diagnostic nuclear medicine are the evaluation of structure and function. For example, a structural study of the liver could reveal the presence of tumors or abscesses and a functional study of blood flow can indicate abnormalities in heart function. Since disease frequently can be defined at a molecular level, radiopharmaceuticals can play an important role in defining the nature of disease and repeated use on a given patient might be beneficial in monitoring the course of treatment. Consequently, the field of nuclear medicine is currently experiencing remarkable growth and development.

This review concentrates on ^{99m}Tc radiopharmaceuticals. Radiopharmaceuticals containing other inorganic species are treated in a separate chapter. The chemistry and structure of technetium compounds are described in two later chapters. Radiopharmaceuticals containing ^{99m}Tc are currently the most widely

[1] Current address: Department of Chemistry, Emory University, Atlanta, Georgia 30322.

0-8412-0588-4/80/47-140-091$05.00/0

used diagnostic studies of bone, brain, liver, lung, renal function, cardiac function and hepatobiliary function. Radiopharmaceuticals based on ^{99m}Tc hold such a prominent position in nuclear medicine for many reasons. ^{99m}Tc is readily available in generator form as $Na^{99m}TcO_4$ (^{99}Mo on an alumina column which decays with a half-life of 2.7 days to ^{99m}Tc is shipped to hospitals where $^{99m}TcO_4^-$ can be eluted daily for one week). The ^{99m}Tc half-life of 6 hours results in a patient receiving a very low internal dose of radiation. The absence of beta radiation and low gamma emission (140 keV) further accentuates this point. The particular advantages of ^{99m}Tc, in addition to its chemical properties which make it useful for incorporation in various radiopharmaceuticals, are therefore also inherent in its physical properties and versatile chemistry.

Technetium Radiopharmaceuticals

The ^{99m}Tc-radiopharmaceuticals which have been developed to date can be categorized as belonging to two broad classes. The first class includes ^{99m}Tc-tagged radiopharmaceuticals. In these agents, the ^{99m}Tc functions as a label since the normal biodistributions are essentially unchanged by the presence of the radionuclide, i.e. the tracer principle is obeyed. The second class, ^{99}Tc-essential radiopharmaceuticals, includes those agents whose biodistribution is, to a large extent, dependent on properties of the complex, as dictated, in large measure, by the technetium itself(1).

^{99m}Tc-tagged radiopharmaceuticals are mainly large substances (e.g. proteins, particles and cells) in which binding of ^{99m}Tc results in a relatively minor perturbation of the overall biological, chemical and physical properties of the substance. Colloids, which are members of one subgroup of this class (Table I), perhaps best exemplify this class. When colloids are injected intravenously, they are rapidly removed from the circulation by the reticuloendothelial system with approximately 80% of the colloidal material localizing in the liver. ^{99m}Tc-sulfur colloid is a clinically useful agent for the evaluation of liver disease. This colloid, prepared by any of several methods(2,3) localizes in the liver regardless of whether or not ^{99m}Tc is bound to it. A typical ^{99m}Tc-colloid scan is shown in Figure 1.

Other members of the ^{99m}Tc-tagged class of radiopharmaceuticals are listed in Table I. Note that with the exception of the bone imaging agents, all members of this class are composed of large substances. The placement of the bone agents in the class of tagged radiopharmaceuticals is somewhat questionable. It has been known for a long time that phosphate and phosphonates localize in the bone(4). This observation was the original basis for labeling these molecules with ^{99m}Tc(5). The finding that the biodistributions of ^{99m}Tc-phosphate and phosphate itself are similar suggests that ^{99m}Tc is, in this case, functioning as a

TABLE I

^{99m}Tc-Radiopharmaceuticals

Technetium-Tagged Radiopharmaceuticals

1. Particles and colloids

   Tc-macroaggregated albumin, Tc-albumin microspheres, Tc-albumin minimicrospheres, Tc-ferric hydroxide aggregates, Tc-sulfur colloid, Tc-antimony colloid, Tc-phytate

2. Proteins

   Albumin, streptokinase, urokinase, fibrinogen

3. Cells

   Erythrocytes, leukocytes, platelets, lymphocytes

4. Small molecules

   Bone agents, e.g., polyphosphates, pyrophosphates, diphosphonates, iminodiphosphonates

Technetium Essential Radiopharmaceuticals

1. Kidney function agents

   Tc-DTPA, Tc-EDTA, Tc-MIDA (methyliminodiacetic acid), Tc-citrate

2. Kidney structure agents

   Tc-gluconate, Tc-glucoheptonate, Tc-Fe-ascorbate, Tc-inulin, Tc-mannitol, Tc-dimercaptosuccinic acid

3. Infarct avid agents

   Tc-pyrophosphate, Tc-glucoheptonate, Tc-tetracycline and Tc-HEDP (hydroxy-ethylidene) diphosphonate

4. Hepatobiliary agents

   Tc-dihydrothioctic acid, Tc-HIDA, Tc-isomercaptobutyric acid and Tc-pyridoxylideneglutamate

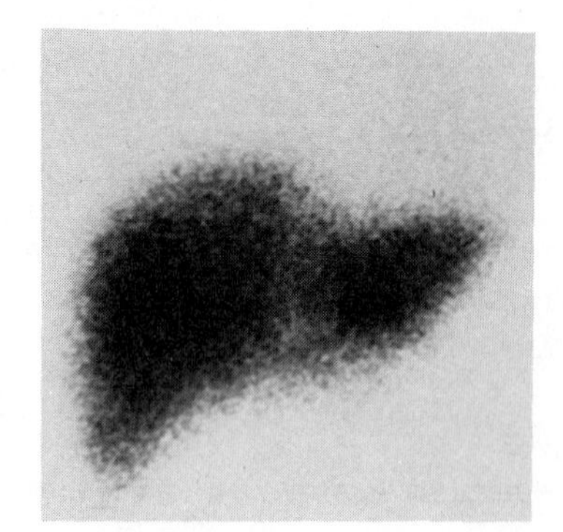

Figure 1. A ^{99m}Tc–sulfur colloid scan of a normal human liver

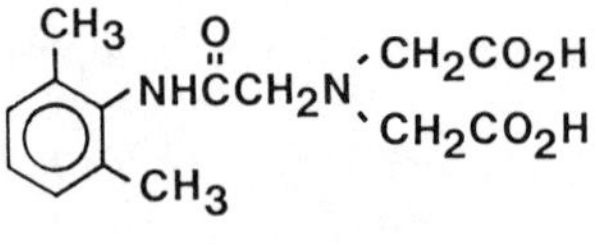

Figure 2. The structure of HIDA (N-(2,6-dimethylphenylcarbamoylmethyl)iminodiacetic acid)

tag for the phosphate molecule. However, the finding that the biodistributions of ^{99m}Tc-HEDP and ^{14}C-HEDP are somewhat different coupled with several reports that complexes in this subgroup are reactive, undergoing rapid exchange with thiol and carboxylate ligands, suggests that the localization in the bone may be related to the reactivity of the complex and that ^{99m}Tc is not functioning as a tag for the phosphate and phosphonate ions(6). Since the evidence supporting this mechanism is only speculative at this point and the similarities in distribution of phosphate and ^{99m}Tc-phosphate are clear, we somewhat arbitrarily assign the bone agents to the ^{99m}Tc-tagged class, until the true behavior of these compounds is established.

The second class, ^{99m}Tc-essential radiopharmaceuticals, contains those ^{99m}Tc-labeled substances for which the Tc atom itself greatly influences the overall physical and chemical properties of the labeled substance. The biodistribution of the ^{99m}Tc-complex of these substances is often markedly different than the biodistribution of the substance when it is not labeled with ^{99m}Tc. The best understood example of this class of compounds is ^{99m}Tc-$(HIDA)_2$.

The structure of HIDA is shown in Figure 2. This compound was originally synthesized by Loberg(7) as a lidocaine analog capable of complexing ^{99m}Tc. When the ^{99m}Tc-complex of HIDA was prepared and its biodistribution determined in mice, it was found to rapidly leave the circulation and accumulate in the liver. From there, ^{99m}Tc$(HIDA)_2$ was excreted into the bile and eventually into the intestines. Loberg also prepared ^{14}C-labeled HIDA and determined its biodistribution. In marked contrast to ^{99m}Tc$(HIDA)_2$, ^{14}C-HIDA is excreted rapidly and almost exclusively by the kidneys(7). The difference in distribution between HIDA and the ^{99m}Tc-complex of HIDA demonstrates that ^{99m}Tc is not functioning as a tag for HIDA and illustrates some of the difficulties encountered when trying to design ^{99m}Tc-radiopharmaceuticals based on small molecules (i.e. Molecular weight < 1000).

Other members of this class of radiopharmaceuticals are listed in Table I. Although the biodistribution of some of these ^{99m}Tc-complexes could be predicted *a priori*, based on a knowledge of the fate of other metal complexes of the same ligands, it is generally difficult to predict the biodistribution of ^{99m}Tc-complexes when one has only a knowledge of the *in vivo* fate of the uncomplexed ligand.

Development of New ^{99m}Tc Radiopharmaceuticals

Although the inorganic chemistry of technetium will undoubtedly yield many new and interesting compounds. The current growing interest in technetium undoubtedly arises from the prospects of preparing either improved diagnostic agents for current uses or in expanding the uses to which Tc compounds can be put. Useful new applications would include detecting tumors and evaluation of cardiac damage with a "cold spot" agent (i.e. one which accumulates in normal, not damaged, tissue).

There are two fundamental areas where knowledge must be improved before greater studies can be expected in the development of new agents.

The first area is the improvement of our knowledge of technetium chemistry. Recent advances in this direction have been made and the future is quite bright(8,9,10). As indicated above, discussions of Tc chemistry appear elsewhere in this volume.

The second area is the improvement of our knowledge of how to predict the biodistribution of metal containing species. Greater insight into this problem would have benefits even beyond radiopharmaceuticals development; it would have application to trace element metabolism, treatment of heavy metal poisoning and to the development of metallodrugs such as new antitumor platinum drugs. A corollary of this problem area is the need to understand how we might introduce a metal species into a small organic molecule which has a desirable biodistribution without significantly modifying that biodistribution.

Factors Influencing the Biodistributions of Small Molecules

Although many new developments can be expected in the area of new radiopharmaceuticals based on larger substances such as proteins, cells, etc., we will confine the remainder of our discussion to the design of small molecules (molecular weight < 1000) for use as radiopharmaceuticals. In attempting to design radiopharmaceuticals of this type, it is useful to consider the general information which is known about factors which influence the biodistribution of small molecules.

When a small molecule is introduced into the circulation, it may attain either an intravascular, extracellular or intracellular volume of distribution(11). The capillary walls are porous and are generally freely permeable to small molecules. Almost all compounds used as drugs are small enough to pass freely through pores in the capillary membrane, thus, in the absence of protein binding (which will be discussed below) small molecules should attain at least an extracellular volume of distribution. The membrane separating the extracellular space of the brain from the intravascular space is somewhat less permeable and is generally considered to be comparable to cellular membranes (with respect to permeability). Thus, many molecules such as pertechnetate ion which are distributed in the extracellular fluid, are unable to penetrate the blood:brain barrier. Penetration of cellular membranes is also related to the size of the molecule, however, in this case the pore size is much smaller, being permeable to only simple ions, such as Na^+, Cl^-, etc. Due to the small size of the pores in cell membranes, it is unlikely that any Tc-complex will be able to diffuse through the pore. Thus, most of these complexes will be limited to an extracellular volume of distribution and will be unable to penetrate the blood: brain barrier via pores.

An alternative to diffusion through pores is passive diffusion through the cell membrane itself. This generally requires that the molecule be uncharged and lipid soluble. Thus, a lipid soluble compound (provided it is not highly protein bound) should attain an intracellular volume of distribution, as a result of passive diffusion through cell membranes. Small molecules which are too large to diffuse through cellular pores and too insoluble in lipids to permit passive diffusion through the cell membrane, may still attain an intracellular volume of distribution as a result of carrier mediated (energy independent) or active (energy dependent) transport. Currently, there are two types of Tc-compounds which are actively taken up by cells, these are:

1) Pertechnetate, which is actively transported (as an iodide analog) into the thyroid.

2) Hepatobiliary agents, such as $^{99m}Tc(HIDA)_2$ which is actively accumulated by hepatocytes.

Many attempts have been made to design ^{99m}Tc-labeled compounds which would be actively accumulated by other than excretory organs, but to date, none of these attempts have been successful.

Molecules which are strongly bound to serum proteins may be limited to an intravascular volume of distribution even if they are small enough to diffuse through membrane pores or lipid soluble enough to diffuse directly through the lipoidal membrane itself. In some cases, protein binding will not completely eliminate distribution in the extracellular and intravascular compartments since it may only delay diffusion into these compartments.

In designing new radiopharmaceuticals, it is essential to consider the effect that molecular size, charge, lipid solubility and protein binding will play in determining the volume of distribution of the radiopharmaceutical. Although it is possible to predict the effect that these factors will have on the biodistribution of ^{99m}Tc-complexes of known structure, charge, lipid solubility, etc., there is still very little known about the relationship of structure to the biodistribution for metal complexes in general or for ^{99m}Tc-compounds in particular. One obvious reason for this is the lack of information on the structures of ^{99m}Tc-radiopharmaceuticals.

As mentioned above, based on work done by our group and by Loberg's group, the structure of $^{99m}Tc\text{-}(HIDA)_2$ can reasonably be represented as shown in Figure 3. The complex contains 2 HIDA ligands(12,13), Tc is in the +3 oxidation state(13) and the complex possesses a net charge of -1(12). While studying the biodistribution of a series of HIDA analogs in mice, we observed a clear relationship between the structure of the complex and the amount of the complex excreted by the biliary system. The total amount of the complex excreted in the bile was found to be directly proportional to the natural log of the molecular weight to charge ratio(12). This finding represents the only known structure:biodistribution study for a series of ^{99m}Tc-complexes. With the rapid accumulation of information on the structure of ^{99m}Tc compounds which is now occurring, it is likely that other structure:biodistribution relationships will be observed. Information such as this should be valuable for the design of new ^{99m}Tc-radiopharmaceuticals.

Design of Bifunctional Tc-radiopharmaceuticals

One approach to the design of new ^{99m}Tc-radiopharmaceuticals which has received considerable attention over the past several years is the design of bifunctional radiopharmaceuticals. In this approach, a bifunctional molecule is synthesized which possesses a chelating functional group (capable of forming a stable complex with ^{99m}Tc) attached to a second functional group which is expected to have a useful biological distribution. The hope is that the functional group which has a useful biodistribution will dictate the biodistribution of ^{99m}Tc-complex. For example, based on work done by Jensen(14), it is known that estradiol, and other compounds which bind with high affinities to estradiol receptors, accumulate in cells which possess estradiol receptors. A ^{99m}Tc-labeled estradiol analog would be useful if the distribution of the Tc-labeled compound was related to the distribution of estradiol receptors. The first step would be the synthesis of an estradiol analog which possessed an appropriate chelating functional group as illustrated in Figure 4. The ^{99m}Tc-complex of this molecule would then be evaluated for localization in tissues possessing estradiol receptors.

For this approach to be successful, it is essential to consider all of the factors discussed above. The initial problem is the choice of a chelating functional group to be incorporated into the molecule. All the important features required of that functional group are not known. However, it is known that the complex must be stable *in vivo*, i.e. the Tc must remain bound to the estradiol analog, at least until the analog reaches and binds to the estrogen receptor. The estrogen receptor is located intracellularly, therefore, it is essential that the Tc-estradiol analog enter the cells which possess the receptor. Since there is no active transport or carrier mediated transport system to facilitate entry of estrogens into cells, and since the Tc-estradiol analog is much too large to diffuse through pores in the cell membrane, it is essential that the complex possess sufficient lipid solubility to permit passive diffusion through the cell membrane.

Equally important is the high affinity binding of the Tc-estradiol analog to the estradiol receptor. The functional group chosen must not interfere with this binding and must be attached to the estradiol molecule at a point which does not interfere with binding to the receptor. The ability of the Tc-compound to

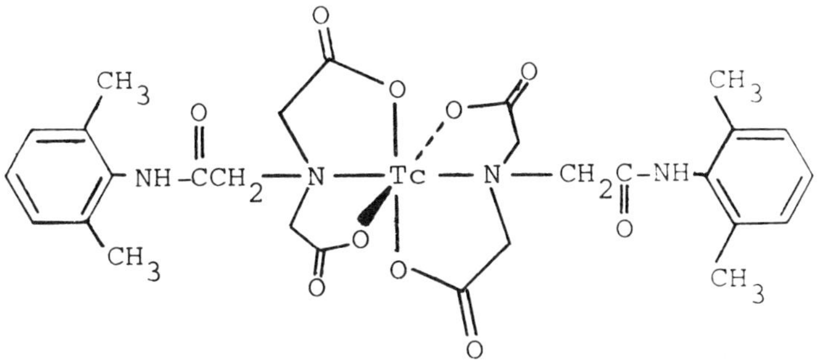

Figure 3. Proposed structure of hepatobiliary agent $Tc(HIDA)_2$

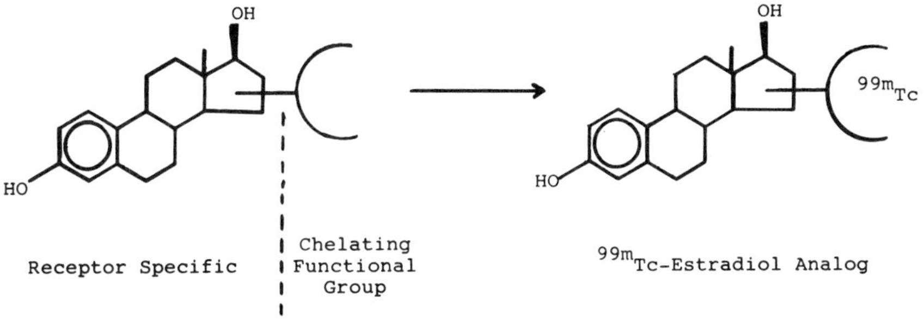

Receptor Specific

Chelating Functional Group

^{99m}Tc-Estradiol Analog

Figure 4. Conceptual representation of a bifunctional ^{99m}Tc radiopharmaceutical for imaging estradiol receptors

bind to the estrogen receptor can be determined *in vitro* by methods described by Eckelman(15). Even if all of these above conditions are met, localization at the receptor could be prevented by excessive plasma protein binding or by extremely rapid removal of this foreign material from the circulation by the excretory organs.

Attempts to design ^{99m}Tc-radiopharmaceuticals of this type in the past have all failed. The past failures were due, at least in part, to the lack of information on the chemistry of Tc. The availability of new knowledge on the chemistry of Tc (such as that in the chapters by Davison and Deutsch) greatly increases the likelihood that new bifunctional Tc-radiopharmaceuticals will be developed.

Abstract

Transition metal complexes containing radioactive isotopes have found widespread use as radiopharmaceuticals in diagnostic nuclear medicine. The most useful transition metal nuclide is technetium-99m and there are many examples of the successful applications of complexes of Tc-99m in both structural and functional studies of patients. Limitations on the design of additional more useful and better technetium radiopharmaceuticals include a) a very incomplete knowledge of the chemistry and structure of technetium compounds and b) inadequate information on the factors which influence the biodistribution and interaction with biomolecules, such as receptors, of transition metal complexes. These factors are discussed in terms of the need for additional metalloradiopharmaceuticals.

Acknowledgments

This work was supported in part by USPHS Grant GM 10548 and by Grant HL 17655.

Literature Cited

1. Burns, H.D., Design of Radiopharmaceuticals; Heindel, N.D.; Burns, H.D.; Honda, T.; Brady, L.W., Eds. "The Chemistry of Radiopharmaceuticals"; Masson Publishing USA, Inc.: New York, 1978.
2. Harper, P.V.; Lathrup, K.; Richards, P. *J. Nucl. Med.*, 1964, *5*, 382.
3. Stern, H.S.; McAfee, J.G.; Subramanian, G. *J. Nucl. Med.*, 1966, *7*, 665.
4. Fels, I.G.; Kaplan, E.; Greco, J.; Veatch, R. *Proc. Soc. Exp. Biol. Med.*, 1959, *100*, 53.
5. Subramanian, G.; McAfee, J.G. *Radiology*, 1966, *99*, 192.
6. Kaye, M.; Silverton, S.; Rosenthall, L. *J. Nucl. Med.*, 1975, *16*, 40.

7. Callery, P.S.; Faith, W.F.; Loberg, M.D.; Fields, A.T.; Harvey, E.B.; Copper, M.D. J. Med. Chem., 1976, 19, 962.
8. DePamphilis, B.V.; Jones, A.G.; Davis, M.A.; Davison, A. J. Am. Chem. Soc., 1978, 100, 5570
9. Smith, J.E.; Byrne, E.F.; Cotton, F.A.; Sekutowski, J.C. J. Am. Chem. Soc., 1978, 100, 5571.
10. Deutsch, E.; Elder, R.C.; Lange, B.A.; Vaal, M.J.; Lay, D.G. Proc. Natl. Acad. Sci. USA, 1976, 738, 4287.
11. LaDu, B.N.; Mandel, H.G.; Way, E.L., Eds. "Fundamentals of Drug Metabolism and Drug Disposition"; The Williams and Wilkins Company: Baltimore, 1971.
12. Burns, H.D.; Worley, P.; Wagner, H.N., Jr.; Marzilli, L.G.; Risch, V.R. Design of Technetium Radiopharmaceuticals. Heindel, N.D.; Burns, H.D.; Honda, T.; Brady, L.W., Eds. "The Chemistry of Radiopharmaceuticals"; Masson Publishing USA, Inc.: New York, 1978.
13. Loberg, M.D.; Fields, A.T. Int. J. Appl. Radiat. Isotopes, 1978, 29, 167.
14. Jensen, E.V.; Jacobson, H.I. Recent Progress in Horm. Res., 1962, 18, 387.
15. Eckelman, W.C.; Reba, R.C.; Gibson, R.E.; Rzeszotarski, W.J.; Vieras, F.; Mazaitis, J.K.; Francis, B. J. Nucl. Med., 1979, 20, 350.

RECEIVED May 12, 1980.

6

Synthetic and Structural Aspects of Technetium Chemistry as Related to Nuclear Medicine

EDWARD DEUTSCH
Department of Chemistry, University of Cincinnati, Cincinnati, OH 45221

B. L. BARNETT
Miami Valley Laboratories, Procter & Gamble Company, Cincinnati, OH 45239

The use and importance of technetium-99m in nuclear medicine has been noted many times (1-4), and is further discussed by Marzilli et al. in this symposium (5). However, realization of the full potential of technetium-99m for diagnostic imaging of internal organs will require a much more extensive and detailed knowledge of technetium chemistry than is now available (1-5). This review covers some recent developments in the synthetic and structural aspects of technetium chemistry that may be relevant to the preparation, use, and understanding of the mode of action, of technetium radiopharmaceuticals.

Synthesis

Current Radiopharmaceutical Synthesis. The aqueous chemistry of technetium is dominated by the oxidizing power of soluble TcO_4^- and the thermodynamic stability of insoluble TcO_2. All technetium-99m radiopharmaceuticals, except pertechnetate itself, are prepared by the aqueous reduction of pertechnetate in the presence of a potential ligand to prevent TcO_2 deposition (2). The most commonly employed reductant is stannous chloride, although many other reductants can, and have, been used (1,2).

$$TcO_4^- + L + \text{excess reductant} \rightarrow \text{Tc–L complex} \qquad (1)$$

While widely used, this procedure is subject to several difficulties and limitations. (a) It allows introduction of only one type of ligand, or one distribution of ligands, into the technetium radiopharmaceutical. (b) It does not allow specific control over the final oxidation state, coordination number, coordination geometry, etc. of the technetium in the Tc-L product. (c) The reductant, especially Sn, is often incorporated into the final product (1,6). (d) The excess reductant is injected into the patient; tin(II) has a long biological half-life (7) and causes several deleterious side effects (8).

In addition, it is very unlikely for the general redox procedure described by eq. 1 to yield a single, well-defined product complex. Since

0-8412-0588-4/80/47-140-103$05.00/0

these preparations are conducted in dilute aqueous solution, in most cases the product radiopharmaceutical will consist of a distribution of Tc-L complexes (in addition to the possible contaminants TcO_2 and TcO_4^-). Figure 1 shows an HPLC chromatogram (9) resulting from separation of $Tc(NaBH_4)$-HEDP, a mixture obtained by $NaBH_4$ reduction of TcO_4^- in the presence of (1-hydroxyethylidene)diphosphonate (HEDP). It is clear from Figure 1 that $Tc(NaBH_4)$-HEDP is not a single species, but rather is a complicated mixture of HEDP-Tc complexes.

Synthesis by Substitution Routes. Many of the limitations inherent in the redox route described above can be avoided by preparation of technetium-99m radiopharmaceuticals by a substitution route, i.e. the classical substitution of ligands onto a pre-reduced and isolated technetium center. Substitution routes allow control over the oxidation state and ligand environment of the technetium product, and permit the synthesis of complexes containing different ligands. By substitution routes it should be possible to prepare series of complexes in which some ligands are held fixed while others are varied in a systematic fashion to affect biological specificity.

Recent work has focused on the use of two specific reduced technetium centers as substrates for substitution reactions: TcX_6^{2-} and $TcOX_4^-$ (X = Cl, Br). The chemistry of the $TcOX_4^-$ system has been developed principally by Davison and co-workers (10). Both of these centers are synthesized from pertechnetate, the starting material for all radiopharmaceutical preparations (2,5), by simple HX reduction: e.g.,

$$TcO_4^- + 9H^+ + 9Br^- \xrightarrow{60^\circ C} Tc^{IV}Br_6^{2-} + 4H_2O + 1.5Br_2 \qquad (2)$$

$$TcO_4^- + 6H^+ + 6Br^- \xrightarrow{0^\circ C} Tc^{V}OBr_4^- + 3H_2O + Br_2 \qquad (3)$$

The only difference between the two preparations is the temperature at which the reduction is conducted; at low temperatures the Tc(V) species $TcOX_4^-$ is kinetically trapped and can be isolated, whereas at higher temperatures the Tc(V) complex suffers further reduction to yield the Tc(IV) species TcX_6^{2-} (10,11). Other potential substances for radiopharmaceutical synthesis by substitution reactions include the undefined, reduced Tc-glucoheptonate complex (12) and the recently reported, lipophilic technetium(V) species $Tc(HBPz_3)Cl_2O$ ($HBPz_3^-$ = hydrotris(1-pyrazolyl)borato ligand) (13).

By substituting HEDP onto $TcBr_6^{2-}$ we have recently been able to generate a radiopharmaceutical with biological properties very similar to those of radiopharmaceuticals prepared by the normal redox route (14). The material prepared by substitution is designated Tc-HEDP, while those prepared by $NaBH_4$ and Sn(II) reduction of pertechnetate in the presence of HEDP are designated $Tc(NaBH_4)$-HEDP and Tc(Sn)-HEDP respectively. Figure 2 compares the biodistributions of these three agents in rats, while Figure 3 compares images of beagle dogs obtained using these agents. It is clear from these figures that all three preparations yield excellent skeletal imaging agents, thus demonstrating that synthesis of technetium radiopharmaceuticals by a substitution route is practicable. This conclusion is

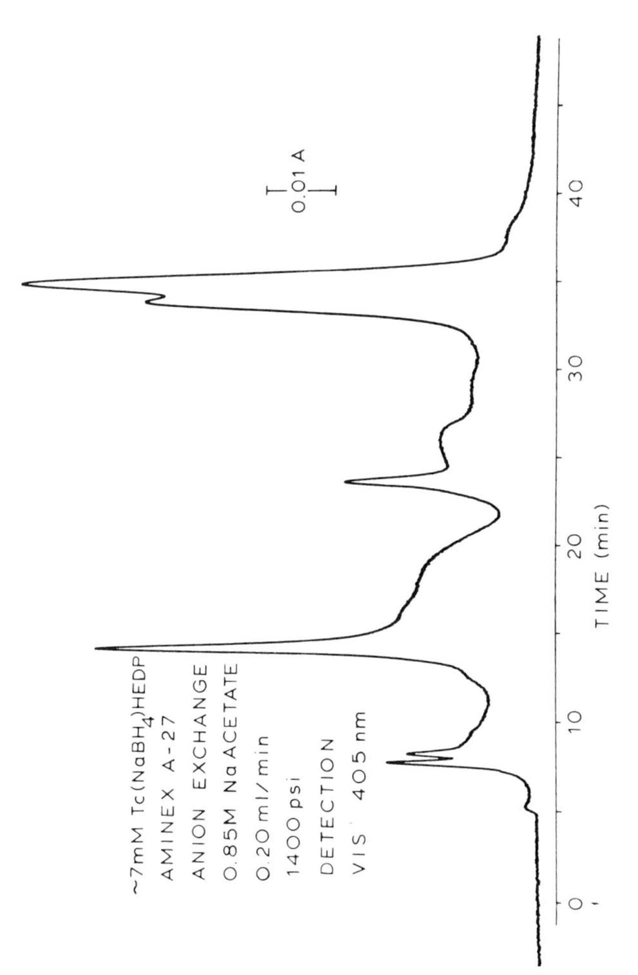

Figure 1. HPLC separation of aqueous $^{99}Tc(NaBH_4)$–HEDP mixture

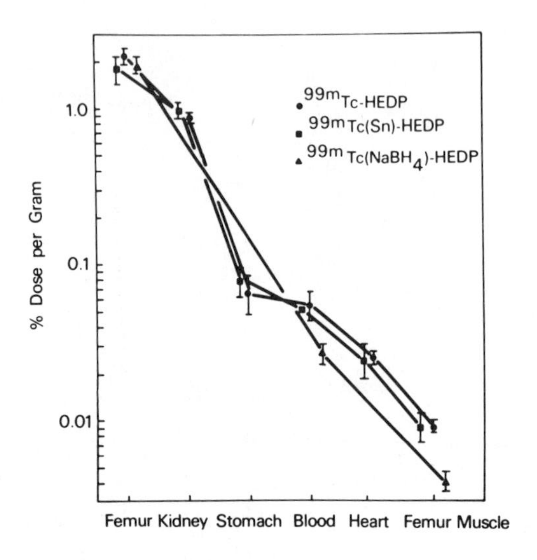

Figure 2. Comparative biodistributions of ^{99m}Tc–HEDP, ^{99m}Tc(Sn)–HEDP, and ^{99m}Tc($NaBH_4$)–HEDP in Sprague Dawley rats at 3 h post iv dose. Each point represents the average of 5 determinations.

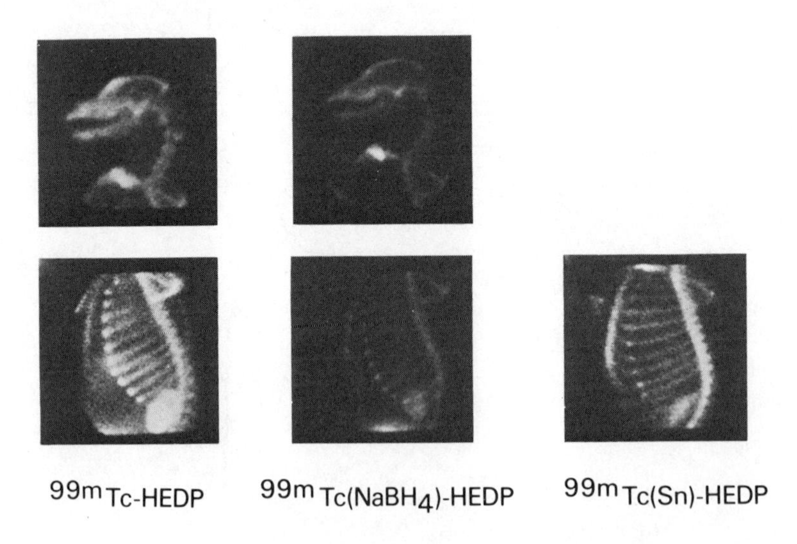

Figure 3. Comparative scintiphotos of beagle dogs imaged with ^{99m}Tc–HEDP, ^{99m}Tc($NaBH_4$)–HEDP, and ^{99m}Tc(Sn)–HEDP at 3 h post iv dose

supported by a recent report (15) in which an efficacious ^{99m}Tc-dimercaptosuccinic acid kidney imaging agent was prepared by ligand substitution onto $TcBr_6^{2-}$.

In summary, substitution routes have the potential of introducing hitherto unattainable flexibility and subtlety into the preparation of technetium radiopharmaceuticals. As currently being developed, these routes should lead to new classes of technetium radiopharmaceuticals, the properties of which will be considerably different and more easily controlled than those of complexes prepared by the standard Sn(II) reduction of pertechnetate.

Structure

Because of the nascent state of technetium chemistry, considerable emphasis is currently being given to the characterization of technetium complexes by single crystal x-ray structure analysis. These analyses provide a firm foundation upon which subsequent development and elaboration of technetium chemistry may be based. Approximately twenty technetium complexes have been characterized by single crystal x-ray methods (1), and several of the resulting structures have considerable relevance to radiopharmaceutical development.

Structures of Tc(V) Complexes Prepared in Aqueous Media. Figure 4 shows the structure of the $TcOCl_4^-$ anion, recently determined by the combined research groups of Cotton, Davison and Day (16), while Figures 5 and 6 (17,18) show the structures of dithiolato complexes which may be prepared by ligand substitution onto the $TcOCl_4^-$ center (10). These structures are dominated by the oxo ligand which induces such a strong structural trans effect (19) that the trans coordination site is vacant, and which is so sterically demanding that the other four ligating atoms are severely bent away from the Tc=O linkage (13). Figures 4-6 also emphasize that there are three distinct types of coordination sites in five-coordinate oxo complexes: the oxo oxygen atom is tightly bound and inert to substitution, the site trans to the oxo group has only weak ligand affinity and is very labile, while the four planar sites have intermediate metal-ligand bond strength and intermediate substitution lability. Figure 7 shows the structure of $Tc(HBPz_3)Cl_2O$ (13) which can be prepared by substitution of $HBPz_3^-$ onto $TcOCl_4^-$ (11). Again, the oxo group dominates the structural description of this complex. The nitrogen atom trans to the Tc=O linkage is 0.17Å further from the technetium center than are the other two nitrogen atoms (which are trans to chloride ligands), showing that again the oxo group induces a large structural trans effect even though the tridentate $HBPz_3^-$ ligand suppresses five-coordination. Also, the large steric requirements of the oxo ligand cause the cis ligands to bend away from the Tc=O linkage and towards the trans pyrazolyl ring (13). Figure 8 shows the structure of $Tc(dmg)_2(SnCl_3)(OH)$ (dmg = dimethylglyoxime in unknown protonation state), which contains a seven-coordinate technetium(V) center connected to a tin(IV) center through an oxygen atom bridge (6).

The observed five-, six- and seven-coordinate complexes of Figures

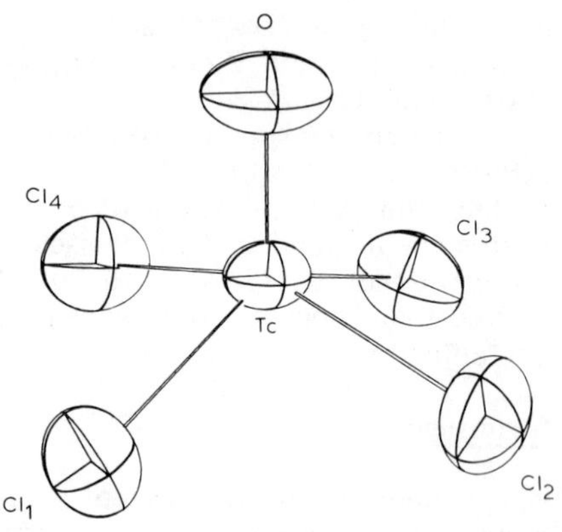

Figure 4. A perspective view of the $TcOCl_4^-$ anion

Inorganic Chemistry

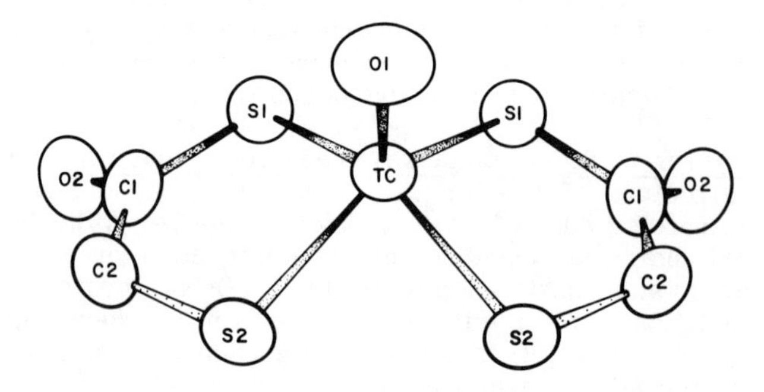

Journal of the American Chemical Society

Figure 5. A perspective view of the $TcO(SCH_2C(O)S)_2^-$ anion (17)

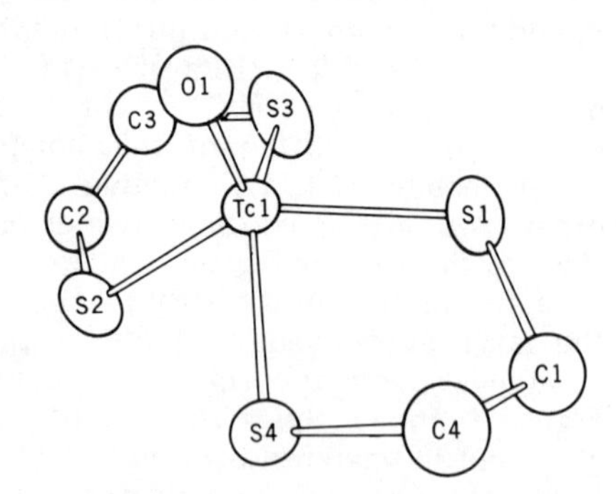

Figure 6. A perspective view of the $TcO(SCH_2CH_2S)_2^-$ anion (18)

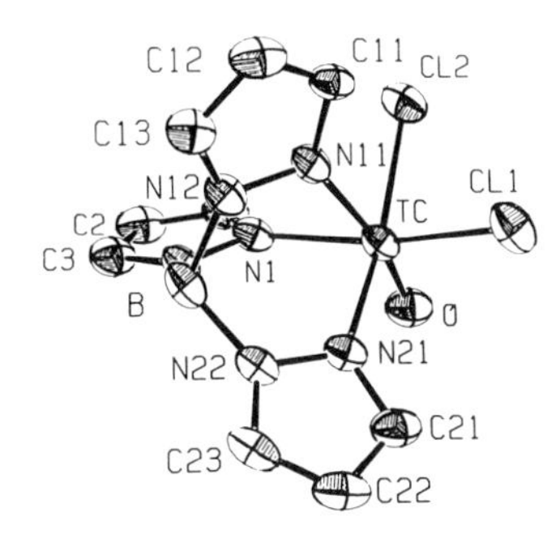

Journal of the American Chemical Society

Figure 7. A perspective view of Tc-$(HBPz_3)Cl_2O$ where $HBPz_3^-$ represents the hydrotris(1-pyrazolyl)borato ligand (13)

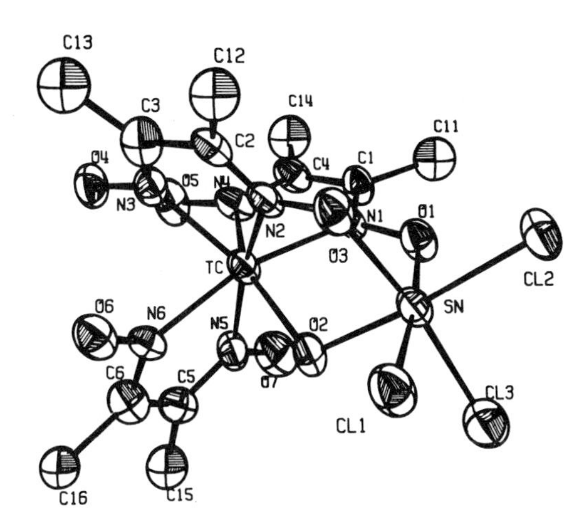

Figure 8. A perspective view of $Tc(dmg)_3(SnCl_3)(OH)$ where dmg represents dimethylglyoxime in an unknown protonation state (6)

4-8 dramatically illustrate that reduced technetium complexes are not restricted to six-coordinate, octahedral structures as is often assumed in the radiopharmaceutical literature. All these complexes were prepared in aqueous, aerobic media, all contain technetium in the +5 oxidation state, and all contain either an oxo ligand or, in the case of $Tc(dmg)_3(SnCl_3)(OH)$, a bridging oxygen atom which may reasonably be assumed to be derived from a Tc=O linkage (6). It is therefore likely that the Tc^V=O moiety will be a predominant feature in the chemistry of technetium radiopharmaceuticals; the different character of the equatorial and axial ligation sites surrounding this moiety must, therefore, be taken into consideration in the design and synthesis of new radiopharmaceuticals.

Structures of Other Relevant Technetium Complexes. It was noted above that technetium(V) can exhibit five-, six-, and seven-coordination. Moreover, in 1960 Fergusson and Nyholm reported (20) the preparation, and indirect characterization, of the eight-coordinate technetium(V) complex, $[Tc(diars)_2Cl_4]^+$, which was synthesized by oxidation of $[Tc(diars)_2Cl_2]^+$ with molecular chlorine (diars = o-phenylenebis [dimethylarsine]). Structural characterization of $[Tc(diars)_2Cl_2]^+$ and $[Tc(diars)_2Cl_4]^+$ (21) shows that the Tc(III) starting material has trans octahedral coordination geometry (Figure 9) and the Tc(V) product has D_{2d} dodecahedral coordination geometry (Figure 10), establishing the preparative reaction as the first known example of oxidative addition from a six-coordinate to an eight-coordinate complex (21). The stability of this particular eight-coordinate species, $[Tc(diars)_2Cl_4]^+$, undoubtedly results in great part from the presence of the diars ligands which are known to promote high coordination numbers (22). However, even for those reactions in which the eight-coordinate products are metastable or unstable, oxidative addition to six-coordinate technetium complexes has great potential as a synthetic route for the interconversion of octahedral technetium complexes and the ultimate synthesis of new technetium radiopharmaceuticals.

Figure 11 shows the structure of $Tc_2Cl_8^{3-}$ (23) which contains a metal-metal bond and which is formed under conditions that are not remote from those used in radiopharmaceutical syntheses. This complex can undergo substitution reactions, e.g. to yield $Tc_2(OOCC(CH_3)_3)_4Cl_2$ shown in Figure 12 (24), and therefore, could be a precursor to a variety of components in radiopharmaceutical mixtures.

Figure 13 shows the structure of tr-$[Tc(NH_3)_4(NO)(OH_2)]^{2+}$ determined by the research group of J. L. Hoard (25). This complex is the first characterized member (26) of what should be a large class of nitrosyl-technetium complexes analogous to the well known nitrosyl-ruthenium complexes. The NO ligand stabilizes low oxidation states and Tc-NO centers should provide suitable templates for synthesis of a variety of radiopharmaceuticals.

Structures of Diphosphonate Complexes. Diphosphonate ligands are widely used to prepare Tc-99m skeletal imaging agents and Tc-99m myocardial infarct imaging agents (1,2). The constitutions, and associated acronyms, of several diphosphonates are shown below along with that of the related ligand pyrophosphate:

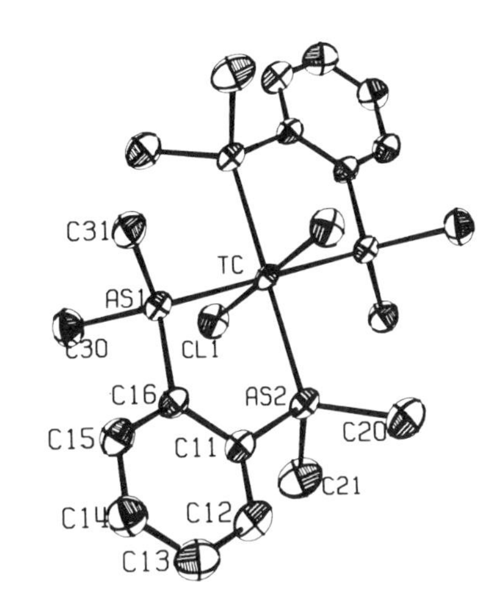

Figure 9. A perspective view of the [Tc-(diars)$_2$Cl$_2$]$^+$ cation where diars represents o-*phenylenebis(dimethylarsine) (21)*

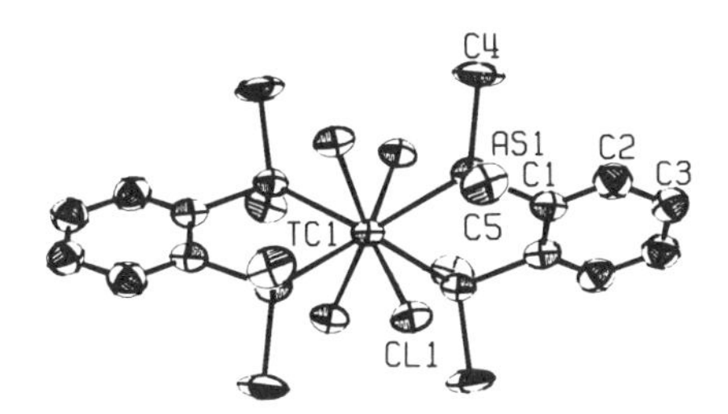

Figure 10. A perspective view of the [Tc(diars)$_2$Cl$_4$]$^+$ cation where diars represents o-phenylenebis(dimethylarsine (21)

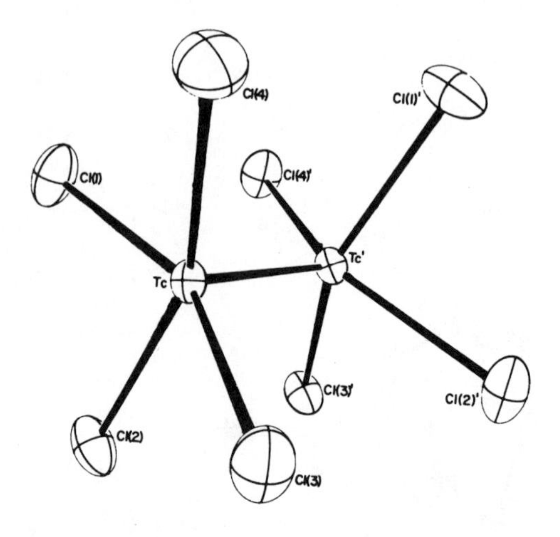

Inorganic Chemistry

Figure 11. A perspective view of the $Tc_2Cl_8^{3-}$ anion (23)

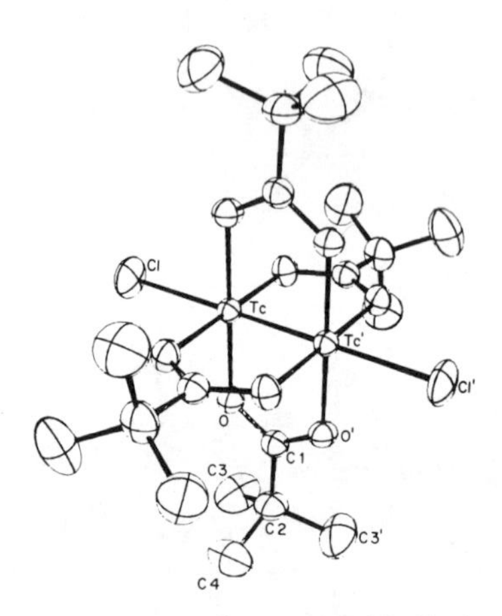

Figure 12. A perspective view of $Tc_2(OOCC(CH_3)_3)_4Cl_2$ (24)

Nouveau Journal De Chemie

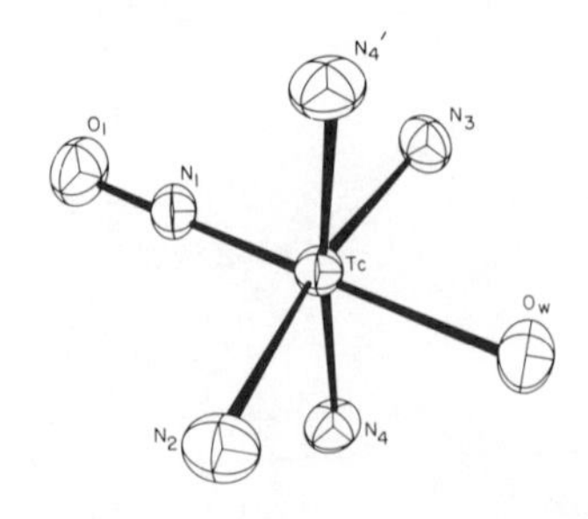

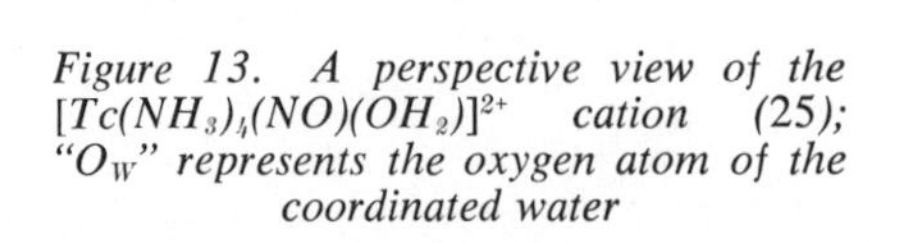

Figure 13. A perspective view of the $[Tc(NH_3)_4(NO)(OH_2)]^{2+}$ cation (25); "O_W" represents the oxygen atom of the coordinated water

PP	$O_3P\text{-}O\text{-}PO_3{}^{4-}$	pyrophosphate
MDP	$O_3P\text{-}CH_2\text{-}PO_3{}^{4-}$	methylenediphosphonate
Cl_2MDP	$O_3P\text{-}C(Cl)_2\text{-}PO_3{}^{4-}$	dichloromethylenediphosphonate
HMDP	$O_3P\text{-}CH(OH)\text{-}PO_3{}^{4-}$	hydroxymethylenediphosphonate
HEDP	$O_3P\text{-}C(OH)(CH_3)\text{-}PO_3{}^{4-}$	(1-hydroxyethylidene)diphosphonate

Clinical applications have focused largely on HEDP and MDP, although considerable attention is currently being given to HMDP. It is generally assumed that technetium complexes of all of these agents are avid bond seekers, and reasonably effective myocardial infarct imaging agents, because the coordinated phosphonate or phosphate ligand retains much of the calcium affinity characteristic of the free ligand. Both bone and myocardial infarcts provide sites of high calcium concentration, and in this context the diphosphonate radiopharmaceuticals are probably best referred to as calcium seeking agents. However, the chemistry of these systems is very complex and no coherent theory explaining the in vivo mechanism(s) of action of technetium diphosphonate radiopharmaceuticals has yet been developed. The evolution of such a theory will require firm structural data as to the possible modes of bonding and interaction between diphosphonate ligands and metal centers. To acquire such data we have conducted structural investigations of several diphosphonate sodium(I) salts (27) (sodium(I) and calcium(II) have similar ionic radii), and of a technetium-MDP complex prepared by substitution of MDP onto $TcBr_6{}^{2-}$ (28).

The solid state structure of the technetium-MDP complex consists of infinite polymeric chains. Each MDP ligand (Figure 14) bridges two symmetry related technetium atoms (Figure 15), and each technetium atom is bound to two symmetry related MDP ligands (Figure 16) -- the MDP/Tc ratio within the polymer is therefore 1/1. The polymeric repeat unit is completed by an oxygen atom (presumably in the form of a hydroxyl ion) that bridges two symmetry related technetium atoms (Figure 15) and by a hydrated lithium cation which neutralizes the charge associated with each repeat unit. In addition, there is a single oxygen atom (presumably in the form of a disordered water molecule) on the three-fold axis of the space group. The molecular formula of the polymeric technetium-MDP complex may thus be represented as $\{[Li(H_2O)_3][Tc^{IV}(OH)(MDP)]\frac{1}{3}H_2O\}_n$ where the indicated protonation states of the bridging and non-coordinated oxygen atoms are chemically reasonable and consistent with an assumed Tc(IV) oxidation state, but are not definitively established by the x-ray diffraction data.

One of the most important structural features of the diphosphonate ligands is the orientation of the $-PO_3$ groups with respect to the P-C-P plane. The "W" configuration, wherein the atoms O2-P1-C-P2-O4 form a planar "W", can easily be seen in Figures 14 and 15. This configuration allows MDP to be doubly bidentate with 01 and 06 on one side of the "W"

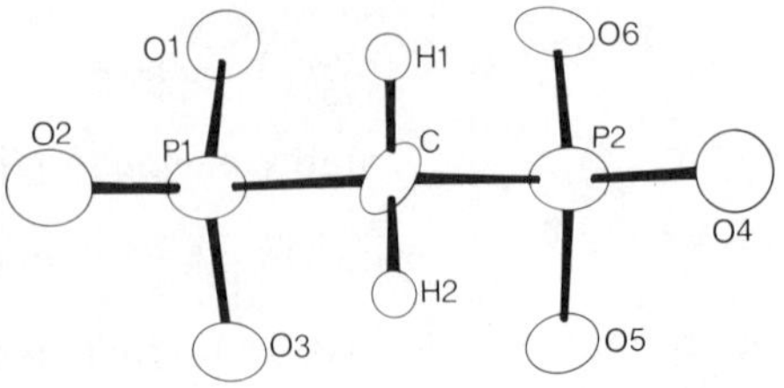

Figure 14. A perspective view of the MDP ligand in the $[Tc(MDP)(OH)^-]_n$ polymer (28)

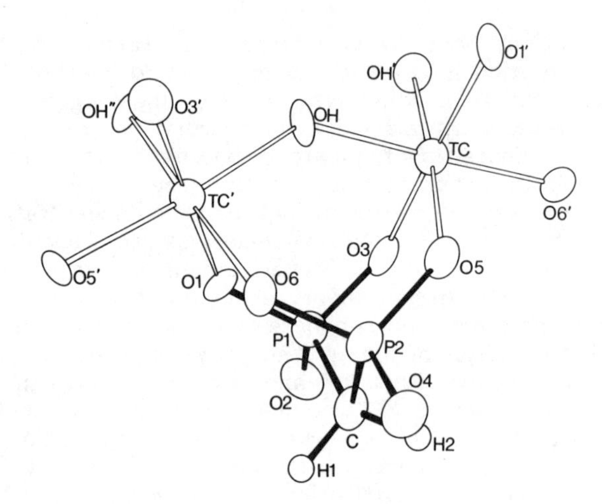

Figure 15. A perspective view of a portion of the $[Tc(MDP)(OH)^-]_n$ polymer showing one MDP ligand bridging two technetium centers (28)

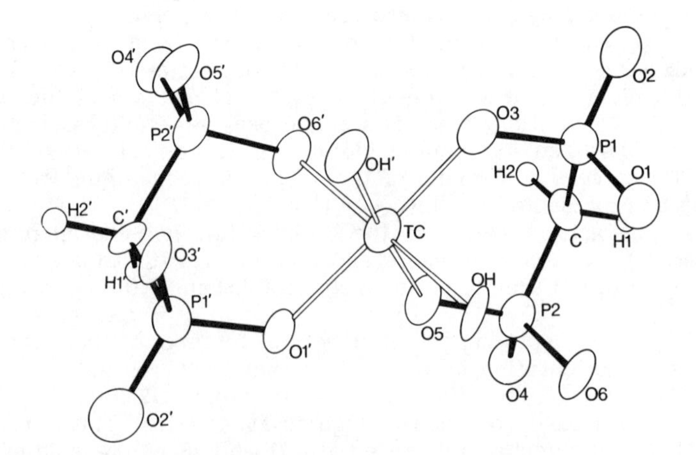

Figure 16. A perspective view of a portion of the $[Tc(MDP)(OH)^-]_n$ polymer showing one technetium center bridging two MDP ligands (28)

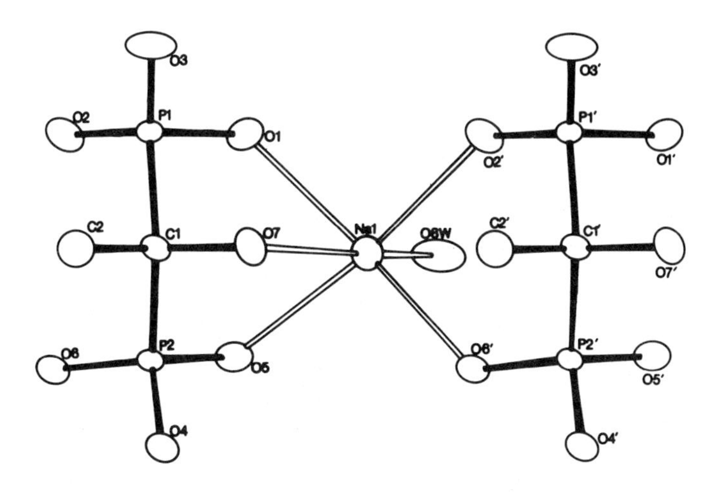

Figure 17. A perspective view of a portion of the polymeric structure of Na_2H_2-HEDP showing one sodium center bridging two HEDP ligands (27); "O8W" represents the oxygen atom of a water molecule coordinated to the sodium center

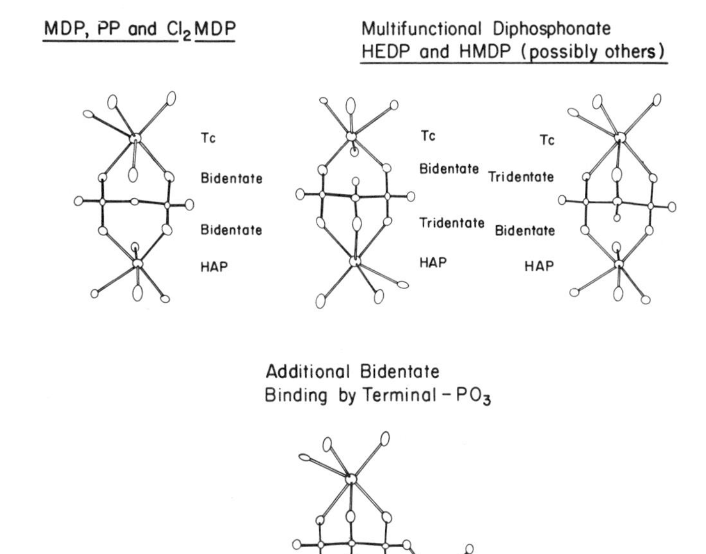

Figure 18. A summary of the established modes by which diphosphonate ligands bridge metal centers. The perspective views are obtained from structural analyses of the respective sodium salts (27), and are interpreted with respect to the hypothesized bridging of technetium to hydroxyapatite (HAP).

Figure 19. A molecular model showing tridentate binding of HMDP to the trigonal face of a calcium center at the surface of hydroxyapatite

coordinating to one metal center, and 03 and 05 on the other side of the "W" coordinating to another metal center (Figure 15). This doubly bidentate character of MDP allows it to bridge metal centers, e.g. Tc-to-Tc in the technetium-MDP polymer, Na-to-Na in Na_2H_2MDP, and Tc-to-Ca in the presumed biological mechanism of action. If the diphosphonate ligand contains a hydroxyl group on the central carbon atom (as in HEDP and HMDP), then the diphosphonate can function as a mixed bidentate-tridentate bridge. Figure 17 shows a portion of the polymeric structure of Na_2H_2HEDP (27) in which each HEDP ligand functions as a bidentate ligand to one sodium center and as a tridentate ligand to another sodium center. This figure illustrates the coordination about one sodium ion, the tridentate and bidentate modes of HEDP coordination being readily apparent. It is therefore clear that by virtue of the extra hydroxyl group, HMDP and HEDP are distinct from those diphosphonates that cannot form mixed bidentate-tridentate bridges (MDP, Cl_2MDP, PP, etc.), and different chemical and biological properties are expected for the two classes of diphosphonate ligands. Figure 18 illustrates the possible modes of bridging between technetium and hydroxyapatite (HAP, the form of calcium most likely encountered in biological systems) by bidentate-bidentate and bidentate-tridentate diphosphonate ligands. The mode wherein tridentate HMDP or HEDP binds to hydroxyapatite is especially intriguing since such tridentate ligation nicely completes the trigonal antiprismatic coordination of calcium at the fastest growing HAP crystal axis. This hypothesized bonding is illustrated more dramatically in Figure 19.

These structural studies emphasize the central role of polymeric metal-diphosphonate complexes in the chemistry of technetium-diphosphonate calcium seeking agents. It is clearly the ability of diphosphonates to bridge metal centers that provides the mechanism for the initial sorption of the radiopharmaceutical onto bone. Mixed metal (technetium, tin, and calcium) diphosphonate polymeric complexes are likely to be the dominant chemical species in clinically used skeletal and myocardial infarct imaging agents. An understanding of the chemistry of these polymeric species will be crucial to an understanding of the mechanisms of action of diphosphonate radiopharmaceuticals and to the development of more efficacious imaging agents.

Acknowledgments

Financial support for this work was provided by the National Institutes of Health (Grant No. HL-21276) and the Procter & Gamble Co. We gratefully acknowledge receiving pre-publication results from Professors J. L. Hoard (Cornell University), A. Davison (MIT), F. A. Cotton (Texas A&M) and V. Day (University of Nebraska), and especially thank our colleagues and students for their many significant contributions to the research discussed in this review.

Abstract

New synthetic routes to technetium-99 complexes and technetium-99m radiopharmaceuticals are based on substitution of ligands onto preformed reduced technetium centers such as Tc(IV) in TcX_6^{2-} and Tc(V) in $TcOX_4^-$ (X = Cl, Br). These procedures avoid many of the difficulties and limitations inherent in the classical pertechnetate reduction route. Single crystal x-ray structural characterizations of several technetium-99 complexes indicate the importance of the Tc(V) oxidation state and the Tc=O linkage in complexes related to radiopharmaceuticals, and dramatically illustrate the variable coordination numbers and geometries exhibited by Tc(V). Structural studies on diphosphonate complexes of sodium and technetium-99 clarify the possible modes of metal-ligand interactions in the calcium seeking Tc-99m diphosphonate radiopharmaceuticals that are widely used as bone and myocardial infarct imaging agents.

Literature Cited

1. Deutsch, E. in "Radiopharmaceuticals II", Society of Nuclear Medicine, New York, N.Y., 1979; pp. 129-146.
2. Siegel, J. A.; Deutsch, E. Ann. Rep. Inorg. Gen. Synth. 1975, 1976, 311-326.
3. Hayes, R. L. in "The Chemistry of Radiopharmaceuticals", Masson, New York, 1978; pp. 155-168.
4. Eckelman, W. C.; Levenson, S. M. Int. J. App. Radiat. Isot., 1977, 28, 67-82.
5. Marzilli, L. G.; Burns, H. D.; Dannals, R. F.; Kramer, A. V.; this symposium.
6. Deutsch, E.; Elder, R. C.; Lange, B. A.; Vaal, J. J.; Lay, D. G. Proc. Natl. Acad. Sci. USA, 1976, 73, 4287-4289.
7. Zalutsky, M. R.; Rayudu, G. V. S.; Friedman, A. M. Int. J. Nucl. Med. Biology, 1977, 4, 224-230.
8. Srivastava, S. C.; Meinken, G.; Smith, T.D.; Richards, P. Int. J. App. Radiat. Isot., 1977, 28, 83-95.
9. Pinkerton, T.; Heineman, W. R.; Deutsch, E. Anal. Chem., in press.
10. Davison, A.; this symposium.
11. Thomas, R.; Davison, A.; Deutsch, E.; unpublished data.
12. Spies, H.; Johannsen, B. Inorg. Chim. Acta, 1979, 33, L113.
13. Thomas, R. W.; Estes, G. W.; Elder, R. C.; Deutsch, E. J. Am. Chem. Soc., 1979, 101, 4581-4585.
14. Deutsch, E.; Libson, K.; Becker, C. B.; Francis, M. D.; Tofe, A. J.; Ferguson, D. L.; McCreary, L. D.; submitted for publication.
15. Kubiatowicz, D. O.; Bolles, T. F.; Nora, J. C.; Ithakissios, D. S. J. Pharm. Sci., 1979, 68, 621-623.
16. Cotton, F. A.; Davison, A.; Day, V. W.; Gage, L. D.; Trop, H. S. Inorg. Chem., 1979, 18, 3024-3029.
17. Davison, A.; DePamphillis, B. V.; Jones, A. G.; Davis, M. A. J. Am. Chem. Soc., 1978, 100, 5570-5571.
18. Smith, J. E.; Byrne, E. F.; Cotton, F. A.; Sekutowlki, J. C. J. Am. Chem. Soc., 1978, 100, 5571-5572.
19. Shustorovich, E. M.; Porai-Koshits, M. A.; Buslaev, Y. A. Coord. Chem. Rev., 1975, 17, 1-98.
20. Fergusson, J. E.; Nyholm, R. S. Chem. Ind., 1960, 347-348.

21. Glavan, K.; Whittle, R.; Johnson, J. F.; Elder, R. C.; Deutsch, E.; submitted for publication.
22. Clark, R. J. H.; Kepert, D. L.; Nyholm, R. S.; Lewis, J. Nature, 1963, 199, 559-562.
23. Cotton, F. A.; Gage, L. D. Inorg. Chem., 1975, 14, 2032-2035.
24. Cotton, F. A.; Gage, L. D. Nouv. J. Chimie, 1977, 1, 441-442.
25. Hoard, J. L.; unpublished data.
26. Armstrong, R. A.; Taube, H. Inorg. Chem., 1976, 15, 1904-1909.
27. Barnett, B. L.; Strickland, L. C.; unpublished data.
28. Deutsch, E.; Barnett, B. L.; Libson, K.; submitted for publication.

RECEIVED April 7, 1980.

7

Radiolabeled Compounds of Biomedical Interest Containing Radioisotopes of Gallium and Indium

M. J. WELCH and S. MOERLEIN

The Edward Mallinckrodt Institute of Radiology, Washington University School of Medicine, 510 South Kingshighway, St. Louis, MO 63110

There has been considerable interest in the application in medicine of the four radioisotopes, ^{67}Ga, ^{68}Ga, ^{111}In, and ^{113m}In, listed in Table 1. Each of the two elements has one isotope with a half-life of 2-4 days which has been used for such applications as tumor scanning [^{67}Ga-citrate (1) and ^{111}In-bleomycin (2)], abscess localization [^{67}Ga-citrate (3) and ^{111}In-labeled white cells (4)], and thrombus detection [^{111}In-labeled platelets (5)]. As shown in Table 1, both indium-111 and gallium-67 are cyclotron-produced, and the most common method of production utilizes proton reactions. It is interesting to note that at the present time (August, 1979) there are at least eight commercial cyclotrons in the United States dedicated to isotope production, and these accelerators produce mainly four radioisotopes for radiopharmaceutical use (^{111}In, ^{67}Ga, ^{123}I, and ^{201}Tl).

TABLE I
RADIONUCLIDES OF INDIUM AND GALLIUM
OF INTEREST IN NUCLEAR MEDICINE

ISOTOPE	^{111}In	^{113m}In	^{67}Ga	^{68}Ga
HALF-LIFE	2.8 days	1.7 hr	3.3 days	68 min
METHOD OF PRODUCTION	^{111}Cd (p,n)	Decay of ^{113}Sn ($t_{1/2}$= 115 days)	^{67}Zn(p,n)	Decay of ^{68}Ge ($t_{1/2}$=275 days)

The short-lived gallium and indium isotopes are produced for medical purposes at the site of use from a radionuclide generator (6). The ^{113}Sn-^{113m}In generator has been displaced in the United States by the ^{99}Mo-^{99m}Tc generator, although it continues to be utilized in countries where there are delivery problems with the much shorter-lived ^{99}Mo-^{99m}Tc system (7). The germanium-68/gallium generator is one of a very limited number of

0-8412-0588-4/80/47-140-121$05.00/0

generator systems that produces a short-lived positron-emitting radionuclide (6). There is currently great interest in positron-emitting radiopharmaceuticals because of the fact that their distribution can be quantitated in vivo utilizing positron emission transaxial tomography. Utilizing this technique, reconstruction of a radionuclide distribution is possible by several techniques to give the true distribution of activity in the source (8,9,10).

The design of radiopharmaceuticals labeled with indium and gallium radionuclides is compounded by the fact that both indium and gallium form very strong chelates with the plasma protein transferrin (11,12). Due to this large stability constant and the larger amount of transferrin in human plasma (0.25 mg/100 ml), one would anticipate indium and gallium compounds being thermodynamically unstable in vivo. It appears, however, that for indium and gallium complexes with stability constants $>10^{20}$, the rate at which the equilibrium is reached is slow compared to the rate of many biological events (11), such as the 1.6 hour glomerular filtration rate of ^{111}InDTPA (13). It should be noted, however, that the injection of weak chelates of indium and gallium leads to very similar biodistributions. Because of this effect of exchange with transferrin, one of the major goals of research in this area is the developement of strongly-binding bifunctional chelates. The first of these was 1-(p-benzenediazonium)-ethylenediamine-N,N',N'-tetra-acetic acid (azo-Ø-EDTA) shown in Figure 1, which was developed by Sundberg, et al (14). This compound forms a link between the metal ETDA complex and a protein by means of the diazo group. Human serum albumin labeled with ^{111}In in such a manner was found to have a biological half-life of 7 days and to lose less than 5% of its activity to transferrin when incubated with serum for 2 weeks (15). This and other approaches (16) have extended the number of available indium and gallium radiopharmaceuticals. The following is a discussion of the major uses of each of the four isotopes.

Indium-113m

The tin-indium generator was introduced in 1966 by Stern et al (17). The tin-113 which is produced by the $^{112}Sn(n,\gamma)^{113}Sn$ reaction in a nuclear reactor is retained in a hydrated zirconium oxide column eluted with 0.05M hydrochloric acid. The generator eluate has been used directly as a blood pool scanning agent (18, 19, 20). The generator eluate, when injected directly, leads to the formation of ^{113m}In-transferrin, which remains in the blood pool for several ^{113m}In half-lives. Increasing the pH of the generator eluate leads to colloidal formations which have been used for the visualization of the liver, spleen, and bone marrow (21, 22, 23). Larger particles where the radioactive indium is associated with iron hydroxide (22, 23) or with macroaggregates

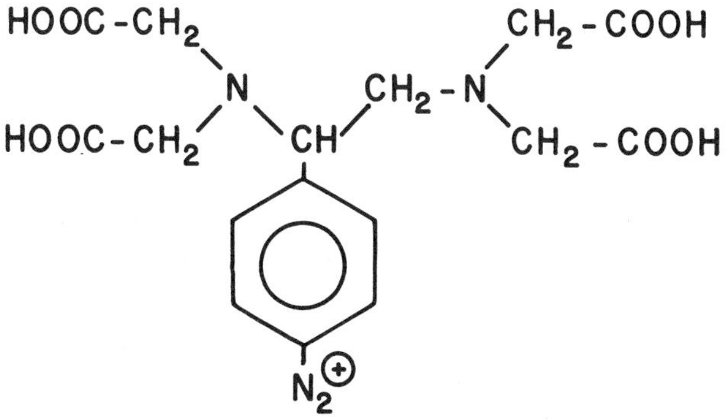

Figure 1. Metal binding molecule that forms a link between the In–EDTA and the protein by means of a diazo bond

of albumin, have been utilized for lung scanning (24,25).

As discussed previously, only chelates with a slow exchange rate remain stable in vivo. Indium-113m chelates with EDTA and DTPA have been utilized for the detection of brain tumors and for the study of renal funcions (26,27). Indium-113m chelates with ethylenediamine tetra(methylene phosphonic acid) (EDTMP) and diethylenetriamine penta(methylene phosphonic acid) (DTPMP) have been utilized to study bone tumors (28, 29). These agents also have promise for the detection of myocardial infarcts (30).

It can be seen from the above discussion that the simple compounds of indium-113m that have been prepared to date can be used to study many organs of the human in a non-invasive manner. Although the 393 keV decay energy and 1.7 hour half-life of ^{113m}In make it a less ideal nuclide than ^{99m}Tc, the long half-life (118 days) of its parent ^{113}Sn make it very useful in developing countries or isolated regions where delivery of radioisotope generators is difficult. The tin-113/indium-113m generator may be eluted several times a day (1.7 hours generates 50% of the equilibrium activity) and need be replaced only twice a year.

Indium-111

The major uses of indium-111 in medicine are listed in Table 2. Indium-111 labeled DTPA is the preferred agent for the study of cerebral spinal fluid kinetics (cisternography)(31). Indium-labeled bleomycin has been used for tumor scanning (2), although ^{67}Ga citrate has achieved greater clinical use. It appears that indium bleomycin is in fact a weak chelate and the in vivo distribution is very similar to that of indium transferrin.

TABLE II
INDIUM-111 RADIOPHARMACEUTICALS

Radiopharmaceutical	Application	Reference
$^{111}InCl_3$	Tumor and Bone Marrow Imaging	32,134
^{111}In Citrate	Bone Marrow Imaging	135
^{111}In-DTPA	Cisternography	31,32,136
^{111}In-EDTA	Cisternography	31,32,136
^{111}In-EDTMP	Bone Imaging	137
^{111}In-HMDTP	Bone Imaging	137
^{111}In-DTPMP	Bone Imaging	137
^{111}In-$Fe(OH)_3$ Colloid	Lymph Node Scanning	138
	Lung Scintigraphy	139
^{111}In-Bleomycin	Tumor Scanning	2,32
^{111}In-HSA	Cisternography	15,32,136
^{111}In-Transferrin	Cisternography and Bone Marrow Imaging	15,32,136
^{111}In-Fibrinogen	Thrombus Imaging	32
^{111}In-RBC's	Cardiac and Spleen Imaging	32

Table II continued

^{111}In-Platelets	Thrombus Imaging	34,37,38, 137
^{111}In-Leukocytes	Abscess and Inflammatory Site Imaging	39
^{111}In-Lymphocytes	Lymph Node Imaging	40,140
	Lymphocyte Kinetics	141

The applications of indium-111 that are currently being investigated include studies with bifunctional chelates and the labeling of blood cells. The bifunctional chelating group (Figure 1) has been utilized to attach ^{111}In to albumin, fibrinogen, and bleomycin (32). Using this bifunctional technique it is possible to prepare a stable indium bleomycin chelate which has great potential for tumor localization.

In recent years the most exciting application has been the utilization of indium-111 labeled 8-hydroxyquinoline to label blood cells. It has been shown that when the 8-hydroxyquinoline complex is mixed with cells separated from plasma, the indium becomes firmly bound inside the cell (33,34). Studies to evaluate the mechanism of uptake suggest that the lipophilic chelate diffuses inside the cell and that there are intracellular binding sites to which the indium exchanges (35, 36). Studies utilizing tritiated 8-hydroxy-quinoline have shown that the 8-hydroxyquinoline is not retained in the cell but is partitioned between the lipophilic cell and the aqueous suspension media. Other studies utilizing both labeled white cells (35) and platelets (36) have shown that when the cells are lysed the activity is attached to proteins. As the indium is attached inside the blood cell a stable label results for reinjection into a patient because the cell membrane prohibits plasma transferrin access to the labeled protein. Labeled platelets (34, 37, 38), labeled white cells (39), and labeled lymphocytes (40) have all been studied extensively. Platelets have been shown in a series of normal volunteers to behave in the same manner as unlabeled platelets (38), and in patients with thrombosis or atherosclerosis (37) to localize at or visualize the site of the lesions. Labeled white cells accumulate in abscesses (33) and have been used for abscess detection in humans (39). This ability of ^{111}In-8-hydroxyquinoline to label blood cells combined with the good imaging characteristics of indium-111 allows this valuable application of this nuclide. It should be noted that the kinetics of uptake of the labeled cells are such that the half-life of ^{113m}In is too short for many applications as a cell label.

Gallium-68

As discussed previously, the germanium-68/gallium-68

generator is of particular interest because it is a convenient generator to produce a positron-emitting radionuclide. A commercially available generator is based on the system initially described by Greene and Tucker (41). In this system the germanium-68 is loaded onto an activated alumina column and the gallium-68 is extracted with 0.005M EDTA. The gallium-EDTA solution at the time of elution contains less than 10^{-2}% of ^{68}Ge as a contaminant. Owing to the large differences in the half-lives of the daughter and parent, the breakthrough of the parent must be very low, as the radiation dose to a patient from ^{68}Ge is many orders of magnitude greater than that from ^{68}Ga. Although ^{68}Ga-EDTA can be used directly for brain or renal scanning (42), the production of any other compound requires one to initially decompose the EDTA complex. Although several methods have been used to accomplish this (43), they are all time consuming and lose a significant fraction of the 68-minute half-lived gallium-68. Because of this problem, there has recently been considerable effort to develop a generator producing the gallium-68 in either an ionic form or as a weak chelate. Both solvent extraction and column systems have been developed to accomplish this. In the solvent extraction technique (44), gallium-68 is extracted from an aqueous solution into chloroform or methylene chloride as the gallium 8-hydroxyquinoline complex. After evaporation of the solvent the ^{68}Ga-8-hydroxyquinoline can either be used directly for cell labeling (45) or exchanged with stronger ligands to form other gallium-labeled radiopharmaceuticals. This type of generator has recently been automated (46) to produce the gallium 8-hydroxyquinoline without operator manipulation. In a new chromatographic generator system (47), the carrier-free germanium-68 is adsorbed on polyantimonic acid in sodium oxalate solution at pH 5-10. Gallium-68 can be eluted as the gallium oxalate over a pH range of 7 to 11, and the germanium-68 breakthrough is less than 0.06%. Other approaches to the production of a generator for ionic gallium-68 have been described by Neirinckx and Davis (48), who have described two systems. In one of these, gallium-68 is eluted with dilute hydrofluoric acid from a strongly basic anion exchange resin, Bio-Rad AG1-X8, onto which ^{68}Ge is strongly adsorbed. The distribution coefficients for germanium and gallium were determined and at the optimum conditions yields of gallium-68 of >95% with germanium breakthrough of less than 10^{-3}% were obtained. When the concentration of hydrofluoric acid was limited to 0.01N HF to decrease the probability of fluoride toxicity, gallium-68 yields of 90% with germanium-68 breakthrough of <0.01% were still obtained (49). Another system described by these investigators uses a chelate resin synthesized by co-polymerization of formaldehyde and pyrogallol. It was found that upon elution with dilute hydrochloric acid good yields of ^{68}Ga were obtained with low levels of ^{68}Ge breakthrough. At this time, however, further investigation of column radiolytic stability and toxicology of the organic resin appears to be needed.

Many gallium-68 radiopharmaceuticals have been prepared, and in several cases the procedures (6) are simply modifications of those already discussed for the indium radionuclides. As examples, ^{68}Ga-colloid has been prepared and used in conjunction with a transverse section imaging device (50) to image the liver of animals and humans. Figure 2 shows the detail obtainable in seven sections of a dog liver and spleen obtained in this manner. Blood components have also been labeled with gallium-68 (47) in a manner analogous to that carried out with indium-111. Labeled red cells have been used to visualize the vascular space and platelets have been used to visualize various areas of platelet deposition. Figure 3 is an example of one of those, in which gallium-68 labeled platelets are accumulating in a performed pulmonary embolus.

In the area of bifunctional chelates, the method of Krejcarek and Tucker (16) has been utilized to attach gallium-68 to proteins (51). In this approach DTPA is coupled to proteins by the formation of an amide bond. It has been shown that ^{68}Ga-proteins can be formed and that the labeled protein is stable for a time period of several hours. The same type of linkage has also been used to attach gallium-68 to human serum microspheres (51).

Whereas most positron-emitting radionuclides in nuclear medicine are cyclotron-produced, gallium-68 has the unique advantage of being readily-available in a generator system. The 275-day half-life of the parent germanium-68 is large to avoid delivery problems, and the 68-minute half-life of gallium-68 is convenient for radiopharmaceutical synthesis without subjecting the patient to excessive absorbed radiation dose. With continued clinical use of positron tomography, these features will make gallium-68 a valuable addition to the nuclear medicine armamentarium.

Gallium-67

Gallium-67 is widely used for the detection of tumors and abscesses. Because most inorganic gallium compounds are hydrolyzed at physiological pH to form the insoluble hydroxide, the soluble gallium-67 citrate is the compound most often used in humans. Gallium-67 bleomycin has not shown the tumor specificity or blood clearance of the citrate salt and hence has not found the widespread application of gallium-67 citrate. Examples of the type of image obtained using gallium-67 citrate are shown in Figures 4 and 5.

When gallium-67 is injected into the bloodstream as either chloride or citrate, it rapidly becomes bound to serum proteins, especially transferrin (52,53,54). Although the gallium-transferrin interaction is weaker than that of iron and

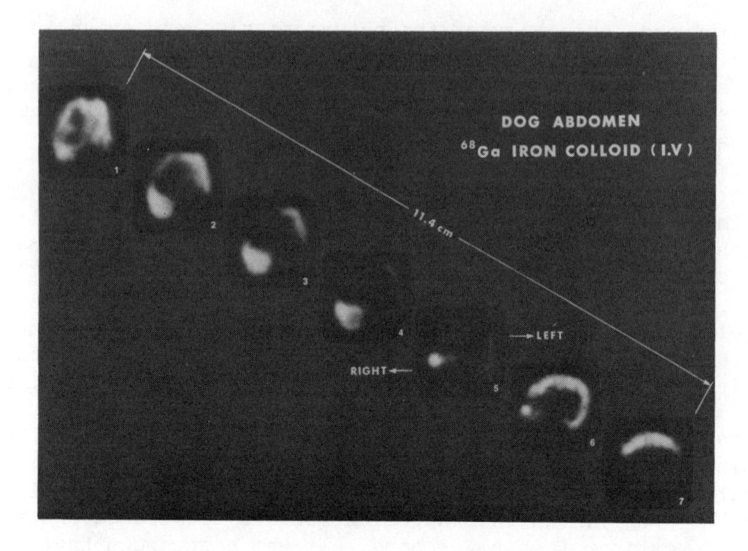

Figure 2. Tomographic images of a dog abdomen obtained with ^{68}Ga–Fe colloid. Slices 1–5 show primarily the liver whereas slices 6 and 7 show the outline of the spleen.

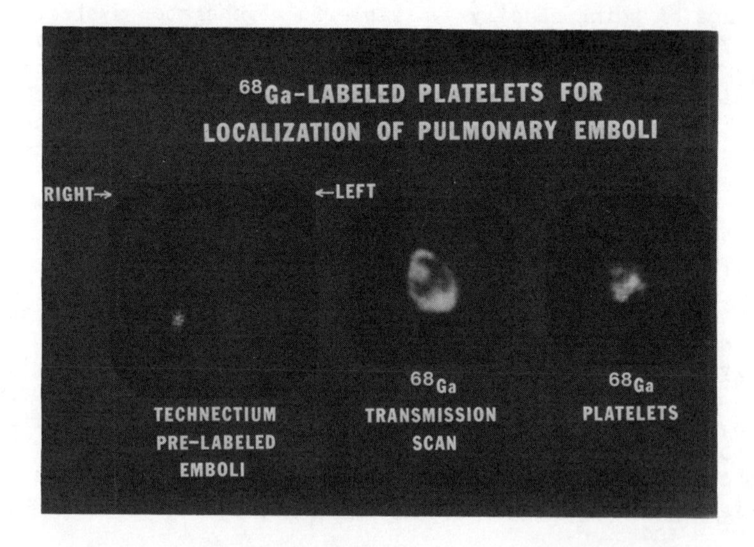

*Figure 3. Tomographic section of ^{68}Ga platelets localizing in an experimentally induced pulmonary embolus. The left panel shows the prelabeled emboli visualized on an Anger camera; the emboli have been prelabeled with technetium-99*m*–sulfur colloid. The middle panel shows the transmission scan used to correct the image for attenuation in various parts of the lung. The right panel shows the localization of the ^{68}Ga platelets in the two emboli shown in the left panel.*

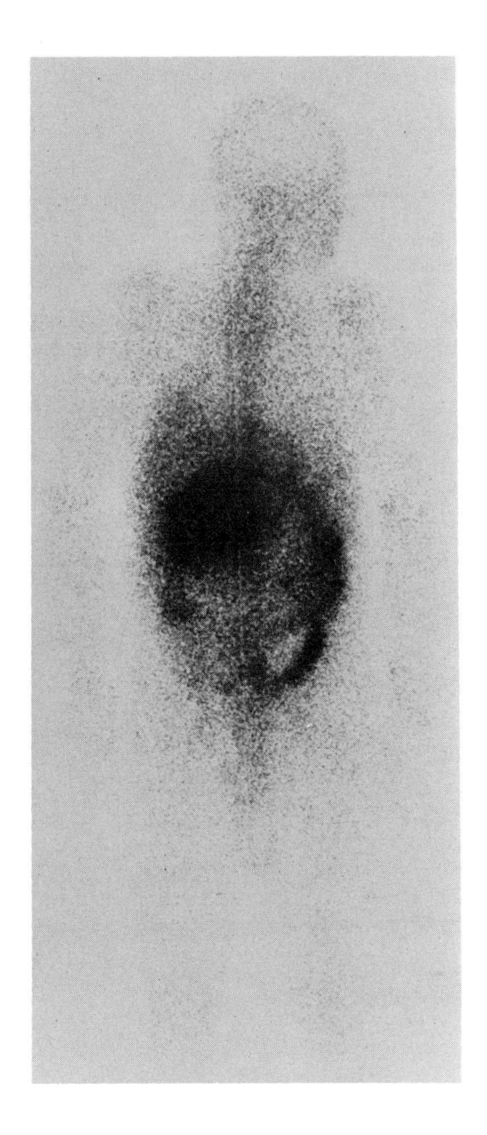

Figure 4. An example of ^{67}Ga localizing in an abscess. The large hot area of activity is the area of the abscess.

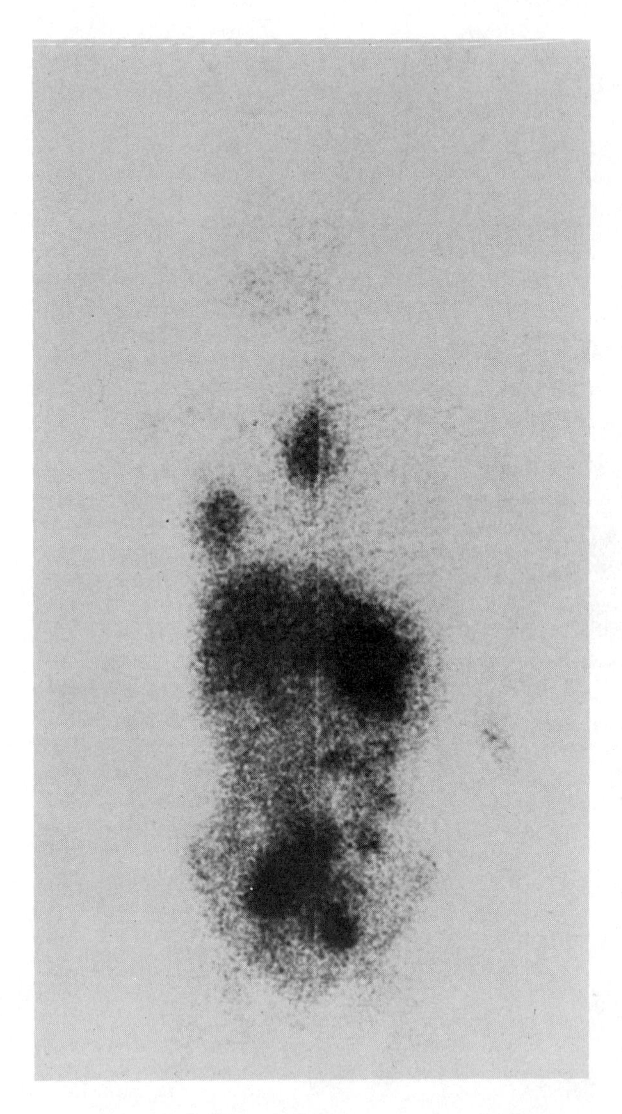

Figure 5. An example of ^{67}Ga localizing in a tumor. This is a patient with lymphoma. As observed, there are multiple areas of increased activity in the lymph nodes.

transferrin, over 90% of serum ^{67}Ga is in the form of ^{67}Ga-transferrin (55). Gallium-transferrin distributes throughout the transferrin pool and the gallium concentrates in tumor, liver, and kidney and is secreted by the salivary glands, mammary glands, and the bowel (56). This tissue distribution of gallium radionuclides is affected by such factors as extent of neoplasm (1), inflammatory processes (57), sex hormone status (58), age (59), lactation (60), radiotherapy and chemotherapy treatment (61,62), amount of carrier gallium administered (63,64), anionic form (63), and administration of compounds which block plasma protein binding of gallium (64,65). The tissue deposition is particularly affected by the specific activity of the gallium, for use of ^{72}Ga (produced by the (n, γ) reaction) shows high bone uptake (66,67,68), whereas application of essentially carrier-free ^{67}Ga (cyclotron-produced) has found widespread clinical use for imaging a variety of epithelial and lymphoreticular neoplasms (69-76) and inflammatory lesions (77,78,79,80). At this time, however, the mechanisms whereby gallium-67 localizes in tumors or inflammatory processes is not definitely established.

The uptake of ^{67}Ga into abscesses has been less well sudied than tumor deposition processes. It has been suggested that the localization of gallium in these lesions is due to its concentration by polymorphonuclear leukocytes (PMN) at the site of inflammation (79,80), but this explanation has been disputed because of ^{67}Ga accumulation in lesions of agranulocytic patients and because less than 1% of injected gallium binds to the cellular fraction (81). Rather, it has been proposed that localization of ^{67}Ga is due to leakage of protein-bound gallium from capillaries with increased permeability resulting from the inflammation (82,83). Once the radionuclide has extravasated into the lesion, it binds preferentially to non-viable PMN (which have an attenuated permeability barrier) and to a lesser extent to bacteria, viable PMN, and extracellular proteins. This general mechanism is also applicable to other nuclides, but ^{67}Ga characteristics make it particularly useful, such as long biological half-life allowing lesion accumulation, low blood levels after twenty-four hours, and long physical half-life allowing imaging two to three days post injection.

A similar approach to tumor uptake mechanism has been presented (84). In this case, the gallium-67 is assumed to accumulate in tumor cells in an unknown manner because of increased deposition of the radionuclide at the tumor site in an inflammatory response. Because acute inflammation is associated with increased capillary blood flow and permeability (85) as well as capillary sprouting (86), these changes allow more ^{67}Ga to be delivered to the tumor site. This simple mechanism offers the advantage of explaining why gallium accumulates only in viable (not necrotic) neoplasms (63) and that it may operate concurrently with the alternative mechanisms which will now be discussed.

The similarity between gallium and the alkaline earth elements has been developed in a different explanation for gallium tumor uptake (87,88,89). Because the lack of tumor specificity in ^{67}Ga uptake suggests that a simple mechanism is involved, perhaps a simple competition between gallium and magnesium or calcium for macromolecular ligands occurs. Since the ionic radii of Ga^{3+} (0.62) and Mg^{2+} (0.65) are similar, the exchange should occur and favor gallium complexation because of its higher valence and hence greater complex stability. Exchange with Ca^{2+} (ionic radius 0.99) is also possible because the macromolecules involved are able to generate steric fitting to accommodate cation size in accordance with the radius-ratio principle. These magnesium and calcium ligands include RNA, DNA, proteins, acid aminoglycans, and phospholipids, all of which show active metabolic synthesis during calcification (90,91). It is not suggested that gallium substitutes for the alkaline earth metals throughout their metabolic pathways, but rather than ^{67}Ga remains in the soluble fraction of tumor cells, bound to the respective macromolecules. The higher tumor content of calcium and magnesium (92,93) implies greater concentration of the metal-binding ligands, and therefore greater cellular uptake of gallium-67 via competitive ionic exchange.

In a follow-up study of the similarities of calcium and gallium biokinetics, a comparative investigation of the uptake of ^{45}Ca and ^{67}Ga in lactating dogs showed that similar subcellular distributions in mammary gland cells (94). However, the uptake of the two radionuclides did not correlate when transmissible venereal tumor was used. It was concluded that although lactating mammary gland uptake of calcium and gallium shows similar characteristics, there is no similarity in the mechanism of uptake of these two elements by tumor tissue.

Using cultured mammalian sarcoma cells, it has been found that transferrin is necessary in the growth medium for gallium-67 uptake to occur (95,96,97). A "transferrin receptor" on EMT-6 sarcoma cells for ^{125}I-labeled transferrin was characterized by Scatchard analysis to have an average association constant $K = 4.54 \times 10^6$ l/mole and approximately (with variation) 500,000 receptors per cell (95). It was proposed that tumor accumulation of gallium-67 can occur only if the metal is complexed with transferrin so that it can interact with the receptors of tumor, as well as non-malignant cells (98). The complex then enters the cell via an "adsorptive endocytosis" process (95,96,97,98,99) similar to the manner in which iron is taken up by reticulocytes and bone marrow cells (100,101). These transferrin receptors are saturable (that is, a plot of ^{125}I-transferrin uptake versus extracellular transferrin concentration reaches a peak (at about 200 μg/ml) as more carrier transferrin is added to the medium) (95). Since uptake is also proportional to the fraction of gallium which is bound to transferrin, this mechanism predicts that total cellular uptake is proportional to the "calculated cellular uptake" (the fraction of ^{67}Ga as ^{67}Ga transferrin times the fraction of total transferrin which is cell-bound)(96). Once the gallium-transferrin is inside the cell, it is deposited in the lysosomes initially and then distributed to other regions of the cells (some gallium is stored in ferritin, but most of the element is deposited in the microvesicles and rough endoplasmic reticulum). The intracellular gallium must be irreversibly bound to macromolecules to prevent it from diffusing into the extracellular space (102). Only cells with a "transferrin receptor" as well as "intracellular receptor" will accumulate and retain ^{67}Ga.

The above hypothesis offers a simple description of gallium-67 uptake since gallium complexes rapidly exchange with plasma proteins which results in primarily gallium-transferrin (52,53,54). It also explains how prelabeled transferrin will give a higher tumor uptake of gallium (102), how administration of scandium will increase the tumor/blood ratio by competitively displacing the gallium from serum transferrin (58), and the correlation between unsaturated iron binding capacity (UIBC) and tumor uptake of gallium (103). However, it does not offer reasons for the discrepancy between ^{67}Ga and ^{59}Fe distribution characteristics (104), the reported inhibitory effects of transferrin on tumor uptake (105), or why gallium accumulation in tumors of iron-deficient animals is not greater than that of animals fed a normal iron diet (106).

Doubting that endocytosis of foreign material was the primary uptake mechanism for tumors (59), other workers chose to examine the intramolecular distribution of gallium in an attempt to elucidate the intracellular receptors involved in the process. Early reports from autoradiographic (107), zonal ultracentrifugal and enzymatic (108), and conventional (109) techniques show that gallium-67 localizes in the lysosomes of both liver and tumor tissue. Large amounts of gallium-67 in the soluble portion of tissue homogenates (84,110) are attributed to the disruption of these organelles during homogenation (111) because of the large amount of ^{67}Ga-acid phosphatase (a lysosomal enzyme) in the preparations (59). More refined methods have shown that gallium-67 binds to a microsomal fraction which probably represents rough endoplasmic reticulum (111). In normal rat liver, most of the gallium localizes in lysosomal particles (64), but the hepatomas sequestered the majority of their ^{67}Ga in the smaller organelles (111).

It has been further shown that a majority (approximately 60%) of the extractable ^{67}Ga (about 70% of the cellular gallium) from tumor and liver cells of the rat is associated with two macromolecular fractions of molar weight 1-1.2 x 10^5 Daltons and 4-5 x 10^4 Daltons (64). The 1-1.2 x 10^5 D band is found in both liver and tumor cells, whereas the 4-5 x 10^4 D band is found primarily in tumor cells, although at minute concentrations only (25 μg of carrier gallium will saturate the binding of this component). Most of the liver uptake is associated with the heavy macromolecule, but 50% of the gallium-67 extracted from tumor cells associates with the low molecular weight fraction (64). This difference in complexation may represent an altered physiology between tumor and normal tissue. Both molecules are glycoproteins, and unstable to heat and alkalinity. Because plasma-bound ^{67}Ga is stable at pH 8.0 for several hours (54), the pH lability of these complexes is evidence that there are intracellular receptors for gallium which differ from the plasma proteins that bind to this metal (64). This pH-dependent process may represent the dissociation of a complex or the change of an ionic species [$Ga(OH)_3 \rightleftarrows Ga(OH)_4^-$] within the intracellular space.

Lactoferrin, with a molecular weight of 8.5-9.0 x 10^4 and a structure similar to transferrin (112) has been suggested as an alternative intracellular gallium-binding agent (113). Lactoferrin binds iron with a greater affinity than transferrin (114), and is found in tissues and secretions (especially milk) which localize gallium-67 (115,116,117,118,119). It was proposed that ^{67}Ga labeled to transferrin and other plasma proteins is transferred to cellular lactoferrin due to the latter's greater chelating ability (113-120). Such a transfer of gallium-67 has been demonstrated in vitro (121), and increased concentrations of the protein has been found in tumors (122,123,124).

This mechanism has been criticized since lactation not only produces lactoferrin (hence breast uptake of gallium-67) but also increased lysosomal activity which may account for increased radiogallium uptake in breast, milk, and tumor (125). However, it offers several advantages as well, one of which is elimination of the "dimerization" of the 4-5 x 10^5 MW molecule to explain the 83,000 MW results when tumor cell homogenates were analyzed using SDS:polyacrylamide gel electrophoresis (126). The lactoferrin hypothesis agrees with reports that (only) about one-third of the gallium-67 in tumors is associated with ferritin (52) and that gallium-67 was associated with 85,000-90,000 MW "fragments of degraded intracellular transferrin" (127). The author of this mechanism emphasized however, that lactoferrin levels are not elevated in all tumors, and ferritin may act as an alternative pathway for binding (128).

Perhaps the most exciting aspect of this proposed mechanism is the manner in which it correlates with recently-isolated ion-binding molecules called siderophores (129). The primary function of lactoferrin is to diminish the amount of extracellular free iron and thereby inhibit bacterial growth (130). Lactoferrin deposited by polymorphonuclear leukocytes is attached to the surface of monocytes and macrophages in inflammatory responses (130). Siderophores are synthesized by bacterial cells to sequester iron needed for growth, and therefore compete with lactoferrin (131). Upon the recent report of a siderophore-like substance isolated from the iron-deficient growth medium of a mammalian tumor tissue (132), it is likely that lactoferrin is deposited as a reactive response during competition for available extracellular iron. Since both lactoferrin and siderophores bind gallium readily (133), the gallium-67 uptake is expected to be higher in the region of a neoplasm.

In conclusion, one may see that the tissue distribution of gallium is well known and clinically useful, but the problem of subcellular localization and uptake mechanism remains to be solved. Studies to date seem to be in agreement on some points and in opposition on others. This state of affairs may merely reflect the variation in the pathology of the different neoplasms, or it may be showing us the complexity of the gallium uptake mechanism. The pharmacology of the gallium ion may be so ubiquitous that there is no single uptake mechanism for any given tumor type. This situation may also hold for other radiometals (such as ^{111}In or the radiolanthanides) which follow a large number of biochemical pathways. In any case more work is needed to find the various intracellular distributions of gallium and the possible uptake mechanism(s) in the hope that a mechanism can be isolated which will be used to optimize radiopharmaceuticals to selectively and rapidly partition the radionuclide from the plasma into the tumor cell.

Acknowledgements

This authors thank their many colleagues who were involved in various phases of this work, particularly Drs. M.M. Ter-Pogossian, B.A. Siegel, B. Kumar, and G. Ehrhardt.

This work was supported by DOE Contract DE-AS02-77EV04318.

Literature Cited

1. Edward, C.L.; Hayes, R.L. J. Nucl. Med., 1969, 10, 103.
2. Thakur, M.L.; Merrick, M.V.; Ganasekera, S.W. in "Radiopharmaceuticals and Labelled Compounds", Vol 2, pp 183-193, IAEA, Vienna, 1973.
3. Lavender, J.P.; Lowe, J.; Barker, J.R.; et al. Br. J. Radiol., 1971, 44, 361.
4. Thakur, M.L.; Coleman, R.E.; Welch, M.J.; et al. Radiol., 1976, 119, 731.
5. Thakur, M.L.; Welch, M.J.; Joist, J.H.; et al. Thromb. Res., 1976, 9, 345.
6. Hnatowich, D.J. Int. J. Appl. Radiat. & Isotopes, 1977, 28, 169.
7. Alverez, J. in "Radiopharmaceuticals", eds. Subramanian, Rhodes, Cooper, & Sodd, pp 102-107, Society of Nuclear Medicine, New York, 1975.

8. Burnham, C.A.; Brownell, G.L. IEEE Trans. Nucl. Sci., 1972, 19, 201.
9. Ter-Pogossian, M.M.; Phelps, M.; Hoffman, E.; et al. Radiol., 1975, 114, 89.
10. Harper, P.V. Int. J. Appl. Radiation & Isotopes, 1977, 28, 5.
11. Welch, M.J.; Welch, T.J. in "Radiopharmaceuticals" eds. Subramanian, Rhodes, Cooper, & Sodd, pp 73-79, Society of Nuclear Medicine, New York, 1975.
12. Hnatowich, D.J. personal communication.
13. McAfee, J.G.; Gagne, G.; Atkins, H.L.; et al. J. Nucl. Med., 1979, 20, 1273.
14. Sundberg, M.J.; Meares, C.F.; Goodwin, D.A.; et al. Nature, 1974, 250, 587.
15. Meares, C.F.; Goodwin, D.A.; Leung, C.S.H.; et al. PNAS,USA, 1976, 73, 3803.
16. Krejcarek, G.E.; Tucker, K.L. Biochem. Biophys. Res. Comm., 1977, 77, 581.
17. Stern, H.S.; Goodwin, D.A.; Wagner, H.N., Jr.; et al. Nucleonics, 1966, 24, 57.
18. Wochner, R.; Adatepe, M.; Van Amburg, A.; et al. J. Lab. Clin. Med., 1970, 75, 711.
19. Mahan, D.F.; Subramanian, G.; McAfee, J.G. J. Nucl. Med., 1973, 14, 651.
20. Stern, H.S.; Goodwin, D.A.; Scheffel, U.; et al. Nucleonics, 1967, 25, 62.
21. Potchen, E.J.; Adatepe, M.; Welch, M.J.; et al. JAMA, 1968, 205, 208.
22. Adatepe, M.; Welch, M.J.; Archer, E.; et al. J. Nucl. Med., 1968, 9, 426.
23. Budine, J.A. Radiology, 1969, 93, 605.
24. Rodriquez, J.; MacDonald, N.S.; Taplin, G.V. J. Nucl. Med., 1969, 10, 368.
25. Alvarez, J.; Maass, R.; Arriago, C. J. Nucl. Med., 1972, 13, 409.
26. O'Mara, R.E.; Subramanian, G.; McAfee, J.G.; et al. J. Nucl. Med., 1969, 10, 18.
27. Hosain, F.; Reba, R.C.; Wagner, H.N. Radiology, 1969, 93, 1135.
28. Subramanian, G.; McAfee, J.G.; Rosenstreich, M.; et al. J. Nucl. Med., 1975, 16, 1080.
29. Jones, A.G.; Davis, M.A.; Dewanjee, M.K. Radiology, 1975, 117, 727.
30. Dewanjee, M.K.; Kahn, P.C. Radiology, 1975, 117, 723.
31. Goodwin, D.A.; Song, C.H.; Finston, R.; Matin, P. Radiology, 1973, 108, 91.
32. Goodwin, D.A.; Sundberg, M.W.; Diamanti, C.I.; et al. in "Radiopharmaceuticals", eds. Subramanan, Rhodes, Cooper, Sodd, pp 80-101, Society of Nuclear Medicine, New York, 1975.
33. Thakur, M.L.; Coleman, R.E.; Welch, M.J. J. Lab. Clin. Med., 1977, 89,217.

34. Thakur, M.L.; Welch, M.J.; Joist, J.H.; et al. Thrombosis Res., 1976, 9, 325.
35. Thakur, M.L.; Segal, A.W.; Louis, L.; et al. J. Nucl. Med., 1977, 18, 1022.
36. Mathias, C.J.; Welch, M.J. J. Nucl. Med., 1979, 20, 659.
37. Davis, H.H.; Heaton, W.A.; Siegel, B.A.; et al. Lancet, 1978, I, 1185.
38. Heaton, W.A.; Davis, H.H.; Welch, M.J.; et al. Brit. J. Haematol., 1979, 42, 613.
39. Doherty, M.B.; Bushberg, J.T.; Lipton, M.J.; et al. Clinical Nucl. Med., 1978, 3, 108.
40. Rannie, G.H.; Thakur, M.L.; Ford, W.L. Clin. Exp. Immunol., 1977, 29, 509.
41. Greene, M.W.; Tucker, W.D. Int. J. Appl. Radiat. & Isotopes, 1961, 12, 62.
42. Anger, H.O.; Gottschalk, A. J. Nucl. Med., 1963, 4, 326.
43. Yano, Y. in "Radiopharmaceuticals from Generator-Produced Radionuclides", pp 117-125, IAEA, Vienna, 1970.
44. Ehrhardt, G.J.; Welch, M.J. J. Nucl. Med., 1978, 19, 925.
45. Welch, M.J.; Thakur, M.L.; Coleman, R.E.; et al. J. Nucl. Med., 1977, 18, 558.
46. Ehrhardt, G.J.; Head, R.; Djordjevic, L.; et al. J. Nucl. Med., 1979, 20, 682.
47. Arino, H..; Skraba, W.J.; Kramer, H.H. Int. J. Appl. Radiat. & Isotopes, 1978, 29, 117.
48. Neirinckx, R.D.; Davis, M.A. J. Nucl. Med., 1979, 20, 681.
49. Neirinckx, R.D.; Davis, M.A. J. Nucl. Med., 1980, 21, 81.
50. Ter-Pogossian, M.M.; Mullani, N.A.; Hood, J.; et al. Radiology, 1978, 128, 477.
51. Wagner, S.J.; Welch, M.J. J. Nucl. Med., 1979, 20, 428.
52. Clausen, J.; Edeling, C.J.; Fogh, J. Cancer Res., 1974, 34, 1931.
53. Gunasekera, S.W.; King, L.J.; Lavender, P.J. Clin. Chim. Acta, 1972, 39, 401.
54. Hartman, R.E.; Hayes, R.L. J. Pharmacol. Exp. Ther., 1969, 168, 193.
55. Larson, S.M. Sem. Nucl. Med., 1978, 8, 193.
56. Nelson, B.; Hayes, R.L.; Edwards, C.L.; et al. J. Nucl. Med., 1972, 13, 92.
57. Ito, Y.; Okuyama, S.; Takahashi, K.; et al. Radiology, 1971, 101, 355.
58. Hayes, R.L. in "Tumor Localization with Radioactive Agents", IAEA-MG-50/14, pp 29-45, IAEA, Vienna, 1976.
59. Hayes, R.L.; Brown, D.H. in "Nuklearmedizin: Forschritte der Nuklearmedizin in klinischer und technologischer Sicht", eds. Pabst, H.W.; Schmidt, H.D., pp 837-848, Schattauer Verlag, New York, 1975.
60. Fogh, J. Exp. Biol. Med., 1971, 138, 1086.
61. Edwards, C.L.; Hayes, R.L.; JAMA, 1970, 212, 1182.
62. Swartzendruber, D.C.; Hübner, K.F. Radiat. Res., 1973, 55, 457.

63. Hayes, R.L.; Edwards, C.L. in "Medical Radioisotope Scintigraphy 1972", Vol. 2, IAEA-SM-164/306, pp 531-552, IAEA, Vienna, 1973.
64. Hayes, R.L.; Carlton, J.E. Cancer Res, 1973, 33, 3265.
65. Hayes, R.L. in "Symposium on the Chemistry of Radiopharmaceuticals", eds. Heindel, N.D.; Burns, N.D.; Honda, T., pp 155-168, Symposium on the Chemistry of Radiopharmaceuticals, Masson, New York, 1978.
66. Andrews, G.A.; Root, S.W.; Kerman, H.D. Radiology, 1953, 61, 570.
67. King, E.R.; Brady, L.W.; Dudley, H.C. Arch. Intern. Med., 1952, 90, 785.
68. Dudley, H.C.; Maddox, G.E.; LaRue, H.C. J. Pharmacol. Exp. Ther., 1949, 96, 135.
69. Vaidya, S.G.; Chandri, M.A.; Morrison, R.; et al. Lancet, 1970, 2, 911.
70. Winchell, H.S.; Sanchez, P.D.; Watanabe, C.K.; et al. J. Nucl. Med., 1970, 11, 459.
71. Pinsky, S.M.; Hoffer, P.B.; Turner, D.A.; et al. J. Nucl. Med., 1971, 12, 385.
72. Langhammer, H.; Glaubitt, G.; Greve, S.F.; et al. J. Nucl. Med., 1972, 13, 25.
73. Johnston, G.S.; Benua, R.S.; Teates, C.D.; et al. J. Nucl. Med., 1974, 15, 399.
74. Greenlaw, R.H.; Weinstein, M.B.; Brill, A.B.; et al. J. Nucl. Med., 1974, 15, 404.
75. DeLand, F.H.; Sauerbrunn, B.J.L.; Boyd, C.; et al. J. Nucl. Med., 1974, 15, 408.
76. Johnston, G.S.; Go, M.G.; Benua, R.S.; et al. J. Nucl. Med., 1977, 18, 692.
77. Lomas, F.; Wagner, H.N. Radiology, 1972, 105, 689.
78. Littenberg, R.L.; Taketa, R.M.; Alazraki, N.P.; et al. Ann. Intern. Med., 1973, 79, 403.
79. Gelrud, L.G.; Arseneau, J.L.; Milder, M.S.; et al. J. Lab. Clin. Med., 1974, 83, 489.
80. Burleson, R.L.; Johnson, M.C.; Head, H. Ann. Surg., 1973, 178, 446.
81. Tsan, M.F.; Chen, W.F.; Scheffel, U.; et al. J. Nucl. Med., 1978, 19, 36.
82. Tsan, M.F.; Camargo, E.E.; Wagner, H.N. in "Second International Congress of the World Federation of Nuclear Medicine and Biology", p 83, Washington DC, 1978 (abst).
83. Tsan, M.F.; Scheffel, U. J. Nucl. Med., 1979, 20, 173.
84. Ito, Y.; Okuyama, S.; Sato, K.; et al. Radiology, 1971, 100, 357.
85. Anderson, W.A.D.; McCutcheon, M. in "Pathology", ed. Anderson, W.A.D., pp 13-51, Mosby, St. Louis, 1966.
86. Büchner, F. in "Allgemeine Pathologie", pp 313-369, Urban & Schwarzenberg, Munich, 1956.
87. Anghileri, L.J.; Heidbreder, M. Oncology, 1977, 34, 74.
88. Anghileri, L.J. J. Nucl. Biol. Med., 1973, 17, 177.

89. Anghileri, L.J. Strahlentherapie, 1973, 146, 359.
90. Eisenberg, E.; Wuthier, R.E.; Frank, R.B.; et al. Calc. Tiss. Res., 1970, 6, 32.
91. Seifert, G. Clin. Orthop., 1970, 69, 146.
92. Anghileri, L.J. Z. Krebsforsch, 1974, 81, 109.
93. Hickie, R.A.; Kalant, H. Cancer Res., 1967, 27, 1053.
94. Paterson, A.H.G.; Yoxull, A.; Smith, I.; et al. Cancer Res., 1976, 36 452.
95. Larson, S.M.; Rasey, J.S.; Allen, D.R.; et al. J. Nucl. Med., 1979, 20, 837.
96. Sephton, R.G.; Harris, A.W. J. Natl. Canc. Inst., 1974, 54, 1263.
97. Harris, A.W., Sephton, R.G. Cancer Res., 1977, 37, 3624.
98. Aulbert, E.; Gebhardt, A.; Schulz, E.; et al. Nuklearmedizin, 1976, 15, 185.
99. Hemmaplardh, D.; Morgan, E.H. Br. J. Haemotol., 1977, 36, 85.
100. Brockxmeer, F.M.; Morgan, E.H. Biochem. Biophys. Acta , 1977, 468, 437.
101. Larson, S.M.; Rasey, J.S.; Allen, D.R.; et al. J. Nucl. Med., 1979, 20, 843.
102. Bradley, W.P.; Alderson, P.O.; Eckelman, E.C.; et al. J. Nucl. Med., 1977, 18, 602.
103. Bradley, W.P.; Alderson, P.O.; Eckelman, E.C.; et al. J. Nucl. Med., 1978; 19, 204.
104. Sephton, R.G.; Hodgson, G.S., DeAbrew, S.; et al. J. Nucl. Med., 1978, 19, 930.
105. Gams, R.A.; Webb, J.; Glickson, J.D. Cancer Res., 1975, 35, 1422.
106. Bradley, W.P.; Alderson, P.O.; Weiss, J.F. J. Nucl. Med., 1979, 20, 243.
107. Swartzendruber, D.C.; Nelson, B.; Hayes, R.L. J. Natl. Cancer Inst., 1971, 46, 941.
108. Brown, D.H.; Swartzendruber, D.C.; Carlton, J.E.; et al. Cancer Res., 1973, 33, 2063.
109. Aulbert, E.; Hanbolt. U. Nucl. Med., 1974, 13, 72.
110. Orji, H. Strahlentherapie, 1972, 144, 192.
111. Brown, D.H.; Byrd, B.L.; Carlton, J.E.; et al. Cancer Res., 1976, 36, 956.
112. Groves, M.L. in "Milk Proteins, Chemistry and Molecular Biology", pp 367-376, Academic Press, New York, 1971.
113. Hoffer, P.B.; Huberty, J.; Khayam-Bashi, H. J. Nucl. Med., 1977, 18, 713.
114. Nagasawa, T.; Kiyosawa, I.; Takase, M. J. Dairy Sci., 1974, 57, 1159.
115. Groves, M.L. JACS, 1960, 82, 3345.
116. Masson, P.L.; Heremans, J.F.; Schonne, E.; et al. Protides Biol. Fluids Proc. Colloq., 1969, 16, 633.
117. Masson, P.L.; Heremans, J.F.; Dive, C.H. Clin. Chim. Acta, 1966, 14, 735.

118. Masson, P.L.; Heremans, J.F.; Schonne, E. J. Exp. Med., 1969, 130, 643.
119. Larson, S.M.; Milder, M.S.; Jonston, G.S. J. Nucl. Med., 1973, 14, 208.
120. Winchell, H.S. Sem. Nucl. Med., 1976, 6, 371.
121. Hoffer, P.B.; Huberty, J.P.; Khayam-Bashi, H. J. Nucl. Med., 1977, 18, 619.
122. Hoffer, P.B.; Miller-Catchpole, R.; Turner, D.A. J. Nucl. Med., 1979, 20, 424.
123. Loisillier, F.; Got, R.; Burtin, P.; et al. Protides Biol. Fluids, 1966, 14, 133.
124. de Sousa, M.; Smithyman, A.; Tan, C. Am. J. Pathol., 1978, 90, 497.
125. Hayes, R.L. J. Nucl. Med., 1977, 18, 740.
126. Lawless, D.; Brown, D.H.; Hübner, K.F.; et al. Cancer Res, 1978, 38, 4440.
127. Aulbert, E.; Gebhardt, A.; Schulz, E.; et al. Nucl. Med., 1976, 15, 185.
128. Hoffer, P.B. J. Nucl. Med., 1979, 19, 1082.
129. Bullen, J.J.; Rogers, H.J.; Leigh, L. Br. Med. J., 1972, 1, 69.
130. Van Snick, J.L.; Masson, P.L.; Heremans, J.F.: J. Exp. Med., 1974, 140, 1068.
131. Neilands, J.B. J. Am. Chem. Soc., 1952, 74, 4846.
132. Fernandez-Pol, J.A. FEBS Lett., 1978, 88, 345.
133. Emery, J. Biochemistry, 1971, 10, 1483.
134. McIntyre, P.A. in "Radiopharmaceuticals", ed. Subramanian, Rhodes, Cooper, Sodd, pp 343-348, Society of Nuclear Medicine, New York, 1975.
135. Glaubitt, D.H.M.; Schluter, I.H.; Hoberland, K.U.R. J. Nucl. Med., 1975, 16, 769.
136. Bell, E.G.; Maher, B.; McAfee, J.G.; et al. in "Radiopharmaceuticals", eds. Subramanian, Rhodes, Cooper, Sodd, pp 399-410, Society of Nuclear Medicine, New York, 1975.
137. Thakur, M.L. in "Radiopharmaceuticals and Other Compounds Labelled with Short-Lived Radionuclides", ed. Welch, M.J., pp 183-202, Pergamon Press, Elmsford, NY, 1977.
138. Goodwin, D.A.; Fiston, R.H.; Colombetti, L.G.; et al. Radiology, 1970, 94, 175.
139. Goodwin, D.A. J. Nucl. Med., 1971, 12, 580.
140. Frost, H.; Frost, P.; Wilcox, C.; et al. Int. J. Nucl. Med. and Biol., 1979, 6, 60.
141. Lavender, J.P.; Goldman, J.M.; Arnot, R.N.; et al. Brit. Med. J., 1977, 2, 797.

RECEIVED April 7, 1980.

ANTICANCER ACTIVITY OF METAL COMPLEXES

8

Introduction to Metal Complexes and the Treatment of Cancers

BARNETT ROSENBERG

Department of Biophysics, Michigan State University, East Lansing, MI 48824

It is somewhat surprising that the classic warhorse of metal coordination chemistry, Peryone's Chloride, a simple, inorganic platinum complex has proven to be one of the most potent of anticancer drugs. This chemical-cis-dichlorodiammineplatinum (II) (cisplatin) has now been approved in most countries of the world for the treatment of advanced, metastatic, testicular and ovarian cancers. It is now considered, when used in appropriate combination chemotherapy, to be curative for these cancers. Many recent clinical advances suggest that it will also be of significant utility in the treatment of other solid cancers such as those of the bladder, prostate, lung, head and neck, certain cancers in children, and finally, in other genitourinary cancers. These clinical trials have been underway for eight years now and are still continuing in an effort to increase the efficacy and decrease the toxicity of cisplatin, as well as broadening the spectrum of responsive cancers.

In the meantime, hundreds of other metal complexes have been shown to be active against various animal-tumor screens. These include predominantly analog structures of the parent drug, but also a scattering of complexes of metals other than platinum. Active pursuit of these areas may be highly rewarding.

It is curious, however, that while the value of the cisplatin drug is well established, knowledge of it's relevant chemistry and mechanisms of action remains still in a fairly primitive state. For example, if cisplatin is simply dissolved in water, the subsequent aquation reactions are numerous and complex. The diammine ligands are not likely to be exchanged under these conditions, but the chloride ligands are. These are sequentially exchanged for water or hydroxyl ligands. The extent of the exchanges is primarily controlled by the chloride concentration in the solution. However, further reactions of the aquated species are known to occur, forming a variety of oligomeric species, including a dimer, a cyclic trimer and a tetramer. The relative concentrations at equilibrium of all of these products is markedly dependent on pH and temperature. Thus the relatively

0-8412-0588-4/80/47-140-143$05.00/0

innocuous action of dissolving cisplatin in water already introduces a large variety of products. This makes any simple interpretation of the biological reactions dubious at the very least.

Neverthless, some guiding interpretations for further experimental work are necessary, even if these be, admittedly, simple minded. We will assume that upon injection of the cisplatin into the extracellular fluid of an animal the high chloride concentration (on the order of 100 milliequivalents per liter) limits any aquation reaction, and the drug remains intact as it courses through the body. Additional reactions of cisplatin with molecules in the blood will have to be left to the pharmacokineticists to sort out. There is evidence to suggest that the intact drug passively permeates cell membranes (viz, no active carrier is necessary). Usually, the intracellular chloride concentration is lower than the extracellular value. In some cells of the body, such as muscle cells, the chloride concentration may be as low as 10 milliequivalents per liter, while in other cells, such as the epithelial cells lining the stomach and intestines, the chloride concentration is approximately the same as in the extracellular fluid. In those cells where the chloride concentration is low, the aquation reaction will occur to some degree. This is the only step necessary to activate the drug. The aquated species are now able to react with various intracellular molecules, and in particular, the nucleic acids. Evidence from both _in vitro_ and _in vivo_ studies strongly imply that the primary target molecule leading to the significant biologic actions is the cellular DNA.

The reactions of the aquated cisplatin species with DNA again provides a plethora of reaction sites, one or more of which may be the significant ligand exchange leading to the anticancer activity. It is the task of those attempting to understand the mechanisms of drug action to unravel the results of these different reactions. A higher order of complexity is added to this system in that the different aquated species may each produce a different set of reaction products with DNA.

It is quite clear by now that drug action of the dichlorodiammineplatinum (II) is stereoselective. For anticancer activity to occur the drug must be in the _cis_ configuration. The _trans_ isomer is inactive. This is true for all the other analogs tested so far. We may try to use this stereoselectivity as a tool to pry out the significant ligand exchanges. We first will limit our considerations to those reactions of aquated cisplatin which the _cis_ isomer is capable of but the _trans_ isomer is not. Of this sub-class of reactions, one is most interesting. This is the formation of a closed ring chelate of cisplatin with the N-7 and O-6 nucleophilic sites of guanine. It had been shown previously that the cisplatin reacted primarily with the GC rich regions of DNA. It has also been suggested that the tertiary structure of DNA is probably too plastic to exhibit the necessary stereoselectivity. It is also known that the cisplatin does not intercalate. And it is believed that reactions with the phosphate-sugar chains are not relevant.

Both intra-and interstrand crosslinkings of DNA by the aquated cisplatin and it's trans isomer are known to occur. Crosslinking has been traditionally invoked to account for cytotoxicity. In this case, however, evidence exists that equal numbers of trans complex crosslinks as cis complex crosslinks cause the same degree of cytotoxicity-but the trans is not an active anticancer agent. I emphasize the importance of separating cytotoxic and anticancer effects. All cellular poisons are not anticancer agents. Somehow a degree of selectivity is necessary in order to have large tumor masses disappear with little or no toxicity to the normal cells of the body.

What is lacking at the present time is hard chemical evidence for the existence of the N-7, O-6 chelate complex with guanine. Reactions of aquated cisplatin with the various nucleophilic sites of guanine produces a multiplicity of reaction products. These are the kinetically allowed species. They form within 48 hours of reaction time. With longer incubation times, however, the number of reaction products decreases. Those remaining are likely to represent the more thermodynamically stable products. These are also the ones that usually end up in the crystallographers hands. It must be pointed out, however, that the significant lesion need exist for no more than one to two days within the cell (one replication of the DNA), and therefore, this lesion may well be one of the less thermodynamically stable products. Therefore, in the court of last resort, crystallography, we still may not obtain a final judgement.

One of the aspects of the chelate complex that makes it particularly attractive to me is the involvement of the O-6 site of guanine. There is an intriguing story developing in the field of carcinogenesis by alkylating agents, which now implicates the alkylation at the O-6 site as possibibly the most relevant in causing mutations in somatic cells. This is considered to be a necessary, but not sufficient step, in the transformation of the cell into a cancer cell. Sufficiency occurs when the lesion is not repaired prior to DNA replication and leads to a mispairing with thymine instead of the correct pairing with cytosine. This then leads on further replication to the replacement of the original GC pair by an AT pair-a base substitution mutation. Whether such mutations need to occur in particular regions of the DNA is not yet clear.

The hypothesis of the O-6 guanine involvement does provide an extra bonus in that a mechanism of action can be postulated, and tested, which allows an explanation of the selective destruction of cancer cells. If the cancer cell becomes so because of its inability to repair the O-6 guanine lesion caused by a carcinogen then it may also be unable to repair the cisplatin induced damage. But the normal cells have intact repair mechanisms and can repair the damage prior to DNA replication and thus survive. Tests of this mechanism of selectivity are in progress.

The discovery of a new class of anticancer drugs based on metal complexes affords us a new opportunity to reexamine the problems of cancer chemotherapy. It is obvious that interactions of metals with DNA is largely an undeveloped field of study, but is too important to remain so.

RECEIVED April 7, 1980.

9

Binding of a Platinum Antitumor Drug to its Likely Biological Targets

STEPHEN J. LIPPARD

Department of Chemistry, Columbia University, New York, NY 10027

Cis-dichlorodiammineplatinum(II) (cis-DDP) is currently being used to treat a wide variety of cancers (1). Ever since the initial discovery that cis-DDP has unusual biological activity (2) not shared by the trans isomer, coordination chemists and biologists have strived to understand the mechanism of action

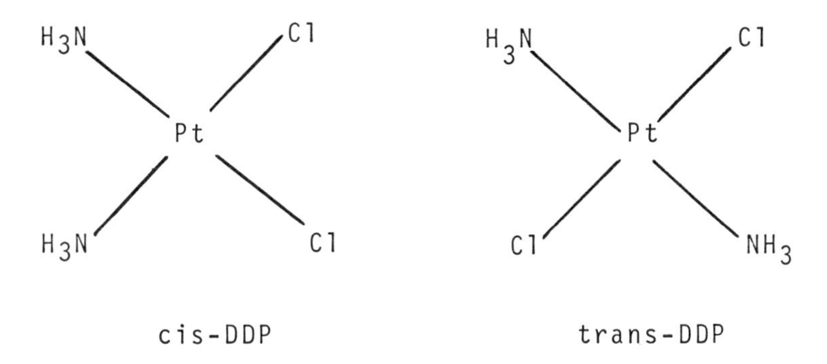

and the reason for the selectivity of the cis isomer. Numerous studies have been carried out that strongly implicate DNA as the target of drug action (3).

The reactivity of cis-DDP in a biological milieu is critically dependent upon the chloride ion concentration (4). Hydrolysis reactions of the kind shown in eq. 1 produce aquo complexes that are kinetically more reactive than the chloro or hydroxo complexes (5). When the drug is administered, there-

0-8412-0588-4/80/47-140-147$05.00/0

$$PtCl_2(NH_3)_2 \underset{+Cl^-}{\overset{-Cl^-}{\rightleftharpoons}} Pt(NH_3)_2Cl(H_2O)^+ \underset{+Cl^-}{\overset{-Cl^-}{\rightleftharpoons}} Pt(NH_3)_2(OH_2)_2^{+2} \tag{1}$$

$$Pt(NH_3)_2Cl(H_2O)^+ \underset{+H^+}{\overset{-H^+}{\rightleftharpoons}} Pt(NH_3)_2Cl(OH) \qquad Pt(NH_3)_2(OH_2)_2^{+2} \underset{+H^+}{\overset{-H^+}{\rightleftharpoons}} Pt(NH_3)_2(OH_2)(OH)^+$$

fore, the high chloride ion concentration (∿0.1 M) in blood suppresses its reactivity. After diffusion across the cytoplasmic membrane, the drug encounters a chloride concentration of 4 mM and reactions with biological targets can take place.

Since DNA is most likely the target of drug action, our discussion will focus on its reactions with cis- and trans-DDP. We shall not attempt to rationalize the ability of cis-DDP to kill cancer cells before destroying normal cells, a requirement of any useful carcinostatic reagent. This specificity cannot be understood simply by examining the chemistry of a drug with its likely biological target. Stimulating discussions of this aspect of the problem are available (3,6).

Binding to the Nucleosome Core Particle

Each chromosome of a eukaryotic cell contains DNA that is probably a single molecule of several centimeters in length if laid out straight (7). Within the nucleus, however, the DNA is folded into a highly compact form having several levels of structural organization. As shown in Figure 1, the simplest building block of chromatin (8,9), the name given to the extractable chromosomal material, is the nucleosome core particle. These particles can be obtained following digestion of chromatin with micrococcal nuclease. They consist of ∿146 base pairs of DNA wrapped in a shallow superhelix about an aggregate of eight histone proteins (Figure 2). During cell division the nuclear DNA is replicated. It is probable that the antitumor activity of cis-DDP involves attack on the nucleosomal DNA.

Studies of the binding of cis- and trans-DDP to

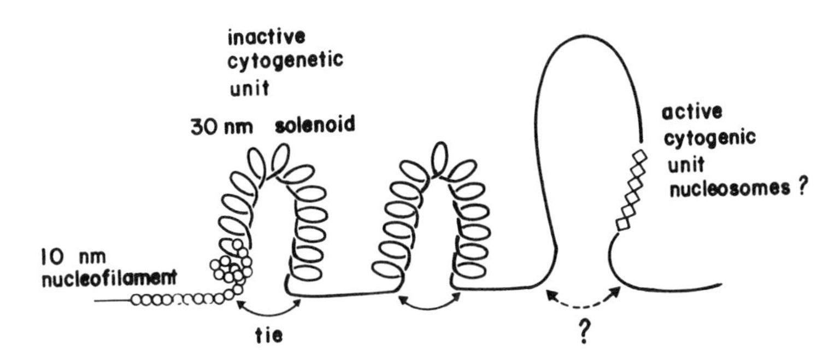

Figure 1. Schematic of the possible levels of organization in chromatin (7)

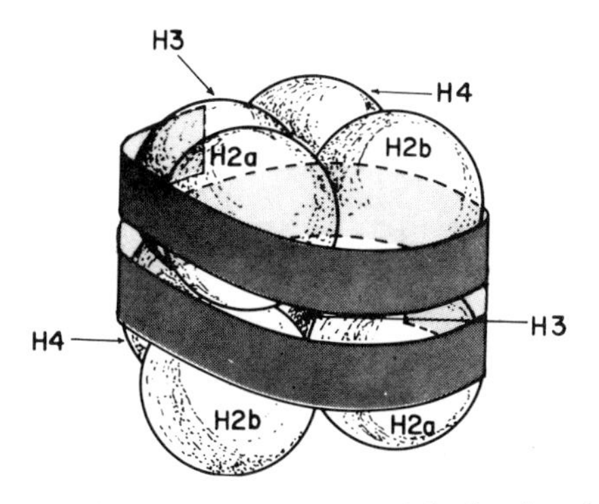

Figure 2. Structure of the nucleosome core particle showing the probable (but not established) relative positions of the histone proteins and the surrounding DNA (23). The labels H2a, H2b, H3, and H4 refer to the four different histone proteins present as two copies each in the nucleosome core particle. The DNA is wrapped around the outside of the protein cluster in a shallow superhelix of 1¾ turns. The approximate dimensions of the particle are 110 × 110 × 57Å. A twofold symmetry axis passes through the particle.

nucleosome core particles have revealed striking differences in reactivity (10). At low ratios of bound platinum atoms per core particle (∿5 - 10), the cis isomer binds mainly to the DNA while the trans complex forms DNA-protein and protein-protein crosslinks. The greater ability of *trans*-DDP to crosslink biopolymers is reasonable since non-bonded steric repulsions between macromolecules will be less than for *cis*-DDP induced crosslinks. *Cis*-DDP forms analogous crosslinks but only at much greater binding ratios. The crosslinks occur because the platinum atom coordinates to protein amino acids and/or DNA bases belonging to two or more different constituents of the nucleosome core particle. This feature was demonstrated through gel electrophoresis studies using ^{195m}Pt radiolabeled *cis*- and *trans*-DDP. In the case of *trans*-DDP, specific crosslinking of histone protein pairs H3/H2a and H2b/H4 was identified. The band containing the crosslinked histones was sliced out of a polyacrylamide gel and soaked in cyanide solution to remove the platinum as $[Pt(CN)_4]^{2-}$ (11). The resulting gel slice was then run in a second dimension to determine the histone proteins specifically crosslinked.

The differences revealed in the gel electrophoretic pattern of the nucleosome core particle after binding *cis*- and *trans*-DDP (10) could possibly provide a simple *in vitro* screen for platinum antitumor drug activity. Preliminary studies (12) have shown that nucleosome cores incubated with either dichloroethylenediamineplatinum(II) or *cis*-dichlorobis(isopropylamine)-platinum(II), two known antitumor drugs, exhibit gel electrophoretic patterns very similar to those of nucleosome cores incubated with *cis*-DDP. Incubation with [(terpy)PtCl]Cl, an inactive compound, gave very different gel patterns. Further work is in progress to evaluate the utility of this assay.

The technique of DNA alkaline elution has demonstrated the greater ability of *trans*- than *cis*-DDP to induce protein-DNA crosslinking in L1210 mouse leukemia cultured cells (13). The extent of DNA-protein crosslinking did not correlate with the cytotoxicity of *cis*- and *trans*-DDP. There was a correlation of intrastrand DNA crosslinking with cytotoxicity, however, with the cis isomer being more effective. Intrastrand crosslinks are probably responsible for the retardation and spreading of the nucleosomal core particle DNA on gels after reaction with *cis*-DDP (10).

In summary, the studies of the binding of *cis*- and *trans*-DDP with the nucleosome core particle have shown a clear difference in reactivity for the two isomers, consistent with their molecular structures. The relevance of this discovery to the cytotoxicity and greater antitumor drug action of the cis isomer is not obvious. Perhaps *trans*-DDP, the more reactive crosslinking reagent, is scavenged *in vivo* before it can reach the nuclear DNA.

Binding to Closed Circular DNA

Closed circular DNAs isolated from *Escherichia coli* strain K12 W677 containing the plasmid pSM1 (14) have been used to monitor the binding of *cis*- and *trans*-DDP. Closed circular DNAs are more useful than linear DNAs for such studies because small changes in the structure of the DNA, for example unwinding of the duplex, produce large effects in the hydrodynamic properties that are readily measured. Covalent binding of both platinum complexes to pSM1 DNAs changes the degree of supercoiling, a result that was attributed to disruption and unwinding of the double helix (15). Electron micrographs showed the platinated DNAs to be shortened by up to 50% of their original length. As shown in Figure 3, a similar shortening occurs upon binding of *cis*-DDP to pM2 DNA (16).

The unwinding and shortening of the DNA double helix in the presence of *cis*-DDP most likely involves disruption of the base pairs accompanied by coordination of the platinum atom to one or more of the heterocyclic nitrogen atoms of the bases. Since *cis*-DDP is biologically active at very low levels ($<10^{-3}$) of bound platinum per DNA phosphate (3), it is possible that the drug recognizes a specific sequence of bases in the DNA chain. The most likely candidates are those rich in guanine-cytosine base pairs. Earlier studies revealed that *cis*-DDP binds more strongly to DNAs with high (G + C) content, and especially to poly(dG)·poly (dC) (17,18). Moreover, intrastrand crosslinking of GpG was proposed (19) to account for the very smeared gel electrophoretic pattern of λ DNA digested by the restriction endonuclease Bam H1 in the presence of [(en)$PtCl_2$], a reagent similar in structure and drug activity to *cis*-DDP. The Bam H1 enzyme recognizes the following sequence, cutting the two polynucleotide

$$\begin{array}{c} \downarrow \\ -G\,G\,A\,T\,C\,C- \\ -C\,C\,T\,A\,G\,G- \\ \uparrow \end{array}$$

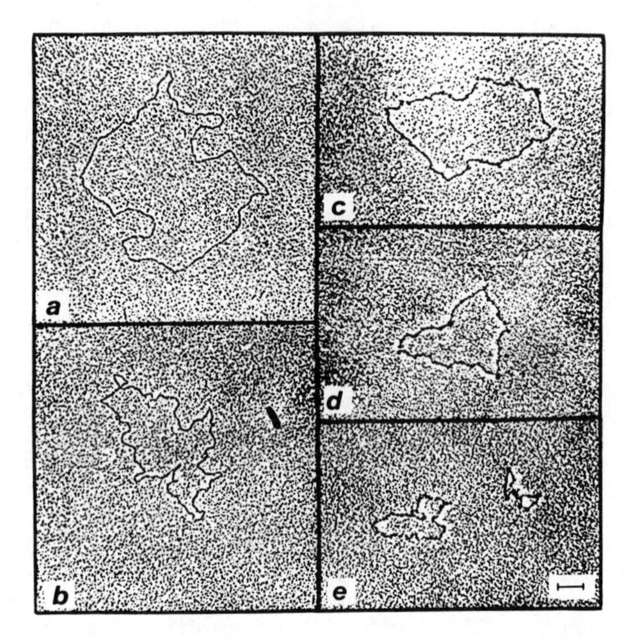

Biochimie

Figure 3. Electron micrographs of the PM2 DNA molecule containing: (a) 0, (b) 0.0125, (c) 0.075, (d) 0.10, and (e) 0.30 mol bound cis-*DDP per DNA phosphate (16)*

chains where shown by the arrows. The addition of cyanide ion to the incubation mixture sharpened the bands. These results (19) were interpreted to imply that $[(en)PtCl_2]$ crosslinks the GpG unit in λ DNA. Unfortunately, this interpretation is not uniquely justified by the data. Since retardation and smearing of the gel pattern of the 146 base pair nucleosome core particle DNA occurs after treatment with *cis*-DDP (10), the smearing of the fragments in the *Bam HI* digest gel could have resulted from platinum binding in a region remote from the cutting site. Addition of cyanide would remove platinum as $[Pt(CN)_4]^{2-}$ (11) and produce the observed sharpening of the bands.

In a different experiment, the closed circular pSM1 DNA was incubated with *cis*-DDP and then treated with the restriction enzyme *Pst* I (20) after removing unbound platinum. As shown in Figure 4, this enzyme normally cleaves the DNA into four fragments, recognizing the

- C T G C A↓G -
- G↑A C G T C -

site. Platinum binding was followed by removing aliquots from the incubation mixture, separating out free platinum by spin dialysis (21), and analyzing for bound platinum by atomic absorption spectroscopy. When 0.05 moles of platinum were bound per mole of DNA phosphate, the enzyme cutting was fully inhibited. Below that ratio (r), gel electrophoretic bands corresponding to fragments A - D (Figure 4) disappeared while partial fragments AC, CD, DB, BA, ACD, CDB, DBA, and BAC arose with increasing platinum binding. Interestingly, the BD partial was the first to appear, concomitant with the early removal of the B and D fragments from the gel at r ∿0.004. This behavior cannot be the result of selective binding of *cis*-DDP to bases within the restriction enzyme recognition site or all four cutting sites would have been equally protected. Rather, a sequence adjacent to the D-B junction was proposed to be responsible for the selective inhibition of the Pst I enzyme in that region. As shown in Fig. 4, a unique $(dG)_4$—$(dC)_4$ sequence occurs on the right flank of the D-B junction. The binding of *cis*-DDP to adjacent guanine or cytosine bases on one DNA strand is stereochemically feasible, as shown by a model building study (20). Interestingly, models reveal that *trans*-DDP cannot bind the N-7 atoms of both guanine bases in the (GpG) fragment (18,20). At comparably low binding levels, *trans*-DDP did not produce the selective inhibition of enzyme cleavage observed for the cis isomer.

Additional work in progress should enable the

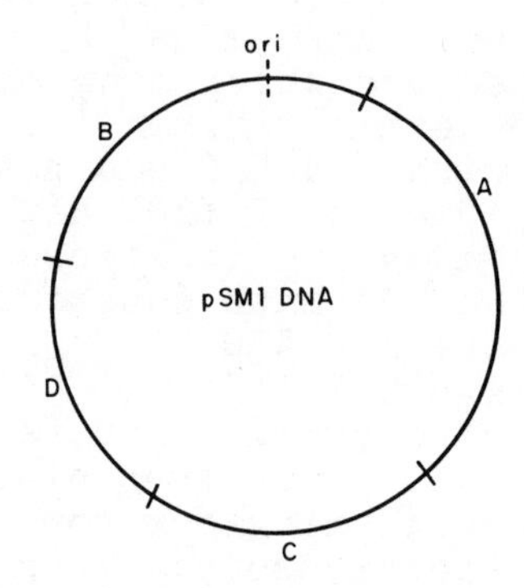

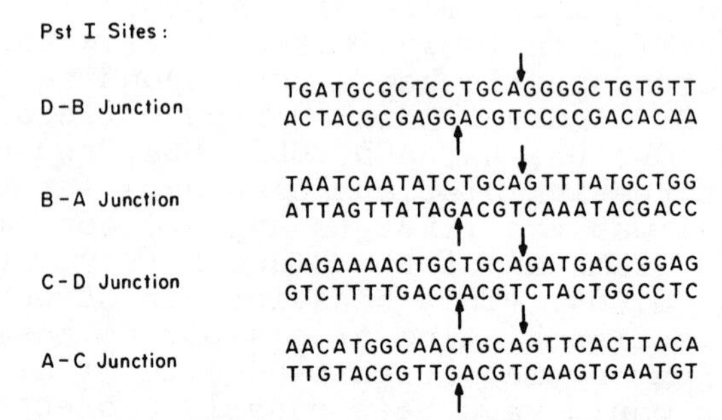

Figure 4. Map of bacterial plasmid pSMl DNA showing the origin of replication (ori) and cleavage pattern by the restriction endonuclease Pst I. Fragments A, B, C, and D are 1.80, 1.60, 1.19, and 1.09 kilobases in length, respectively. The sequence of bases at each of the four cutting sites also is shown.

specific bases that bind platinum to be identified. In the meantime, it is a reasonable hypothesis that the antitumor drug activity of cis-DDP involves recognition of the $(dG)_n(dC)_n$, $n \leq 4$, sequence in DNA. More specifically, the mechanism probably involves intrastrand crosslinking of adjacent guanine bases, as proposed previously (3,22), or possibly adjacent cytosine bases. This supposition, if confirmed, could lead to the rational design and development of more effective drugs.

Acknowledgment

The work cited here was supported by NIH grant CA-15826 and by a Senior International Fellowship from the Fogarty International Center (F06-TW00252). I also wish to thank my co-workers and collaborators, especially Dr. W.R. Bauer, and Engelhard Industries for generous loans of platinum.

Abstract

The binding of cis-dichlorodiammineplatinum(II), a powerful anticancer drug, to nucleosomes and to DNA is reviewed. The drug preferentially attacks the DNA of the nucleosome core particle, the basic building block of the chromosomes comprised of 146 base pairs of DNA wrapped in a shallow superhelix about an octameric aggregate of histone proteins. By contrast, the biologically ineffective trans isomer forms DNA-histone and histone-histone crosslinks. The unwinding of DNA by cis-dichlorodiammineplatinum and its recognition of a specific base pair sequence support a mechanism for cytotoxicity involving intrastrand crosslinking of adjacent guanine residues.

Literature Cited

1. Burchenal, J.H. Biochimie, 1978, 60, 915.
2. Rosenberg, B.; Van Camp, L.; Krigas, T. Nature (London), 1965, 205, 698.
3. Roberts, J.J.; Thompson, A.J. Prog. Nucleic Acid Res. Mol. Biol., 1979, 22, 7, and references cited therein.
4. Lim, M.C.; Martin, R.B. J. Inorg. Nucl. Chem., 1976, 38, 1911.
5. Johnson, N.P.; Hoeschele, J.D.; Rahn, R.O. Chem.-Biol. Interact., in press.

6. Rosenberg, B. Interdisc. Sci. Rev., 1978, 3, 134.
7. Klug, A. Phil. Trans. R. Soc. Lond. B, 1978, 283, 233.
8. Kornberg, R.D. Ann. Rev. Biochem., 1977, 46, 931.
9. Felsenfeld, G. Nature, 1978, 271, 115.
10. Lippard, S.J.; Hoeschele, J.D. Proc. Natl. Acad. Sci. U.S.A., 1979, 76, 6091.
11. Bauer, W.; Gonias, S.L.; Kam, S.K.; Wu, K.C.; Lippard, S.J. Biochemistry, 1978, 17, 1060.
12. Lippard, S.J., unpublished results.
13. Zwelling, L.A.; Anderson, T.; Kohn, K.W. Cancer Research, 1979, 39, 365.
14. Mickel, S.; Bauer, W.R. J. Bacteriol., 1976, 127, 644.
15. Cohen, G.L.; Bauer, W.R.; Barton, J.K.; Lippard, S.J. Science, 1979, 203, 1014.
16. Macquet, J.-P.; Butour, J.-L. Biochimie, 1978, 60, 901.
17. Stone, P.J.; Kelman, A.D.; Sinex, F.M. Nature, 1974, 251, 736.
18. Stone, P.J.; Kelman, A.D.; Sinex, F.M.; Bhargava, M.M.; Halvorson, H.O. J. Mol. Biol., 1976, 104, 793.
19. Kelman, A.D.; Buchbinder, M. Biochimie, 1978, 60, 893.
20. Cohen, G.L.; Ledner, J.; Bauer, W.R.; Ushay, H.M.; Caravana, C.; Lippard, S.J. J. Am. Chem. Soc., 1980, 102, 2487.
21. Neal, M.; Florini, J. Anal. Biochem., 1973, 55, 328.
22. Kelman, A.D. Peresie, H.J.; Stone, P.J. J. Clin. Hematol. Oncol., 1977, 7, 440.
23. Finch, J.T.; Lutter, L.C.; Rhodes, D.; Brown, R.S.; Rushton, B.; Levitt, M.; Klug, A. Nature, 1977, 269, 29.

RECEIVED April 18, 1980.

10

The Potential of Ruthenium in Anticancer Pharmaceuticals

MICHAEL J. CLARKE

Department of Chemistry, Boston College, Chestnut Hill, MA 02167

Pertinent aspects of the chemistry of Ru(II) and Ru(III) complexes are briefly outlined with regard to the utility of these ions in anticancer pharmaceuticals. Of particular interest is that Ru(III) ammine complexes containing hard ligands may be activated by reduction *in vivo* to lose the acido groups and subsequently add a nitrogen base, which can then be firmly coordinated to the metal in either oxidation state. Work with nucleosides and nucleic acids indicates that the coordination of Ru(II) to such ligands is similar to that of the Pt(II) anticancer drugs, so that analogous effects might be exerted on nucleic acid metabolism. Ammineruthenium(III) ions, on the other hand, form unusual cytosinato and adenosinato species. Collaborative studies demonstrate that a number of Ru compounds serve as bacterial mutagens, and so indicate that at least some Ru complexes are capable of damaging genetic material. The *in vivo* production of the more easily substituted Ru(II) aquoammine species from the Ru(III) prodrug should be favored in the relatively reducing and hypoxic environment provided by the interior of many tumors. *In vitro* experiments utilizing subcellular components as electron-transfer catalysts provide support for this. Screening studies on a series of Ru compounds show that many complexes, which would be thought to function by an activation-by-reduction mechanism, do exhibit anti-tumor activity. Tissue distribution studies by other workers reveal significant concentrations of ruthenium, injected as cis-$[Cl_2(NH_3)_4Ru]Cl_2$, in tumor tissue. The potential of ruthenium isotopes for incorporation into radiodiagnostic pharmaceuticals for imaging tumors is also briefly discussed.

0-8412-0588-4/80/47-140-157$06.00/0

Possible applications of ruthenium complexes in the treatment of cancer have been recognized by workers in diverse areas and the utilization of this metal has been approached from widely different perspectives. Early interest centered on the therapeutic properties of ruthenium complexes with aromatic chelates (1) and then, following a dormancy, efforts became focused on complexes bearing analogies to cis-$Cl_2(NH_3)_2Pt$ (2,3). Slightly later, the suggestion was made that ^{97}Ru could provide the basis for a family of radiodiagnostic agents for organ imaging (4). This suggestion holds the promise that tumors may be specifically imaged, located and diagnosed with the help of tumor-localizing Ru-containing radiopharmaceuticals (5,6). In addition to radioscintigraphic agents of this type, tumor-specificity is a desired (but not a required) property for chemotherapeutic pharmaceuticals. A final category of anticancer drugs, which has not been addressed with ruthenium-containing compounds, is that of radiotherapeutic pharmaceuticals, which would provide a dose of short range radiation directly at the tumor site. These might be possible with the β-emitting radionuclides, ^{103}Ru or ^{106}Ru, provided that a high degree of tumor-localization could be obtained.

Approaches to the incorporation of ^{97}Ru or ^{103}Ru into radioscintigraphic agents include: 1) The use of ruthenium-red (7) a μ-oxo trimer which is known to bind preferentially to acidic animal mucopolysaccharides. The stroma of many neoplasms are high in these materials and so could conceivably cause the ruthenium dye to concentrate in some tumors. 2) A range of ruthenocene complexes in which the cyclopentadienyl moieties serve as derivatizable units for the attachment of organ specific molecules (8-10). 3) Complexes with biomolecules which are known to concentrate in tumors. Particular interest centers on the bleomycins, a group of tumor specific antibiotics which induce nucleic acid cleavage. Glucose, nucleic acid and protein precursors, DNA and some proteins also tend to concentrate in some types of tumor cells. With many of these biochemicals the relatively high affinity of Ru(II) and Ru(III) ions for nitrogen ligands can be taken to advantage. 4) The use of Ru(III) complexes as prodrugs, which can be transformed by the body into more active species which, in turn, should behave similarly to the platinum chemotherapeutic agents (11).

The last approach depends upon particular aspects of ruthenium in the II and III oxidation states as well as certain differences between tumor and normal tissue metabolism. It is, in concept, applicable to each of the major categories of anticancer pharmaceuticals mentioned above and provides a rational basis for drug design. The remainder of this report will focus upon the chemical and biological hypotheses involved in this approach and will present the results of experiments which these have guided.

Ruthenium-Nucleoside Interactions.

The initial phases of this work were carried out in the laboratory of Henry Taube at Stanford and were initiated to explore the effects of a relatively "hard" metal ion, and a relatively "soft", pi-donor metal ion, on nucleosides and nucleotides (12-14). Because of their simplicity, high affinity for nitrogen heterocycles, and ease of switching oxidation states, the pentammineruthenium (II-III) ions were chosen for study. A primary focus of these studies, which were begun before it was widely known that the activity of the platinum anticancer drugs probably depended upon nucleic acid binding, was to examine novel modes of metal-purine coordination, particularly carbenoid binding (14,15)

Nucleosides offer a number and a variety of metal coordination sites, but many of these can be eliminated by judicious alkylation of the heterocycle. This coupled with unique spectral patterns arising from ligand-to-metal charge transfer (LMCT) transitions and predictable effects of the metal ion on the ionizing ability of ring protons usually eliminates the need for structural assignments by x-ray crystallography. Indeed, the cautious interpretation of chemical and spectroscopic data has provided the correct assignment of the many possible linkage isomers in every case involving ammineruthenium(III) ions that have subsequently been verified by x-ray methods. Moreover, solution conditions can be chosen so that only a single species exists and, owing to the relative inertness to substitution of Ru(II) and Ru(III) ions, these complexes persist for periods of time sufficient for chemical and biochemical study.

The dominant mode of pentaammineruthenium coordination to purine nucleosides with a keto group at the 6-position is at the N(7) site on the imidazole ring (12-14,16). Figure 1 illustrates this mode of coordination and the hydrogen bonds, which further add stability to the complex, that form between coordinated ammines and O(6). Alkylation at N(9), as occurs in nucleosides, prevents metal binding at both the N(9) and N(3) sites. At low pH protonation is preferred over metallation at N(1) and no ruthenium complexes involving N(1) coordination to this type of nucleoside have been isolated or characterized. Attachment at the N(7) of deoxyguanosine, which is thought to be the initial point of attack of the platinum pharmaceuticals on DNA, has been shown to occur for ammineruthenium(II and III) ions (13,17).

Figure 2 indicates that N(9) coordination is indeed

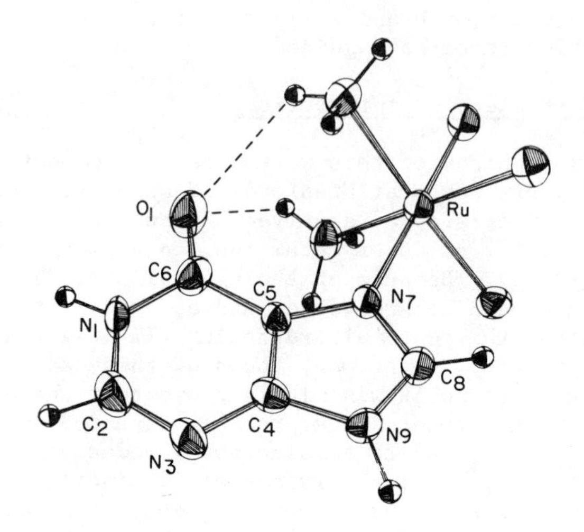

Figure 1. Structure of 7-[(Hyp)$(NH_3)_5$Ru(III)]$^{3+}$ (18)

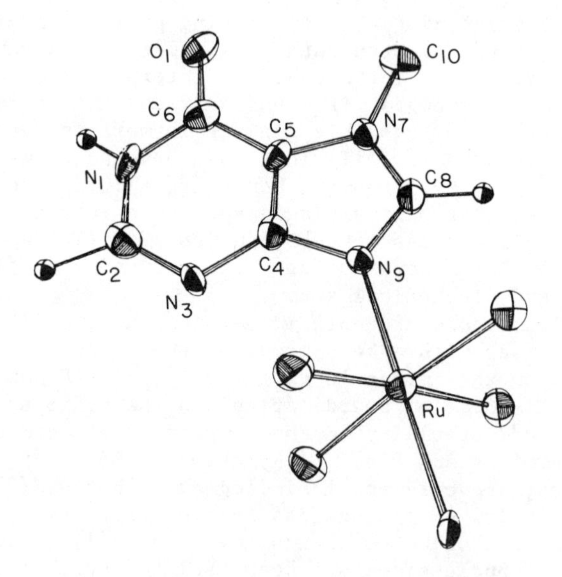

Figure 2. Structure of 9-[(Hyp)$(NH_3)_5$Ru(III)] (18)

possible, when this site is available (18). The corresponding N(3)-bound complex has also been prepared pure in solution (16), but has not been fully characterized as a solid. A novel mode of purine binding which, among true transition metal ions, has only been observed for Ru(II and III) is shown in Figure 3 (14,15). In this case, the caffeine ligand is bound via the C(8) position. Purines other than xanthines have not yielded to attempts to form ylidene or carbenoid complexes of this type.

The $(NH_3)_5Ru(III)$ group is also the only species known to form stable monodentate complexes with cytidine, adenosine and related nucleosides via coordination to the exocyclic nitrogen (Figure 4) (19). The Ru-N(4) bond in the cytidine complex is approximately 0.13 Å shorter than that expected for a Ru(III)-N single bond (20) and indicates that the partially filled $d\pi$-orbital on the metal is accepting some degree of electron density from a $p\pi$-orbital on the nitrogen. While complexes of this type are stable over a wide range of pH, they are most readily formed by redox catalysis at neutral pH. Catalytic synthesis in the presence of a small amount of Ru(II) suggests that initial attack probably occurs by the metal in this oxidation state on the available pyrimidine ring nitrogen (19). Oxidation of the metal ion to Ru(III) should facilitate deprotonation of the exocyclic amine, thus allowing for a subsequent and fairly rapid ring-to-exocyclic nitrogen linkage isomerization. Spectral studies suggest that at low pH, reprotonation occurs at the adjacent ring nitrogen rather than on the exo-N (19).

Since adenine occurs in a variety of water soluble coenzymes, these can also coordinate Ru(III) via the exo-N site (21). Other nucleosides (or near nucleosides) such as riboflavin are capable of forming stable ruthenium adducts and such coordination may severely affect the coenzymic activity (22). Figure 5 illustrates the mode of metal binding and the structural bending induced in a flavin. The pi-retrodative bonding in these species is intense and, interestingly, the Ru(II)-N(5) bond length (1.979 Å) is quite close to that of the Ru(III)-N(4)bond in the cytosinato complex (1.983 Å, Figure 4). The surprisingly similar geometries around the ligand nitrogen in both of these complexes suggest a near equivalency in the mode of binding, even though one would normally be considered to be a Ru(II) complex and the other Ru(III). In fact, the formulation of the flavin complexes of $Ru(II)-Fl_{ox}$ does not always appear to be appropriate and the canonical form $Ru(III)-Fl^{\overline{\cdot}}$ may be preferred for some purposes (23).

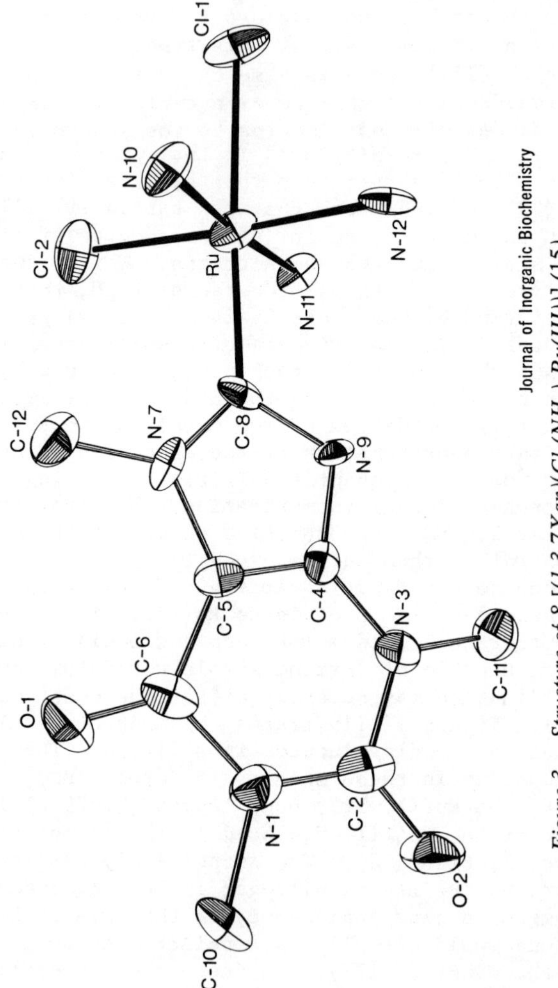

Journal of Inorganic Biochemistry

Figure 3. Structure of 8-[(1,3,7Xan)($Cl_2(NH_3)_3$Ru(III))] (15)

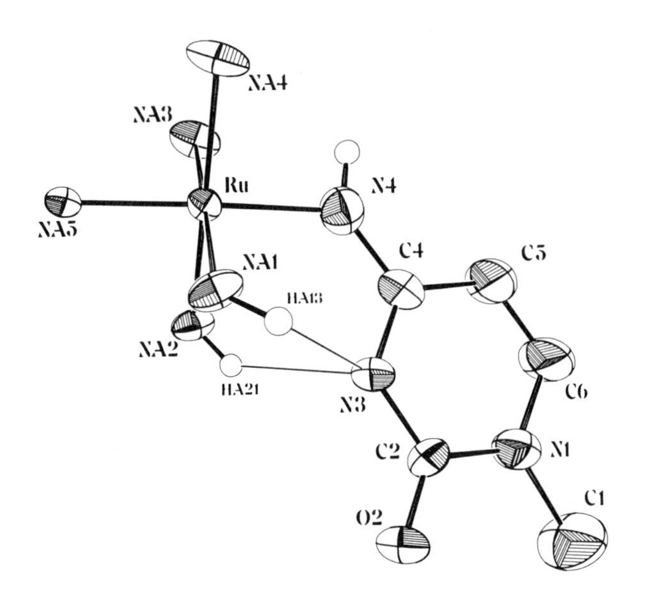

Figure 4. Structure of 4-[(1MeCyt$^-$)$(NH_3)_5$Ru(III)] (20)

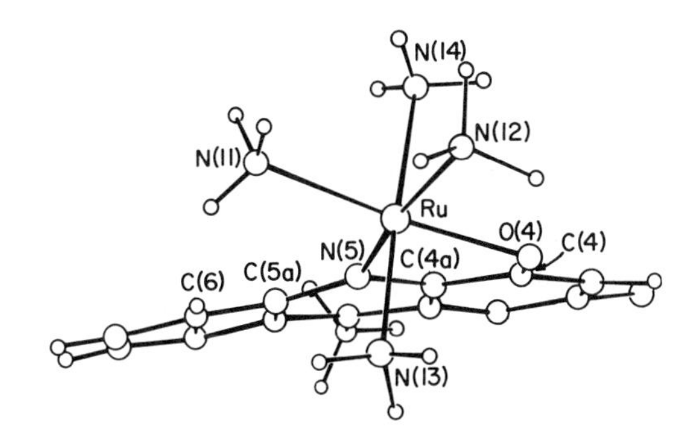

Figure 5. Structure of 4,5-[(10MeIAlo)$NH_3)_4$Ru]$^{2+}$ (22)

The spectra of $(NH_3)_5Ru(III)$-nucleoside complexes invariably exhibit fairly intense LMCT transitions, which are well resolved from the intraligand bands and provide a convenient probe into the nature of these metal-nucleoside interactions (12-14,16,19). Reference to Figure 6 reveals that these absorptions are usually localized in two distinct regions of the spectrum, one in the near UV and the other in the visible. The energy of these bands is primarily dependent upon the particular purine or pyrimidine involved, its protonation state and site of protonation or deprotonation. Their intensities, on the other hand, are largely a function of the metal binding site, but also depend somewhat on the nature of the ligand (16). Bands of this type have not been reported for any other metal-nucleoside adducts and contribute to making the present system one of the most convenient for study.

Application of the maxim that the closer a hard metal ion is to a deprotonation site, the greater the increase in acidity of that site (cf. Figure 7) facilitates assignments of the metal coordination position and separation of the various linkage isomers (see Table I). The latter can usually be accomplished by ion-exchange chromatography since the charge of the sundry isomers varies differently with the pH of the eluant buffer (12-14,16). A particularly drastic change in the acidity of nucleosides is seen in the cases of Ru(III) coordination to cytidine and adenosine in which the proton ionization constant increases by a factor of at least 10^9 over that of the free ligands and the preferred protonation site is altered (19).

The affinity of tr-$(SO_3)(H_2O)(NH_3)_4Ru(II)$ for guanosine is approximately 200 times greater than its affinity for adenosine (24). The lower binding constant for adenosine corresponds well with the relative instability of (Ado) $(NH_3)_5Ru(II)$ at low pH and it is likely that the ligand in both complexes is coordinated at the N(1) position (12,19,24). The selectivity for guanosine may be exploited for the specific labelling of such sites on nucleic acids, so long as the metal is restricted to the lower oxidation state when binding to the macromolecule.

In general, the Ru(II) coordination site is identical to that of Ru(III) since both are normally substitution-inert and have fairly high affinities for most types of nitrogen ligands (12-14,16,25). However, this is not always the case and the reduction of 4-(Ado)$(NH_3)_5Ru(III)$ results in a rapid linkage isomerization reaction ($k=1.6\ sec^{-1}$) with the Ru(II) ion presumably coordinating at the N(1) position (19,26). Similarly, reduction of 7-(1,3-Me_2Xan)$(NH_3)_5Ru(III)$ in acid yields 8-(1,3-Me_2Xan)$(H_2O)(NH_3)_4Ru(II)$ (14,15).

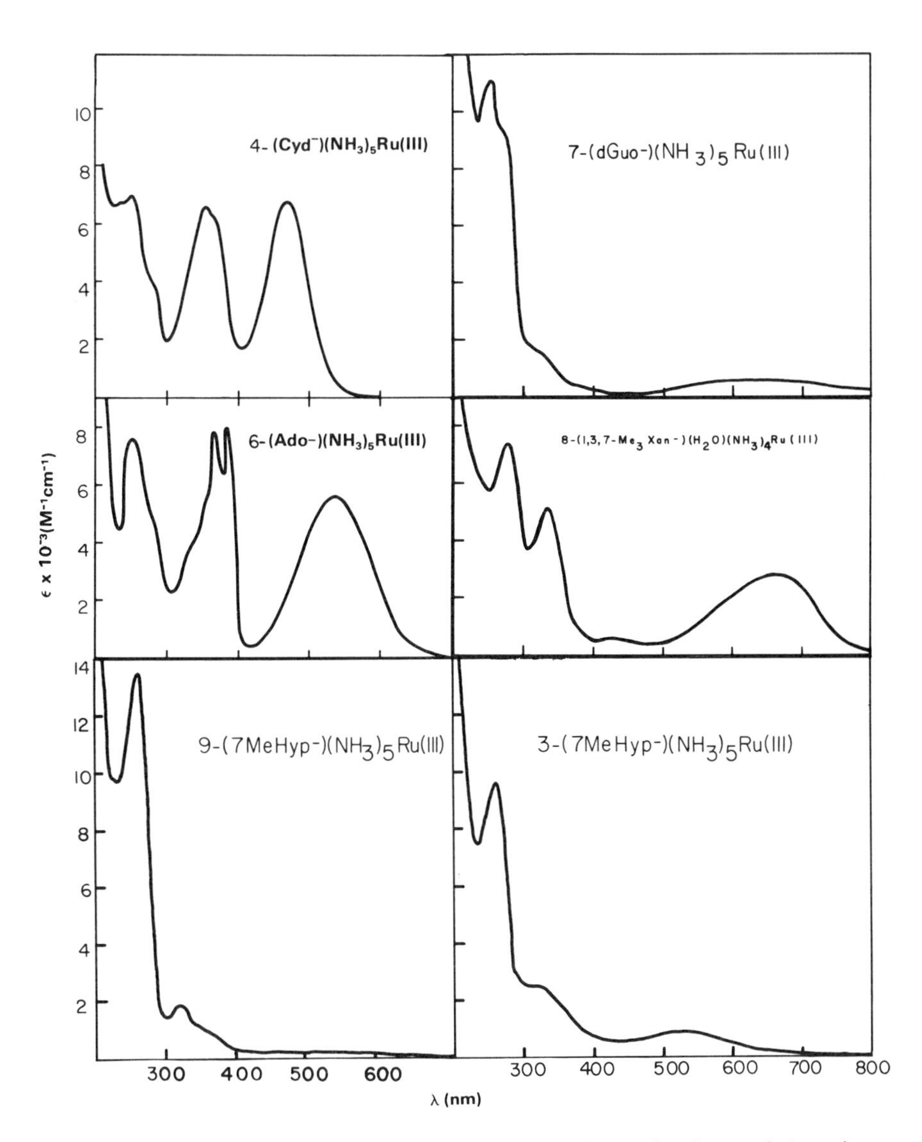

Figure 6. Spectra of various $(NH_3)_5Ru(III)$ complexes showing variations in absorption patterns with ligand and binding site

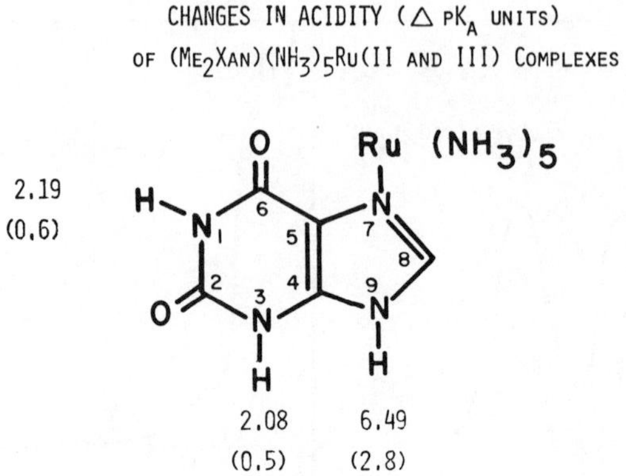

CHANGES IN ACIDITY (Δ pK$_A$ UNITS) OF $(Me_2Xan)(NH_3)_5Ru$(II AND III) COMPLEXES

2.19 (0.6)

2.08 (0.5)

6.49 (2.8)

Journal of the American Chemical Society

Figure 7. ΔpK_a values of isomers of 7-[$(Me_2Xan)(NH_3)_5Ru$(II and III)] (14). Values are reported in ΔpK_a units relative to the free ligand. Numbers in parentheses are for the Ru(II) complexes.

Table I. Changes in Acidity of Hypoxanthine Complexes on Coordination of $(NH_3)_5Ru$(II and III) (*16*)

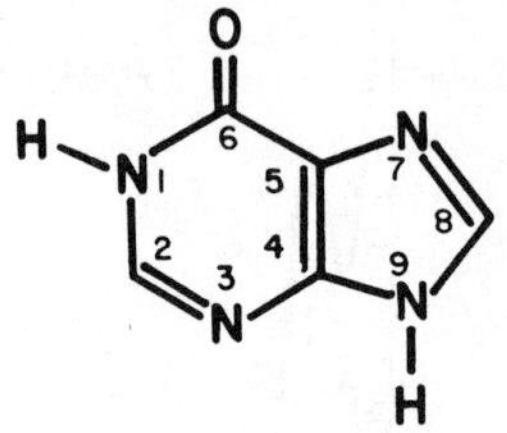

Metal Binding Site	Ligand	Deprotonation Site	ΔpK$_A$ Relative to Free Ligand Ru(II)	Ru(III)
3	7MeHyp	1	0.9	4.18
7	Ino	1	0.4	2.11
9	7MeHyp	1	0.6	1.54
9	1MeHyp	7	1.2	4.7

Inorganic Chemistry

The carbenoid ligand in the latter case serves as a potent trans-labilizer so that the ammine group opposite it quickly exchanges for water. Nor is a change in oxidation state necessary to initiate a change in binding site since protonation of 3-(7-MeHyp$^-$)$(NH_3)_5$Ru(III) in 1 M HCl results in a movement of the metal to the more electron rich N(9) position with an observed half-life of 1.45 hrs. at 37° (16,21).

These various linkage isomerization reactions suggest that even "substitution-inert" metal ions are not always held to a single position once bound to a nucleotide or nucleic acid. In fact, it is possible to envision sequential isomerizations resulting in metal migration over the perimeter of a single base residue or along the chain of a nucleic acid. While certainly speculative, these ideas imply that the primary lesion inflicted on a nucleic acid by metal coordination need not necessarily be the most damaging, and that subsequent metal movement to other coordination positions, particularly those normally on the interior of the nucleic acid, may yield the actual therapeutic or toxic effect. Indeed, such metal migration might be especially effective in producing interstrand crosslinks in DNA.

Ruthenium Interactions with Nucleic Acids.

The spectra of samples of $[(NH_3)_5Ru(III)]_n$-DNA prepared from normal and heat-denatured DNA are shown in Figure 8. Comparison with Figure 6 reveals a coincidence of bands in the visible region suggesting that helical DNA binds Ru(III) primarily at N(7) sites on guanine residues, while the single-stranded DNA coordinates the metal additionally at the exocyclic nitrogens of cytosine and adenine. Subsequent acid hydrolysis of these samples followed by ion-exchange chromatography allows the separation and spectrophotometric identification of the individual $(NH_3)_5$Ru(III)-purine complexes (Figure 9), which substantially confirms this interpretation (17). However, the cytosine complex cannot be isolated by the techniques employed so that the evidence for Ru(III)-cytosine complexation is entirely spectroscopic. Interestingly, the spectra of the Ru-DNA prepared using helical DNA at the higher ruthenium concentrations exhibit similarities to those obtained for the single-stranded samples. This implies some metal-induced uncoiling of the nucleic acid allowing subsequent metal attack on the "interior" adenine and cytosine sites.

Since ammineruthenium ions can coordinate the exo-N sites of cytosine and adenine as well as ring nitrogens, a variety of options for inter- and intrastrand crosslinking of DNA become available (11). However, the stability of these various modes of binding depends both on pH and the oxidation

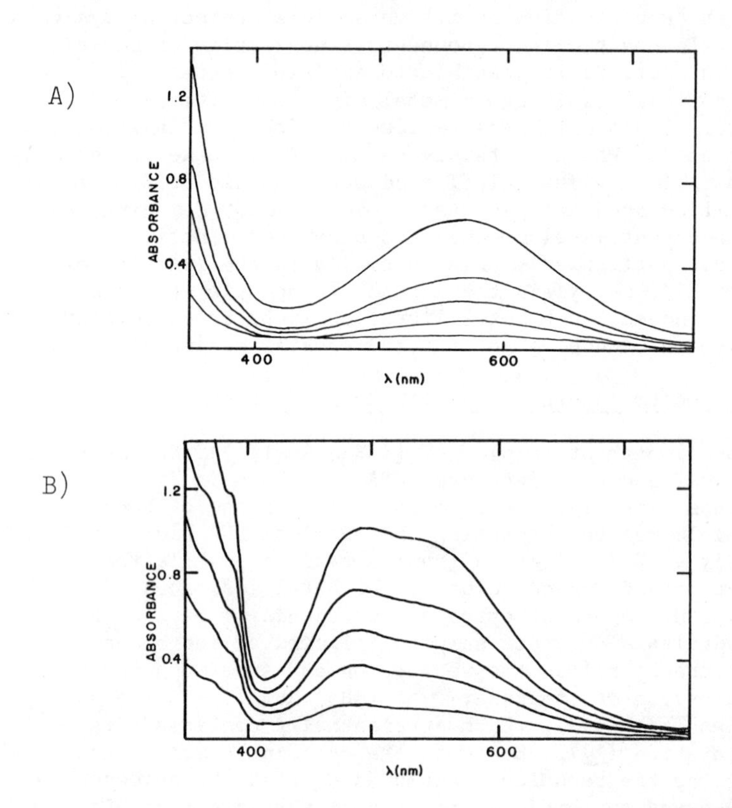

Inorganica Chimica Acta

Figure 8. Spectra of $[(NH_3)_5Ru(III)]_n$*–DNA samples prepared from: a, helical and b, single-stranded DNA with DNA concentration held constant and increasing concentrations of* $(H_2O)(NH_3)_5Ru(II)$ *followed by air oxidation (17)*

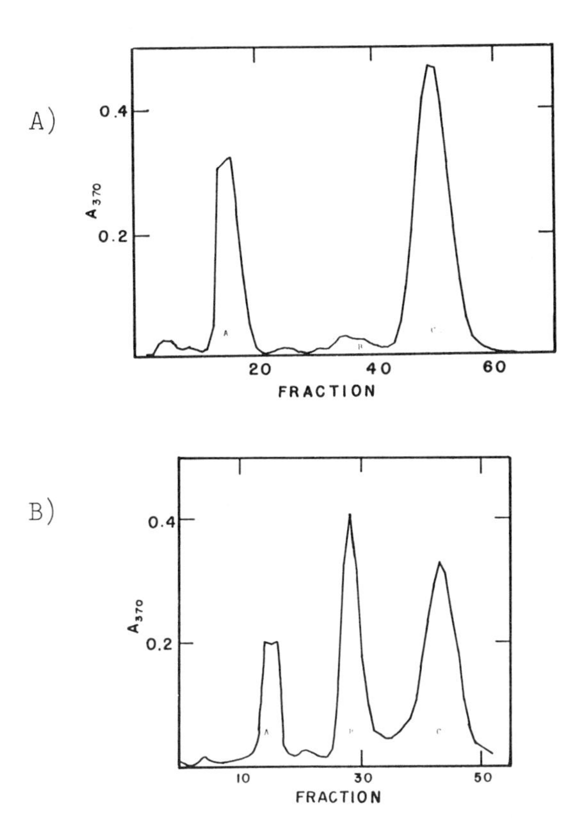

Inorganica Chimica Acta

Figure 9. Chromatography of acid hydrolyzed $[(NH_3)_5Ru(III)]_n$*–DNA samples prepared from: a, helical and b, single-stranded DNA (17)*

state of the metal. Long-lived coordination of Ru(III) occurs with the exo-N sites of adenine and cytosine and the N(7) of guanine, so that crosslinking involving these modes would appear most likely. In the case of Ru(II) only linkages involving the N(7) of guanine are expected to perist for significant periods (13). However, transient linking involving Ru(II) and the exo-N sites of adenine and cytosine and the N(1) position on adenine are possible (19).

Activation by Reduction of Ru(III) Prodrugs.

Significant differences exist between the chemistry of ammine Ru(II) and Ru(III) ions (11,25) which can be taken to advantage in the design of anticancer pharmaceuticals. While both $(NH_3)_5$Ru(II and III) have comparable affinities for imidazole ($K \simeq 2 \times 10^6$), Ru(III) has a five-fold higher affinity for ammonia and the stability constant of Ru(II) for pyridine is 4×10^3 greater than that of Ru(III) (28). In general, Ru(II) ions bind more firmly to those ligands which can serve as good π-acceptors of electron density from metal $d\pi$ -orbitals, while Ru(III) ions exhibit a relative preference for acido ligands such as chloride and carboxylates. Also the substitution rates of water or acido ligands from ammineruthenium(II) ions are usually much more rapid than those involving Ru(III). For example, the rate of aquation of $Cl(NH_3)_5$Ru(II) is approximately 5 sec^{-1}, while that of the analogous Ru(III) complex can be estimated to be a factor of 4×10^6 slower at neutral pH (27-31).

The relative chemical properties of Ru(II) versus Ru(III) suggest that ammineruthenium(III) ions should be far less active toward binding biochemical ligands than analogous Ru(II) complexes. In the case of most nitrogen ligands a wealth of chemical evidence exists in support of this (11). Thus a relatively inactive and so, hopefully, fairly non-toxic Ru(III) complex might be activated toward binding to nitrogen heterocycles by in vivo reduction. Innocuous anionic ligands such as chloride or acetate could be employed as leaving groups to lower the charge of the complex and enhance lipophilicity, so as to facilitate delivery of the Ru(III) prodrug across membrane barriers. Activation of the drug should take place preferentially in reducing environments. Inactivation would be expected to result should reoxidation of the Ru(II) species take place before binding to a nitrogen ligand occurred. This simplistic approach, therefore, predicts greater levels of drug binding in tissues high in reducing power and low in oxygen content.

Recent studies on tumor metabolism indicate very low levels of O_2 to be available, even at very short distances from blood capillaries (31-33). This appears to be due to a high rate of oxygen utilization by tumor cells, so that O_2 is rapidly depleted and largely unavailable to much of the tumor tissue. Gylcolytic metabolism must then be relied upon to generate the major portion of the energy supply for much of the neoplastic tissue with concomitant increase in lactic acid production and lowering of pH (34,35). Such anaerobic metabolism and glycolytic production of NADH should and does provide a more reducing environment than the normal surrounding tissue (36). Therefore, production of the lower oxidation states of metal ions should be particularly favored in many types of neoplastic tissues. Moreover, for those metal ions whose reduction potentials are pH dependent, the more acidic mileu provided by most tumors should additionally favor the reduced species (11,37).

Most organic reductants occurring in vivo, such as NADH or succinate, do not rapidly reduce metal ions from the III to II oxidation states since a two-electron transfer is required for the organic molecule to reach a stable product, while the metal ion requires but a single electron. Owing to this mismatch, some interface, usually supplied by a flavoprotein, is necessary between the organic and inorganic reactants. This is not meant to imply that flavins are the preferred in vivo reductants of Ru(III) ions, but only that such reduction would be expected to occur at or subsequent to the electron-pair splitting process in a biological electron transfer system. Similarly, deactivation of Ru(II) species does not necessarily have to involve O_2; however, few relatively strong biochemical oxidants are available in tissue in its absence.

The results of experiments employing subcellular components to catalyze the reduction of $Cl(NH_3)_5Ru(III)$ and subsequent metal complexation by a nitrogen heterocycle carried out in both the presence and absence of air are illustrated in Figures 10 and 11 (38). These studies show that the NADH reduction of Ru(III) proceeds smoothly under anaerobic conditions when microsomal enzymes are present. The actual reductant is not known but is likely to be NADH- or NADPH-dehydrogenase or cytochrome-b_5, which accepts single electrons from the former enzyme. The cytochrome P-450 enzymes, which apparently serve to reduce chromate (39), probably do not reduce the metal complex in the present system, since addition of metyrapone, a specific inhibitor for these proteins, did not affect the net rate of Ru(II) complexation. In keeping with the activation by reduction hypothesis, the rate of formation of (isonicotinamide)$(NH_3)_5Ru(II)$ significantly decreases when the reaction is run in air. Moreover, when the reaction is run under N_2, creation of the dinitrogen complex does not greatly interfere with binding of the nitrogen heterocycle. Thus, diversion of potential ruthenium-containing

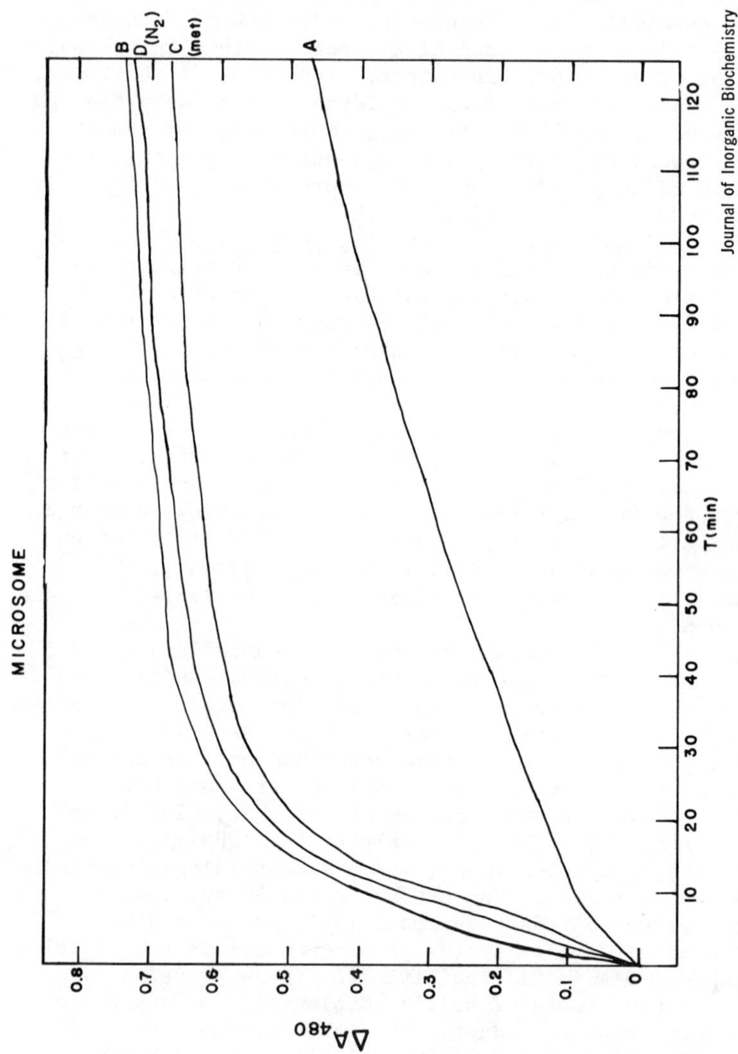

Journal of Inorganic Biochemistry

Figure 10. Microsome-catalyzed NADH reduction of Cl(NH$_3$)$_5$Ru(III) and subsequent formation of (Isonicotinamide)(NH$_3$)$_5$Ru(II): A, reaction run in air; B, reaction run under Ar; C, reaction run under Ar in presence of metyrapone; D, reaction run under N$_2$ (38)

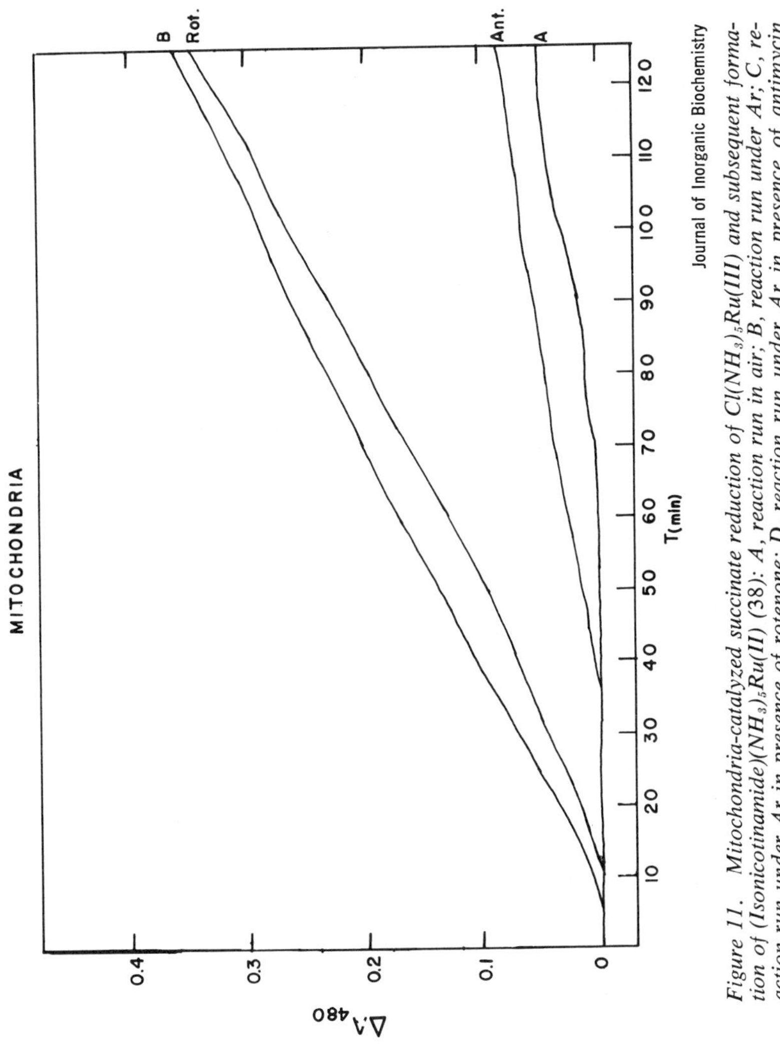

Journal of Inorganic Biochemistry

Figure 11. Mitochondria-catalyzed succinate reduction of $Cl(NH_3)_5Ru(III)$ and subsequent formation of (Isonicotinamide)$(NH_3)_5Ru(II)$ (38): A, reaction run in air; B, reaction run under Ar; C, reaction run under Ar in presence of rotenone; D, reaction run under Ar in presence of antimycin

drugs in this manner is unlikely to present a problem (38).

Analogous experiments were carried out using mitochondria as the electron-transfer catalyst and succinate as the electron source. While large differences between the aerobic and anaerobic rates of metal complexation were also observed in this system, only 1-3% of the metal was coordinated as Ru(II) even under relatively forcing conditions (38). Addition of malonate, a specific inhibitor for succinate dehydrogenase, or antimycin-A, which blocks the respiratory electron transfer chain between cyt-b and cyt-c_1, resulted in severe inhibition of metal complexation (Figure 11). This implies that the actual metal reductant occurs subsequent to cyt-b in the electron-transfer sequence and either cyt-c_1 or cyt-c are likely candidates. Assuming that either cytochrome serves to reduce Ru(III) at least partially explains the low yield of the reaction, since the reduction potentials of these proteins (0.225 and 0.254 V, respectively) are rather high relative to that of the metal complex (-0.042 V) and steric interactions would probably prevent close contact with the reducing heme moiety (40-41).

Biological Screening of Ruthenium Compounds.

A partial summary of the results of biological studies performed in collaboration with other laboratories are summarized in Tables II and III. *In vitro* work on the mutagenic properties of a series of ruthenium compounds has recently been carried out by Yasbin, Miehl and Matthews (42). Kelman, Edmonds and Peresie have studied the inhibition of cellular DNA and protein synthesis and were involved in the submission of a number of ruthenium compounds to the NCI for screening in animal tumor systems (43).

The results of the Ames test for mutagenesis indicate that many ruthenium compounds introduce serious lesions into cellular genetic material so that an error-prone DNA repair mechanism is induced. These results are similar to those obtained for cisplatin (44) and suggest that these complexes probably bind directly to nuclear DNA. In concert with this, many of the ruthenium complexes also inhibit cellular DNA synthesis (11,43), another property also noted for the cis-platinum drugs. Unfortunately, however, there is no correlation between either of these studies and the antitumor activity of ruthenium compounds tested in animal systems.

A high percentage of the compounds tested, which would be expected to function as Ru(III)-prodrugs, have exhibited antitumor activity in rats. An exception to this are those complexes containing bipyridyl or o-phenanthroline ligands which strongly stabilize the lower valent state and which

Table II. Antitumor and Mutagenic Activity of Selected Ruthenium Complexes

Compound	Dose (mg/kg)	%T/C	% Inhib. DNA Synthesis	Ames Test
$Cl_3(NH_3)Ru(III)$	50	189	86	
cis-$[Cl_2(NH_3)_4Ru(III)]Cl$	12.5	157	86	
cis-$[Cl_2(en)_2Ru(III)]Cl$	50	139	26	+++
$[Ox(en)_2Ru][(Ox)_2enRu]$	50	151	33	++
$[Br(NH_3)_5Ru(III)]Br_2$	12.5	120	40	
$[CH_3COO(NH_2)_5Ru](ClO_4)_2$	12.5	149	50	
$[CH_3CH_2COO(NH_3)_5Ru](ClO_4)_2$	12.5	163	0	
$[(Ino)(NH_3)_5Ru(III)]Cl_3$	25	125	96	+++
$[(1,3Me_2Xan)(NH_3)_5Ru]Cl_3$	50	90	35	+++
$[(NH_3)_6Ru(III)]Cl_3$	3.1	113		0
$K_3[Cl_6Ru(III)]$	25	138	6	
$[Isn(NH_3)_5Ru(II)](PF_6)_2$	6.25	104		
$[Ox(bipy)_2Ru(II)]$	3.13	101		+
$[Cl_2(bipy)_2Ru(III)]Cl$	50	97		
$[(NO)(NH_3)_5Ru(II)]$	3.12	127	62	
cis-$Cl_2(DMSO)_4Ru(II)$	565	125		++

Table III. Tissue Distribution of ^{103}Ru Complexes[a] (*49*)

Compound	Time (Hrs)	Tumor	Blood	Muscle	Liver	Kidney	TCI
		(Tumor/Tissue Ratios in Parentheses)					
$RuCl_3$	24	5.42	4.27 (1.27	1.57 (3.45)	6.11 (0.89)	12.24 (0.44)	1.33
	48	6.10	3.20 (1.90	1.15 (5.30)	9.17 (0.66)	10.48 (0.58)	1.86
$[Ru(NH_3)_5Cl]Cl_2$	24	3.29	1.97 (1.67)	0.79 (4.16)	2.37 (1.38)	5.89 (0.56)	1.70
	96	2.30	0.67 (3.43)	0.80 (2.87)	1.63 (1.41)	3.81 (0.60)	1.49
cis-$[Ru(NH_3)_4Cl_2]Cl$	24	5.38	5.18 (1.03)	1.54 (3.49)	5.12 (1.05)	9.23 (0.58)	1.56
	72	5.93	2.10 (2.82)	1.32 (4.49)	4.14 (1.43)	7.21 (0.82)	2.03
cis-$[Ru(en)_2Cl_2]Cl$	24	5.97	5.62 (1.05)	1.69 (3.53)	5.76 (1.03)	11.15 (0.53)	1.70
^{67}Ga-citrate	24	7.09	1.81 (3.92)	0.53 (13.38)	9.30 (0.76)	9.21 (0.77)	1.81
	96	3.54	0.36 (9.83)	0.37 (9.57)	8.19 (0.43)	7.14 (0.49)	1.45

$$\text{TCI (Tumor Concentration Index)} = \frac{\%\ \text{Injected Dose/g of Tumor}}{\%\ \text{Injected Dose/g in Whole Body}}$$

[a] Tissue from EMT-6 sarcoma bearing mice. Date given as % dose/g of tissue.

would significantly decrease the rate of loss of the acido groups from Ru(II). The bulky aromatic ligands in these complexes might also limit DNA binding. Several nitrosylruthenium(II) compounds investigated by Dr. S. Pell (45) and $Cl_2(DMSO)_4Ru(II)$ under study in Italy (44,46) have also shown anticancer activity. The fac-$Cl_3(NH_3)_3Ru$ isomer has been shown to induce filamentous growth in E. Coli (47). However, it should be noted that this is not the same triammine species that is presented in Table I, which is probably a mixture of mer and fac isomers (48).

While the results of biological testing are in large part consistent with the hypothesis of Ru(III) prodrugs being activated by in vivo reduction toward nucleic acid binding, this is by no means proven. Moreover, recent radioactive tracer studies carried out in collaboration with P. Richards and S.C. Srivastava of the Brookhaven National Laboratory working in conjunction with S. M. Larson of the Seattle V. A. Hospital (49), indicate that significant quantities of the metal, injected as ammineruthenium(III) complexes, are retained by several tissues in addition to the tumor (Table III). Most troublesome are relatively high levels remaining in the blood, suggesting that binding to proteins occurs even in this aerated environment. A possible reason for this is the high affinity of both Ru(II) and Ru(III) ammine complexes for RS^- groups which are available on some plasma proteins. Nevertheless, the prototypical complex, cis-$Cl_2(NH_3)_4Ru(III)$, does exhibit localization in the tumor on a level comparable with that of the most widely used tumor-imaging agent ^{67}Ga-citrate.

Pitha (50) has shown that the cellular toxicity of cis-$Cl_2(NH_3)_4Ru(III)$ is considerably higher than the corresponding trans isomer when tested on cells grown in liquid media. Moreover, the toxic effect was essentially the same when the cells were grown in the presence of the complex for only one hour, suggesting that the metal initially binds to the cell surface and then enters through pinocytosis. Since the rate of aquation of this complex is expected to be comparable to that of cis-$Cl_2(NH_3)_2Pt$ (11), direct Ru(III) coordination is possible under physiological conditions. However, redox-induced translocation should remain the dominant effect following entry into the cell and would allow more rapid coordination of intracellular nucleic acids.

Conclusion.

A number of Ru(III) complexes have exhibited antitumor activity and it is likely that redox assisted substitution is involved in causing the metal to interact with genetic material. Unfortunately, however, the toxic doses of most of the compounds tested are not greatly different from those at which the highest level of anticancer activity occurs. Thus, the high degree of antitumor selectivity hoped for in pursuing the "activation-by-reduction" hypothesis has not yet been obtained. Nevertheless, this approach remains encouraging and testing of analogs of those compounds exhibiting the greatest therapeutic index should prove profitable.

The use of ^{97}Ru in radioscintigraphic agents for tumor diagnosis offers several advantages over other isotopes in present use. The half-life of this radionuclide (2.9 days) is sufficient to allow chemical synthesis and purification, but not so long as to provide an intolerable level of radiation to the patient or to be inconvenient for hospital use. Moreover, the radiation from this isotope allows its use in presently available radioscintigraphic equipment. A series of precursor complexes involving Ru(II) and Ru(III), which could be used to coordinate a range of nitrogen-containing biochemical ligands and macromolecules such as proteins and nucleic acids, seems possible. Finally, improvement of the tumor localization, already exhibited by some ruthenium complexes, based on the elevated Ru(II)/Ru(III) ratios hypothesized to occur in tumor relative to normal tissue may make available a new class of tumor-imaging pharmaceuticals.

Acknowledgements.

Grateful acknowledgement is again made to the many collaborators whose contributions have been cited in the text. Funding for work carried out by the author has been provided by grants from the American Cancer Society, Massachusetts Division, the Research Corporation and by NIH grants: GM-13638, CA-18522 and GM-26390.

Literature Cited

1. Dwyer, F. P.; Mayhew, E.; Roe, E.M.F.; Shulman, A. Brit. J. Cancer, 1965, 19, 195.
2. Rosenberg, A.; Renshae, E.; Van Camp, L.; Hartwick, J.; Drobnik, J. J. Bacteriol., 1967, 93, 716.
3. Clear, M. J. Coord. Chem. Rev., 1974, 12, 349.
4. Subramanian, G.; McAfee, J. G. J. Nucl. Med., 1970, 11, 365.
5. Tanabe, M. Radioisotopes, 1976, 25, 44.
6. Mizukawa, K.; Yamamoto, G.; Tamai, T.; Tanabe, M.; Yamato, M. Radioisotopes, 1978, 27, 19.
7. Anghileri, L.J. Strahlentherapie, 1975, 129, 173.
8. Wenzel, M.; Herken, R.; Close, W. Z. Naturforsch, 1977, 32, 473.
9. Wenzel, M.; Subrumanian, N.; Nipper, E. Naturwiss, 1976, 63, 341.

10. Wenzel, M. Strahlentherapie, 1978, 154, 506.
11. Clarke, M. J. in Metal Ions in Biological Systems, Vol.11, Sigel, H., ed., Marcel Dekker, New York, 1979.
12. Clarke, M.J. 1974, Ph.D. Thesis, Stanford University.
13. Clarke, M.J.; Taube, H. J. Am. Chem. Soc., 1974, 96, 5413.
14. Clarke, M.J.; Taube, H. J. Am. Chem. Soc., 1975, 97, 1397.
15. Krentzien, H.; Clarke, M.J.; Taube, H Bioinorganic Chem., 1975, 4, 143.
16. Clarke, M.J. Inorg. Chem., 1977, 16, 738.
17. Clarke, M.J.; Buchbinder, M.; Kelman, A.D. Inorg. Chim. Acta, 1978, 27, 187.
18. Edmonds, S.; Kastner, M.E.; Clarke, M.J.; Eriks, K.; unpublished results.
19. Clarke, M.J. J. Am. Chem. Soc., 1978, 100, 5068.
20. Graves, B.J.; Hodgson, D., J. Am. Chem. Soc., 1979, 101, 5608.
21. Clarke, M.J.; Coffey, K.F.; Kirvan, G. unpublished results.
22. Clarke, M.J.; Dowling, M.G.; Garafalo, A.; Brennan, T. F. J. Am. Chem. Soc., 1979, 101, 223.
23. Clarke, M.J.; Dowling, M.G.; Garafalo, A. R.; Brennan, T.F. J. Biol. Chem., 1980, in press.
24. Brown, G.M.; Sutton, J.E.; Taube, H. J. Am. Chem. Soc., 1978, 100, 2767.
25. Kuehn, C. G.; Taube, H. J. Am. Chem. Soc., 1976, 98, 689.
26. Kirvan, G.; Clarke, M.J. unpublished results.
27. Coleman, G.N.; Gesler, J.W.; Shirley, F.A.; Kuempel, J.R.; Inorg. Chem., 1973, 12, 1036.
28. Taube, H. Survey of Progress in Chemistry, 1973, 6, 1.
29. Marchant, J.A.; Matsubara, T.; Ford, P.C. Inorg. Chem., 1977, 16, 2160.
30. Coleman, G.M.; Gesler, J.W.; Shirley, F.A.; Kuempel, J.R. Inorg. Chem., 1973, 12, 1036.
31. Broomhead, J.A.; Basolo, F.; Pearson, R.G. Inorg. Chem., 1963, 3, 826.
32. Gullino, P.M. Adv. Exp. Biol. Med., 1975, 75, 521.
33. Vaupel, P.; Thews, G. Adv. Exp. Biol. Med., 1976, 75, 547.
34. Vaupel, P. Microvasc. Res., 1977, 13, 399.
35. Weinhaus, S. Z. Krebsforsch, 1976, 87, 115.
36. Cater, D. B.; Phillips, A. F. Nature, 1954, 174, 121.
37. Lim, H.S.; Barclay, P.J.; Anson, F.C. Inorg. Chem., 1972, 11, 460.
38. Clarke, M.J.; Rennert, D.; Buchbinder, M.; Bitler, S. J. Inorg. Biochem., 1980, in press.
39. Gruber, J.E.; Jennette, K.W. Biophys. Biochem. Res. Comm.,
40. Ewalt, R.X.; Bennett, L.E. J. Am. Chem. Soc., 1975, 96, 940.
41. a) McArdle, J.V.; Gray, H.B.; Creutz, C.; Sutin, N. J. Am. Chem. Soc., 1974, 96, 5737.
b) Sutin, N. in "Bioinorganic Chemistry II", Raymond K., ed., Am. Chem. Soc., Washington, D.C., 1972, p.156.

42. Yasbin, R.; Miehl, R.; Matthews, R. C. Departments of Biology and Chemistry, The Pennyslvania State University, unpublished results.
43. Kelman, A.D.; Clarke, M.J.; Edmonds, S.D.; Peresie, H.J. J. Clin. Hemat. Oncol., 1977, 7, 274.
44. Monti-Bragadin, C.; Tamaro, M.; Banfi, E. Chem. Biol. Interactions, 1974, 11, 469.
45. Pell, S. Boston University, private communcations.
46. Giraldi, T.; Save, G.; Bertoli, G.; Mestroni, G.; Zassinovich, G. Canc. Res., 1977, 37, 2662
47. Durig, J.R.; Danneman, J.; Behnke, W.D.; Mercer, E.E. Chem. Biol. Interactions, 1976, 13, 287.
48. Creutz, C. A. Ph.D. Thesis, Stanford University, 1970.
49. Srivastavea, S.C.; Richards, P.; Meinken, G.W.; Som, P.; Atkins, H.L.; Larson, S. M.; Grunbaum, Z.; Dowling, M.G.; Clarke, M.J. Proceedings of the Second Int. Symp. on Radiopharmaceuticals, Seattle, Washington, in press.
50. Pitha, J. private communications. See Pitha, J. Eur. J. Biol., 1978, 82, 285 for experimental details.

RECEIVED April 7, 1980.

11

Correlations of Physico-Chemical and Biological Properties with in Vivo Biodistribution Data for Platinum-195*m*-Labeled Chloroammineplatinum(II) Complexes

JAMES D. HOESCHELE, THOMAS A. BUTLER, and JOHN A. ROBERTS

Nuclear Medicine Technology Group, Health and Safety Research Division, Oak Ridge National Laboratory, Oak Ridge, TN 37830

Cis-Dichlorodiammineplatinum(II), *cis*-[Pt$(NH_3)_2Cl_2$], is a unique metal-based antitumor drug which is most effective in combinational chemotherapy of cancers of genitourinary origin (1,2). Its mechanism of action is presumed to involve the inhibition of DNA synthesis (3,4). *cis*-[Pt$(NH_3)_2Cl_2$] is a member of a family of third-row transition metal complexes known collectively as the chloroammineplatinum(II) complexes. The typical square-planar structure and formulae of these complexes is illustrated in Figure 1. It is significant that *cis*-[Pt$(NH_3)_2Cl_2$] is the only complex in this series which exhibits clinically useful antitumor activity. In contrast, the structurally similar *trans* isomer is inactive (5,6), and the remaining complexes in this series exhibit either marginal or no antitumor activity (7). This striking contrast in biological activity for a series of related complexes, coupled with the fact that this model series of complexes ideally lends itself to the study of structure-activity relationships, prompted us to study the *in vivo* distribution properties of this important class of complexes. The purpose of this study was to obtain biodistribution data of a systematic nature which could be correlated with available physico-chemical and biological data for these complexes and with existing structure-activity criteria for platinum(II) antitumor agents. The type of data available for correlation is tabulated in Table I. Correlations of this nature have the potential of providing additional insight into the mechanism of action of Pt(II) compounds, a better understanding of, and perhaps a basis for predicting the biological fate, distribution, and potential utility of these and related transition metal complexes *in vivo*. These correlations could also stimulate new ideas for the design of more effective and, hopefully, less toxic antitumor agents.

Our study of the *in vivo* distribution of all six chloroammineplatinum(II) complexes has been completed and represents, to the best of our knowledge, the first biodistribution study of a complete series of transition metal complexes. The purposes of this paper are to report (a) a brief description of the microscale synthesis of the ^{195m}Pt-radiolabeled complexes, (b) the empirical

0-8412-0588-4/80/47-140-181$07.00/0

$[Pt(NH_3)_{4-x}Cl_x]^{2-x}$

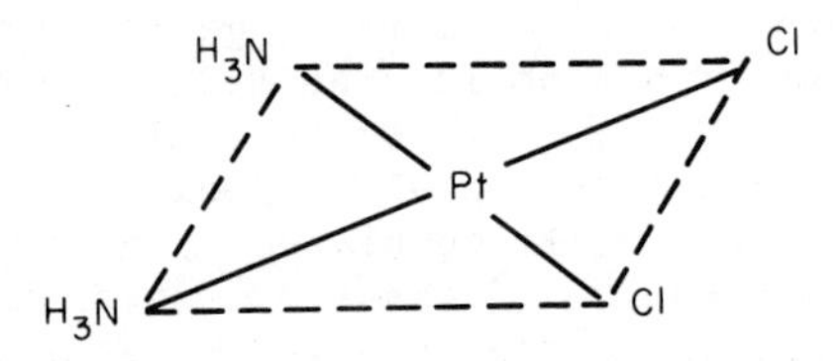

$[PtA_4]^{2+}$, $[PtA_3Cl]^+$, $[PtA_2Cl_2]^\circ$, $[PtACl_2]^-$, $[PtCl_4]^{2-}$

CIS & TRANS

*Figure 1. Square-planar structure and formulae of the chloroammineplatinum(II) complexes. The series includes two cationic complex species $[Pt(NH_3)_4]^{2+}$ and $[Pt(NH_3)_3Cl]^+$, two neutral species (*cis-*$[Pt(NH_3)_2Cl_2]$, polar, and* trans-*$[Pt(NH_3)_2Cl_2]$, nonpolar), and two anionic complex species $[Pt(NH_3)Cl_3]^-$ and $[PtCl_4]^{2-}$. Counter ions are K^+ and Cl^-.* A *denotes NH_3.*

Table I. Physico-chemical and Biological Data Available for Chloroammineplatinum(II) Complexes

	Physico-chemical Properties	Reference
1.	Chemical Kinetic and Thermodynamic Constants for Aquation	16
	a) Reactivity of Cl^- and NH_3 Ligands	
2.	Number and Position of Ligands (e.g., *cis, trans*)	-
3.	Net Charge on Complex (-, 0, +)	-
4.	Dipole Moment, μ (for *cis* and *trans*)	-
5.	Partition Coefficients, K_d (organic/aqueous)	18, *
	Biological Properties	
1.	Antitumor Activity	5, 7
2.	Toxicity	5, 7
3.	Mutagenicity (CHO Cells)	27
4.	Biological Half-time, $T_{1/2}(B)$	23, 24, *
5.	DNA Binding Constants	26

*This Study

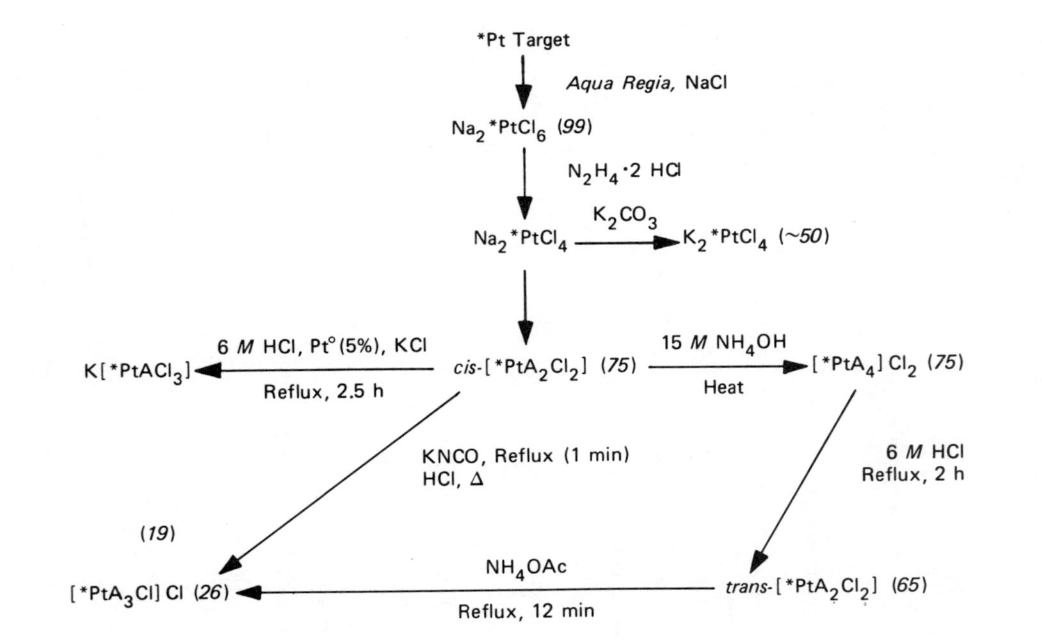

Figure 2. Schematic of the microscale syntheses of ^{195m}Pt-radiolabeled chloroammineplatinum(II) complexes. Reactions are carried out at ambient temperature unless specified otherwise. Yields relative to the Pt target are enclosed in parentheses. NH_3 and ^{195m}Pt are denoted by A *and* *Pt, *respectively.*

correlations and qualitative observations discerned from these biodistribution studies thus far, and (c) the implications of these observations.

Experimental

^{195m}Pt Radiolabel. ^{195m}Pt ($T_{1/2}$ = 4.02 d) is produced by neutron irradiation (n,γ) of enriched ^{194}Pt (97.41%) in the High Flux Isotope Reactor (HFIR) at the Oak Ridge National Laboratory. Platinum-195*m* is an ideal radiolabel since it emits penetrating radiation (γ-photons of 99 and 129 keV) and thus provides the capability of sensitive detection of Pt at low concentrations in biological specimens. The maximum specific activity achievable is $\sim$1 mCi ^{195m}Pt/mg Pt, which is equivalent to $\sim 2 \times 10^3$ disintegrations/min/ng Pt. Also, ^{195m}Pt is ideally suited for tissue localization studies by autoradiographic techniques since approximately three conversion electrons are emitted per event.

Microscale Syntheses. A schematic outline of the microscale synthesis of the ^{195m}Pt-labeled complexes is illustrated in Figure 2. In general, the methods of synthesis employed were essentially adaptations and/or refinements of published procedures (8,13). Syntheses were performed at a scale of 50-100 μmoles Pt and overall yields of a given complex ranged from 40-99%. The radiolabeled complexes were purified by recrystallization from physiological saline and/or by techniques utilizing ion-exchange resins. The purity of these complexes was certified by comparing the ultraviolet and visible absorption spectra with available spectral data (14,15,16) as well as by thin-layer and paper chromatographic techniques employing a radiochromatographic scanning device. A detailed description of the methods for synthesis, purification, and evaluation of the purity of these complexes will be published elsewhere.

At the microscale level, a key stratagem which greatly facilitates the synthesis of these complexes is to accurately determine the specific activity of ^{195m}Pt just after dissolution of the Pt target in *aqua regia* and, where necessary, after the synthesis of *cis*-$[Pt(NH_3)_2Cl_2]$. This permits critical stoichiometric requirements to be met, as in the reduction of Na_2PtCl_6 with $N_2H_4 \cdot 2HCl$, and a rapid determination of final product yields.

The central position of *cis*-$[Pt(NH_3)_2Cl_2]$ in this scheme emphasizes the importance of this complex as the preferred starting material in the synthesis of three of the remaining chloroammineplatinum(II) complexes. It is an ideal starting material because the microscale synthesis has been optimized to a degree which permits synthesis of a pure compound in a 75% overall yield based on the irradiated Pt metal target (17).

The ^{195m}Pt-labeled complexes of the general structure, $[PtA_4]Cl_2$ (A=CH_3NH_2 and *i*-$C_3H_7NH_2$), were synthesized using procedures identical to that for $[Pt(NH_3)_4]Cl_2$, except that NH_3 was replaced by the other amines.

Biodistribution Studies. Normal female Fischer 344 rats, 8-12 weeks old, were used in the *in vivo* distribution studies. Based on a dose of 4 mg/kg of *cis*-$[Pt(NH_3)_2Cl_2]$, a dose which is therapeutically effective in treating rat tumors, equimolar doses of the ^{195m}Pt-labeled complexes were administered intravenously in physiological saline *via* the lateral tail vein. The concentration of each complex in the administered solution (0.7 to 1.0 ml) was 1.0 mg/ml with the exception of *trans*-$[Pt(NH_3)_2Cl_2]$ which was 0.36 mg/ml, i.e., a saturated solution at 37°C. Fifty to 100 μCi of ^{195m}Pt was administered to each rat. Groups of 4 to 5 rats were sacrificed at 0.25, 1, and 4 h and 1 and 7 d post-injection. The organs were excised, rinsed with physiological saline, and the Pt-195*m* radioactivity determined in the whole organ by means of a Packard Auto-Gamma spectrometer. Biodistribution/retention data were obtained as a function of time for all compounds and, in the case of the *cis* isomer only, as a function of the dose of the injected compound. Biological half-times, $T_{1/2}(B)$, were evaluated from the cumulative % dose of Pt excreted in the urine and feces, which were collected in metabolism cages over a period of a week.

Partition coefficients were determined radiometrically using *n*-octanol/saline mixtures at 37°C. Saline solutions, of the respective radiolabeled complexes (10^{-4} *M*) were equilibrated with an equal volume of *n*-octanol for at least 4 h. Aliquots of the organic and aqueous phases were counted and K_d values computed from the ratio of the μCi ^{195m}Pt/ml *n*-octanol to the μCi ^{195m}Pt/ml saline.

Results and Discussion

Biodistribution Data. The distribution and retention of ^{195m}Pt determined in 14 tissues 24 h after administration of ^{195m}Pt-labeled complexes to female Fischer rats is compiled in Tables II, III, and IV. Data in Table II are expressed in terms of the % injected dose/g tissue, which is a measure of the concentration of Pt in the tissues rather than the total amount of Pt retained per organ. Data in Tables III (tissue/blood ratios) and IV (relative tissue distribution data) were computed using the data in Table II. The average uncertainty in these data is on the order of 15%. For purposes of data comparison, an equimolar dose of 13.3 μmoles of complex was administered per kg rat body weight. A retention of 1% of the injected dose/g tissue for a 0.2 kg rat corresponds to 26.6 *n* moles or 5.19 μg Pt/g tissue. Unless stated otherwise, reference to the distribution data should be interpreted as the % injected dose/g tissue. Also, reference to the level or concentration of a complex should be interpreted as the level or concentration of Pt, as measured by ^{195m}Pt, since the nature of the retained species is unknown.

Using the data in Table IV, tissues were ranked for each compound in the order of decreasing concentration of Pt. Rankings were averaged and the average order of decreasing tissue retention was established as follows:

Table II. Tissue Distribution of Chloroammineplatinum(II) Complexes in the Fischer (female) 344 Rat at 24 h Post-Injection (*i.v.*, tail vein) (A = NH_3)

Complex:	$[PtA_4]Cl_2$	$[PtA_3Cl]Cl$	*cis*-$[PtA_2Cl_2]$	*trans*-$[PtA_2Cl_2]$	$K[PtACl_3]$	$K_2[PtCl_4]$
Dose (mg/kg)[a]	4.4	4.2	4.0	2.2	4.8	5.5
Tissue	% Dose/g Tissue					
Blood	0.0070	0.093	0.21	1.33	1.37	0.63
Liver	0.16	0.673	0.37	0.57	0.58	1.10
Spleen	0.066	0.124	0.25	1.04	0.96	0.57
Pancreas	0.047	0.058	0.17	0.32	0.23	0.47
Stomach	0.044	0.077	-	0.092	0.17	0.28
S. Intestine	0.081	0.089	-	0.18	0.20	0.33
Adrenals	0.089	0.160	0.33	0.45	0.26	0.30
Kidneys	0.25	2.13	1.26	3.80	3.11	3.76
Heart	0.053	0.092	0.081	0.48	0.31	0.41
Lungs	0.20	0.166	0.21	0.91	0.64	0.79
Brain	0.047	0.019	0.012	0.029	0.027	0.014
Genitals[b]	0.13	0.106	0.41	0.63	0.66	0.87
Colon	0.17	0.161	-	0.71	0.25	0.31
Skin	0.037	0.062	-	0.18	0.26	0.53
Totals[c]	1.24	3.62	3.44	9.54	8.15	8.91

[a] Equimolar doses except for *trans*-$[PtA_2Cl_2]$ (2.15 instead of 4.00)

[b] Ovaries, uterus and fallopian tubes

[c] Data for stomach, small intestine, colon, and skin are excluded from this total

Table III. Tissue/Blood Ratios[a] for Chloroammineplatinum(II) Complexes in the Fischer 344 Rat at 24 h Post-Injection (*i.v.*, tail vein) (A = NH_3)

Complex:	$[PtA_4]Cl_2$	$[PtA_3Cl]Cl$	*cis*-$[PtA_2Cl_2]$	*trans*-$[PtA_2Cl_2]$	$K[PtACl_3]$	$K_2[PtCl_4]$
Tissue						
Liver	22.9	7.2	1.8	0.36	0.42	1.7
Spleen	9.4	1.3	1.2	0.43	0.70	0.90
Pancreas	6.7	0.62	0.81	0.78	0.17	0.75
Stomach	6.3	0.83	-	0.069	0.12	0.44
S. Intestine	11.6	0.96	-	0.14	0.15	0.52
Adrenals	12.7	1.7	1.6	0.34	0.19	0.48
Kidneys	35.7	22.9	6.0	2.86	2.3	6.0
Heart	7.6	1.0	0.39	0.36	0.23	0.65
Lungs	28.6	1.8	1.0	0.51	0.47	1.3
Brain	6.7	0.20	0.057	0.022	0.020	0.022
Genitals[b]	18.6	1.1	2.0	0.47	0.48	1.4
Colon	24.3	1.7	-	0.53	0.18	0.49
Skin	5.3	0.67	-	0.14	0.19	0.84

[a] Computed from data in Table II

[b] Ovaries, uterus and fallopian tubes

Table IV. Relative Tissue Distribution[a] of ^{195m}Pt-Labeled Chloroammineplatinum(II) Complexes in the Fischer Rat (female)[b] at 24 h Post-Injection (*i.v.*, tail vein) (A = NH_3)

Complex:	$[PtA_4]Cl_2$	$[PtA_3Cl]Cl$	*cis*-$[PtA_2Cl_2]$	*trans*-$[PtA_2Cl_2]$	$K[PtACl_3]$	$K_2[PtCl_4]$
Dose (mg/kg)[c]	4.4	4.2	4.0	2.2	4.8	5.5
Tissue	Ratios of % Dose/g Tissue Relative to the Compound with the Lowest Level in the Same Tissue Type					
Blood	1(*7.0x10^{-3})	13.3	30.0	190.0	†196.0	90.0
Liver	1(*1.6x10^{-1})	4.2	2.3	3.6	3.6	†6.9
Spleen	1(*6.6x10^{-2})	1.9	3.8	15.8	†14.5	8.6
Pancreas	1(*4.7x10^{-2})	1.2	3.6	6.8	4.9	†10.0
Stomach	1(*4.4x10^{-2})	1.8	-	2.1	3.9	†6.4
S. Intestine	1(*8.1x10^{-2})	1.1	-	2.2	2.5	†4.1
Adrenals	1(*8.9x10^{-2})	1.8	†3.7	5.1	2.9	3.4
Kidneys	1(*2.5x10^{-1})	8.5	5.0	15.2	12.4	†15.0
Heart	1(*5.3x10^{-2})	1.7	1.5	9.1	5.8	†7.7
Lungs	1.2	1(*16.6x10^{-2})	1.3	54.8	3.9	†4.8
Brain	†3.9	1.6	1(*1.2x10^{-2})	2.4	2.3	1.2
Genitals[d]	1.2	1(*1.1x10^{-1})	3.9	5.7	6.2	†8.2
Colon	1.1	1(*16.1x10^{-2})	-	4.4	1.6	†1.9
Skin	1(*3.7x10^{-2})	1.7	-	4.9	7.0	†14.3

[a] Computed from the data in Table II; used to determine the average order of tissue retention.

*Compound with lowest level per given tissue; in % dose/g tissue.

†Compound with highest level per given tissue; relative to *.

[b] 8-12 weeks old

[c] ~ 1.0 mg/ml except for *trans*-DDP (0.36 mg/ml).

[d] Ovaries, uterus, and fallopian tubes

kidney > liver > lung > genitals > spleen > bladder > adrenals > colon > heart > pancreas > small intestine > skin > stomach > brain

while the order for the *cis*-[$Pt(NH_3)_2Cl_2$] alone was:

kidney > genitals > liver > adrenals > spleen > bladder > lung > pancreas > heart > brain (only 10 tissues).

Several general observations have been made from a comparison of the tissue distribution after 24 h (Table II) and the average order of decreasing tissue retention. Platinum was detected in every tissue examined [14] for every compound. The highest concentrations were found in the kidney (0.2–4%) and, in general, the lowest in the brain (0.01–0.05%). The genitals (ovary, fallopian tubes, and uterus) exhibited the second highest levels for *cis*-[$Pt(NH_3)_2Cl_2$], and it is noteworthy that ovarian tumors (solid) respond well to *cis*-[$Pt(NH_3)_2Cl_2$] (Cisplatin) chemotherapy (2). However, this association does not imply that high tissue uptake necessarily predicts (and that low uptake necessarily rules out) chemotherapeutic effectiveness toward tumors of the same tissue of origin. A case in point is the testes, which (in the rat) shows low Pt uptake (0.036% for *cis*-[$Pt(NH_3)_2Cl_2$]); however *cis*-[$Pt(NH_3)_2Cl_2$] is most effective clinically against metastatic testicular cancer.

Distribution Profiles as a Function of Time for Selected Tissues. Profiles of the distribution of chloroammineplatinum(II) complexes as a function of time (for periods up to 7 d) in four specific tissues are shown in Figure 3. The distribution patterns are distinct and presumably reflect the unique chemical behavior expected for discrete substitution-inert complexes, i.e., these complexes appear to retain sufficient identity *in vivo* to exhibit characteristic distribution patterns. This can be contrasted with the behavior of simple binary metal salts, which exist as labile cations in solution, and whose distribution properties show little or no difference from one salt to another. All compounds are cleared relatively rapidly from the blood (panel A), especially the cationic species, $[Pt(NH_3)_3Cl]^+$ and $[Pt(NH_3)_4]^{2+}$, for which the levels are on the order of 0.1% and 0.02%, respectively after 4 h. The $[Pt(NH_3)Cl_3]^-$ ion shows the highest level after 7 d (∿1%).

Platinum levels in the kidney (Figure 3B), an organ of considerable clinical interest because of the dose-limiting toxicity exhibited by *cis*-[$Pt(NH_3)_2Cl_2$], are appreciably higher than in the blood. An important feature of the kidney profiles is that while the levels of most complexes steadily decline with time, retention is invariant at 4% (∿21 μg Pt/g tissue) for $PtCl_4^{2-}$ and *trans*-[$Pt(NH_3)_2Cl_2$] from 0.25 h to ⩾7 d, suggesting irreversible and accumulative tissue binding. The levels for *cis*-[$Pt(NH_3)_2Cl_2$] in the kidney 24 h post-injection are relatively low (∿1.3%) but

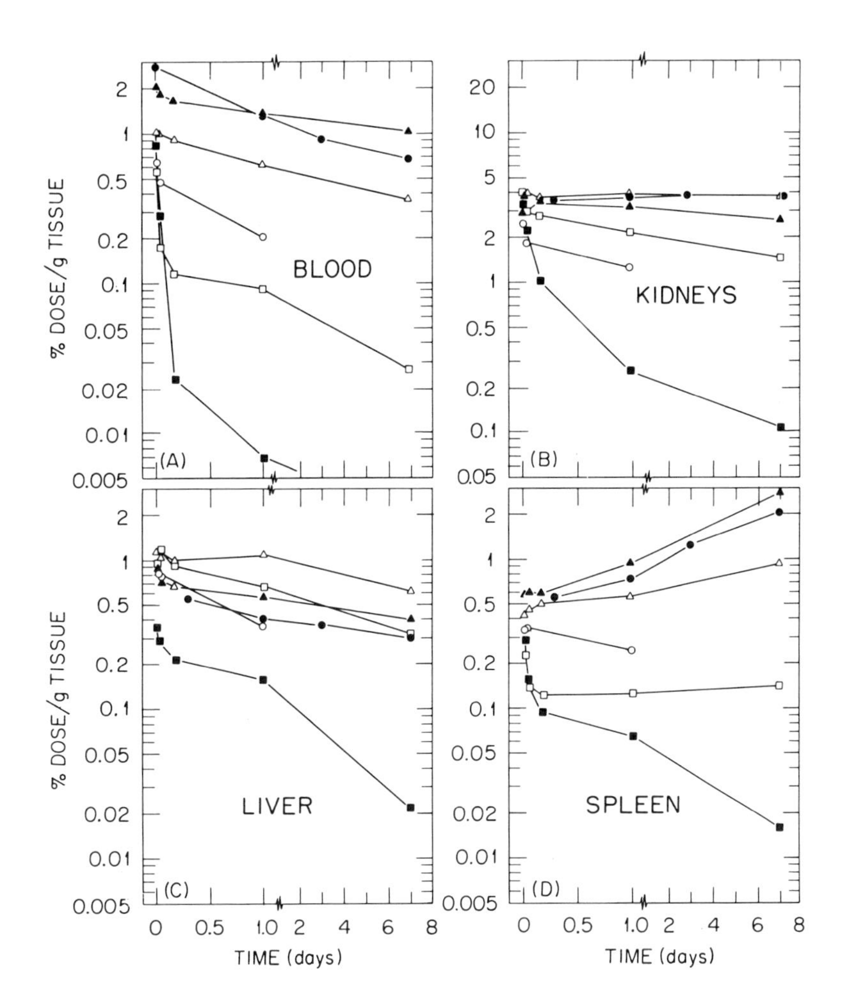

Figure 3. Profiles of the distribution of ^{195m}*Pt-radiolabeled chloroammineplatinum(II) complexes in the blood, kidneys, liver, and spleen as a function of time (5 days). Distribution is in terms of % injected dose/g tissue. Symbols:* (Δ) $[PtCl_4]^{2-}$; (▲) $[Pt(NH_3)Cl_3]^-$; (○) cis-$[Pt(NH_3)_2Cl_2]$; (●) trans-$[Pt(NH_3)_2Cl_2]$; (□) $[Pt(NH_3)_3Cl]^+$; (■) $[Pt(NH_3)_4]^{2+}$. *Panels: (A) blood; (B) kidneys; (C) liver; (D) spleen.*

still ~5X higher than for $[Pt(NH_3)_4]^{2+}$ (0.25%). The spread in Pt concentrations after 7 d is nearly 40-fold.

With the possible exception of the skin envelope, the liver (Figure 3C) shows the highest retention on a whole organ basis; however, liver concentrations of Pt are, in general, substantially lower (2 to 6 times) than in the kidney. After 7 d, the range of Pt concentrations is approximately 30-fold: 0.62 for the $[PtCl_4]^{2-}$ ion compared to 0.022 for the $[Pt(NH_3)_4]^{2+}$ species.

A unique pattern of distribution is observed for the spleen (Figure 3D). With the exception of the $[Pt(NH_3)_4]^{2+}$ species, retention gradually increases rather than decreases as a function of time beyond 0.25 h post-injection. A similar pattern was observed for ^{195m}Pt-labeled *cis*- and *trans*-$[Pt(NH_3)_2Cl_2]$ in the mouse (19) where the apparent increase in retention in the spleen with time was accounted for by a decrease in splenic weight without a proportional decrease in Pt content. Retention of *cis*-$[Pt(NH_3)_2Cl_2]$ gradually declines to 24 h, the last time point available.

Tissue Selectivity. As judged by tissue/blood ratios at 24 h post-injection of ^{195m}Pt-labeled chloroammineplatinum(II) complexes (Table IV), tissue selectivity is optimal for the tetra-ammine complex, $[Pt(NH_3)_4]Cl_2$. For example, values for the kidneys, lungs, colon, liver, and brain are 36, 29, 24, 23 and 7, respectively. A most intriguing result is that the inert, dipositive cationic species exhibited both the highest concentration of Pt in the brain (0.05%) and highest brain/blood ratios (6.7) of any of the chloroammineplatinum(II) complexes. Although the absolute amount of Pt retained is very small, these values are still 3.9 and 118 times the corresponding values for neutral *cis*-$[Pt(NH_3)_2Cl_2]$. It is unlikely that a trace impurity in the ^{195m}Pt-labeled $[Pt(NH_3)_4]Cl_2$ preparation could be responsible for the brain uptake for the following reasons: (a) brain levels of Pt were reproducible for three independent syntheses of the $[Pt(NH_3)_4]Cl_2$ complex; (b) syntheses of $[Pt(NH_3)_4]^{2+}$ were carried out using the purified starting material, *cis*-$[Pt(NH_3)_2Cl_2]$, which exhibits the lowest brain uptake of any chloroammine analog; and (c) radiochromatograms of radiolabeled $[Pt(NH_3)_4]^{2+}$ and precursor showed no detectable extraneous radioactivity. Rectilinear scans of a female rat at both 4 and 24 h post-injection of 0.2 mCi of ^{195m}Pt-labeled analog clearly showed uptake in the brain and other body organs (Figure 4). The higher brain/blood ratio, which suggests better transport through the "brain-blood barrier", is surprising. Moreover, there is a trend of increasing retention and tissue selectivity with increasing positive charge on the chloroammineplatinum(II) complex. If this relationship is a general one, it suggests that multi-charged cationic complexes and/or metal ions may exhibit a selective affinity for brain tissue and, therefore, might have utility as brain imaging agents. In the context of brain tissue affinity, Ga(III) and Bi(III) have potential utility as brain-imaging agents (20) and Al(III) has

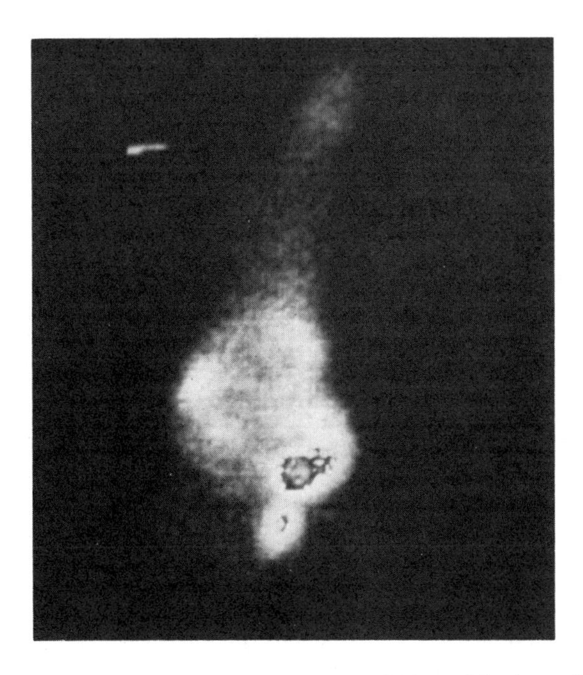

Figure 4. Rectilinear scan of a (female) Fischer 344 rat 4 h after administration (iv) of 0.2 mCi of ^{195m}Pt-labeled $[Pt(NH_3)_4]Cl_2$, demonstrating uptake in the brain and other body organs. $[Pt(NH_3)_4]Cl_2$ was administered in saline at a dose of 4.4 mg/kg. Anterior view; head (top).

been detected in abnormally high levels in autopsied brain tissue of patients afflicted with the degenerative Alzheimer's disease (21).

Localization studies originating out of the observations noted above and designed to evaluate the potential of ^{195m}Pt-labeled $[Pt(NH_3)_4]Cl_2$ as a brain-imaging agent, have shown that (a) uptake is not selective for any one anatomical region of the brain, and (b) ^{195m}Pt is localized (by means of autoradiographic examination of brain slices) principally within the extravascular spaces rather than within the cells. Attempts to enhance brain uptake using homologs of $[Pt(NH_3)_4]Cl_2$, i.e., ^{195m}Pt-labeled-$[PtA_4]Cl_2$ complexes where $A{=}CH_3NH_2$ and *i*-$C_3H_7NH_2$, were unsuccessful, as shown in Table V. Brain uptake after 24 h is reduced by nearly a factor of 10, and organ uptake is generally diminished compared to $[Pt(NH_3)_4]Cl_2$. Increasing the alkyl character of the A substituent of $[PtA_4]Cl_2$ complexes presumably increases the lipophilicity of the complexes but does not lead to increased uptake of such complexes by the rat brain. Although the distribution/retention patterns are unique, uptake is neither sufficiently high (with the exception of the kidney) nor sufficiently selective in any given tissue to warrant consideration of any chloroammineplatinum(II) complex as a potential imaging agent.

Tissue Distribution as a Function of Dose of *cis*-$[Pt(NH_3)_2Cl_2]$ 24 h Post-Injection. As a first step in assessing whether kidney retention (damage) might be altered by varying basic pharmacological parameters, tissue distribution was studied as a function of the dose of administered ^{195m}Pt-$[Pt(NH_3)_2Cl_2]$. The results in Figure 5 indicate that tissue distribution is essentially independent of the injected dose of this drug. This means that a constant fraction of the injected dose is retained in all tissues for a single *i.v.* injection over a six-fold range of doses (1.0 to 6.0 mg/kg). Hence, proportionally greater amounts of Pt would be retained by the kidney at higher doses as in high dose *cis*-$[Pt(NH_3)_2Cl_2]$ therapy. If Pt retention is directly proportional to kidney damage, there would appear to be no benefit gained (as regards reducing kidney retention) in substituting a daily-dose regimen for a single-dose regimen since the same amount of Pt will be retained in either case for the same total dose. This premise would be valid only if a constant fraction of the dose is retained for each successive daily dose; preliminary ^{195m}Pt tracer studies indeed indicate that this is true (22). It would appear that kidney retention is not likely to be diminished by variation of drug concentration unless the pharmacokinetics of uptake are different at very low drug concentrations.

Order of Compound Retention. The complexes were ranked in the order of increasing total retention at 24 h post-injection. The total tissue retention of a compound (listed in Table II under Totals) is the mathematical sum of the tissue retention in ten tissues. Tissues were selected for which data were obtained for

Table V. Tissue Distribution of ^{195m}Pt-labeled $[PtA_4]Cl_2$ Complexes in the Fischer (female) 344 Rat at 24 h Post-Injection (*i.v.*, tail vein)

A:	NH_3	CH_3NH_2	$(CH_3)_2CHNH_2$ [a]
Tissue	10^2 x % Dose/g Tissue		
Brain	4.8	0.50	0.65
Blood	0.66	0.75	1.6
Liver	14	3.5	21
Kidney	25	12	97

[a] Slight impurity in this preparation; probably the $[PtA_3Cl]Cl$ analog.

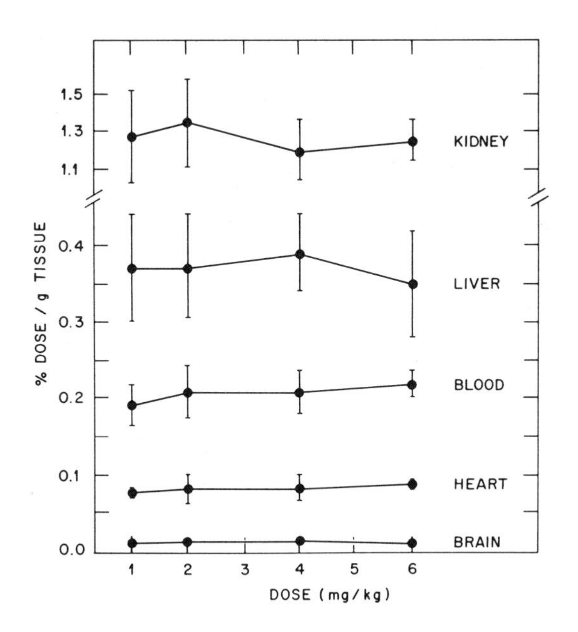

Figure 5. Distribution of ^{195m}Pt in (female) Fischer 344 rat tissues 24 h after administration (iv) of various doses of ^{195m}Pt-labeled cis-$[Pt(NH_3)_2Cl_2]$. *The distribution data are expressed in terms of the % of the injected dose/g tissue; dose of* cis-$[Pt(NH_3)_2Cl_2]$ *is in terms of mg/kg body weight.*

all six complexes. The importance of this order, which is essentially the same at 7 d, is shown below (Table VI) and will be discussed in relation to selected physico-chemical and biological data.

Empirical Correlations of Biodistribution Data with Physico-chemical and Biological Properties

Partition Coefficients. The trend in partition coefficients, K_d, determined for *n*-octanol/saline mixtures is consistent with the polarity of these complexes but only partially consistent with the order of compound retention. As illustrated in Figure 6, *trans*-$[Pt(NH_3)_2Cl_2]$ exhibits the highest K_d value and $[Pt(NH_3)_4]Cl_2$ the lowest, and the tissue distribution of these complexes reflects this contrast. The K_d for *trans*$[Pt(NH_3)_2Cl_2]$, is consistent with its high tissue uptake, the neutral and non-polar nature of this molecule ($\mu = 0$) and its greater tendency toward fecal excretion (vide infra). These results underscore the contention that membrane transport and tissue uptake for these complexes are not limiting factors in the sequence of events responsible for anti-tumor activity. If partition coefficients for *n*-octanol/saline mixtures are indeed a satisfactory measure of membrane transport for these complexes, the relatively high tissue uptake for $K[Pt(NH_3)Cl_3]$ is surprising in view of the low K_d observed for this analog. A partition coefficient for K_2PtCl_4 has not been determined but is predicted to be less than that for $K[Pt(NH_3)Cl_3]$. The K_d data are tabulated in Table VII.

Biological Half-times. Biological half-times, $T_{1/2}(B)$, (Table VII) of chloroammine-platinum(II) complexes also parallel the order of compound retention, i.e., complexes retained to the greatest extent also show the longest half-times. Half-times for *cis* and *trans* could not be determined using the present data but would appear to be $\geq$7 d. Literature estimates of $T_{1/2}(B)$ for *cis* range from 1 d (23) to > 7 d (24) and, based on whole-mouse counting data (*i.p.* injections) (19), the $T_{1/2}(B)$ for *trans* could be 4 to 10 X greater than that for *cis*.

Cumulative urine and fecal excretion profiles (Figure 7) appear to be multiphasic in nature and the order of initial rates of urinary excretion is essentially the reverse of the order of compound retention. It is interesting to note that the amount of ^{195m}Pt excreted *via* the urine compared to that for the feces at 24 h (U/F ratios in Table VII) is approximately 10 X greater for *cis* than for *trans* and 1.5 X greater for *cis* than for the $[Pt(NH_3)_4]^{2+}$ species. The approximately equal amounts of Pt excreted *via* the urine and feces for the $[PtCl_4]^{2-}$ ion, strongly suggests that this charged species is metabolized extensively. This is consistent with the high binding capacity (four replace-able chloride ligands) and high tissue retention for this ion. In sharp contrast, one would expect that the chemically inert and

Table VI. Order of Increasing Compound Retention 24 h Post-Injection[a]

$[PtA_4]^{2+}$ <	$[PtA_3Cl]^+$ <	$[PtA_2Cl_2]°$ <	$[PtACl_3]^-$ <	$[PtCl_4]^{2-}$ <	$[PtA_2Cl_2]°$
I	II	III, *cis*	IV	V	VI, *trans*
(0.36)[b]	(1.05)	(1.00)	(2.37)	(2.60)	(2.78)

[a] Counter cations (K^+) and anions (Cl^-) have been omitted to emphasize the charge on the complexes.

[b] Numbers in parentheses are the total retention normalized to that for *cis*$[Pt(NH_3)_2Cl_2]$.

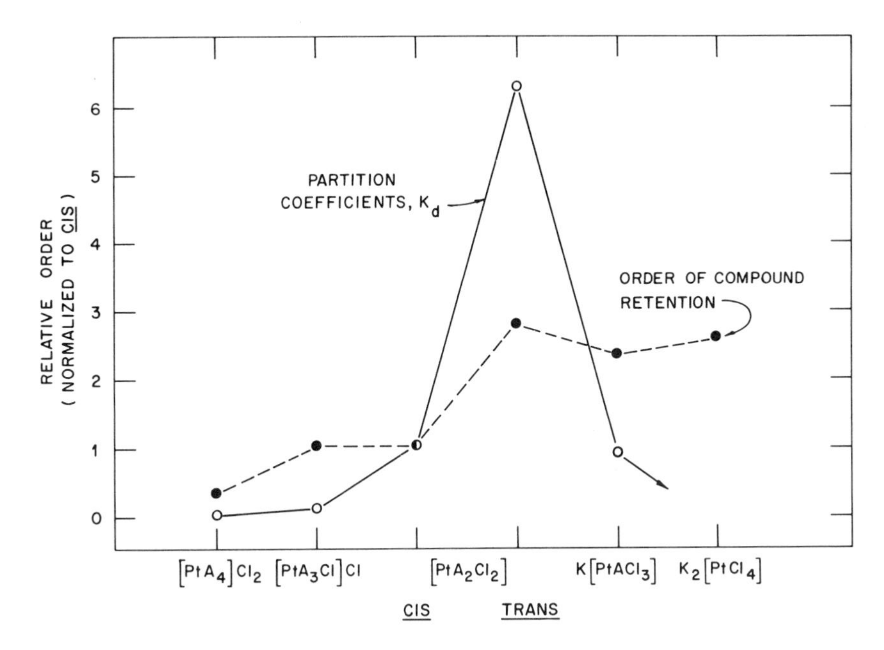

Figure 6. Comparison of partition coefficients, K_d*, with the order of* ^{195m}*Pt-labeled chloroammineplatinum(II) compound retention in the tissues.* K_d *values refer to* n-*octanol/saline mixtures at 37°C. Data is normalized relative to that for* cis-*$[Pt(NH_3)_2Cl_2]$.*

Table VII. Partition Coefficients, K_d, Biological Half-times, $T_{1/2}(B)$, and Ratios of Urinary/Fecal Excretion (U/F) for ^{195m}Pt-Labeled Chloroammineplatinum(II) Complexes.

Complex[a]	K_d[b] (Relative)	$T_{1/2}(B)$[c]	U/F (24 h)
$K_2[PtCl_4]$	--	6.3	0.95
$K[Pt(NH_3)Cl_3]$	0.84	2.8	4.1
cis-$[Pt(NH_3)_2Cl_2]$	1.0	≥7	20[d]
trans-$[Pt(NH_3)_2Cl_2]$	6.2	≥7	2.1
$[Pt(NH_3)_3Cl]Cl$	0.11	0.23	4.7
$[Pt(NH_3)_4]Cl_2$	0.036	0.13	13

[a] $[^{195m}Pt\text{-}Pt(II)] = 10^{-4}M$

[b] Between *n*-octanol/saline at 37°C; K_d (absolute) for *cis* = 7.3×10^{-3}

[c] Determined from cumulative rat urine and fecal excretion

[d] Data from reference 24

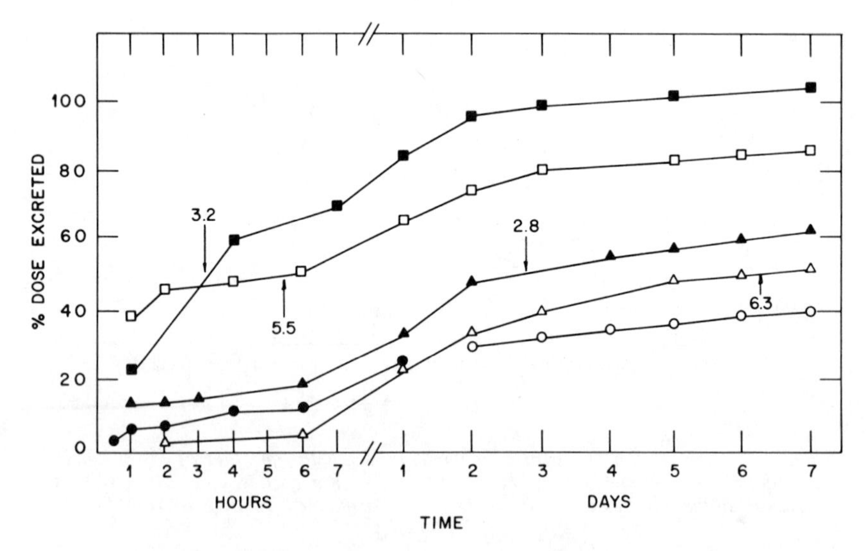

Figure 7. Profiles of the cumulative urine and fecal excretion of ^{195m}Pt as a function of time. Biological halftimes, $T_{1/2}$ (B), are indicated by arrows at the point of 50% dose excretion. Data for cis *(Ref. 24). Symbols: (△) $[PtCl_4]^{2-}$; (▲) $[Pt(NH_3)Cl_3]^-$; (○)* cis-$[Pt(NH_3)_2Cl_2]$; *(●)* trans-$[Pt(NH_3)_2Cl_2]$; *(□) $[Pt(NH_3)_3Cl]^+$; (■) $[Pt(NH_3)_4]^{2+}$.*

biologically inactive $[Pt(NH_3)_4]^{2+}$ species undergoes little or no reaction *in vivo* since virtually 100% of the injected activity is excreted within 3 d.

Charge on the Complex; Number of Potential Binding Sites. With the exception of the *trans* isomer, the order of compound retention exactly parallels trends in the negative charge on the complex and in the number of replaceable chloride ligands (or number of potential binding sites) on the central metal atom, Pt(II). Based only on this correlation, it would appear that gross tissue binding is rather non-selective and is related statistically to the number of available coordination sites. However, a closer comparison of the data and nature of the complex shows that the complexes retained to the greatest extent (IV, V, VI, Table VI) are complexes which contain kinetically reactive chloride ligands, by virtue of the *trans*-effect of mutually opposed chloride groups. For example, the $[PtCl_4]^{2-}$ complex ion contains two pairs of mutually opposed chloride ligands and shows the highest tissue retention in 6/14 comparisons. Conversely, the cationic complexes, I and II, which show the lowest tissue retention (except as noted for I in the brain) do not contain chloride ligands which are mutually opposed. Thus, the order of compound retention appears to be related to kinetic considerations.

Kinetic *vs* Equilibrium Constants for Aquation. The order of compound retention in the tissues more closely follows the trend in rate constants, k_1, rather than the thermodynamic equilibrium constants, K_{eq}, for the first step of the aquation process. The aquation process is likely to be an important first step in the *in vivo* reactions of chloroammineplatinum(II) complexes and can be represented by the following equilibrium equation using *cis*-$[Pt(NH_3)_2Cl_2]$ as an example:

$$cis\text{-}[Pt(NH_3)_2Cl_2] + H_2O \underset{k_2}{\overset{k_1}{\rightleftharpoons}} cis\text{-}[Pt(NH_3)_2Cl(H_2O)]^+ + Cl^- ,$$

where k_1 and k_2 are first-order rate constants (sec^{-1}) for the forward step (aquation) and reverse step (anation), respectively, and the equilibrium constant is numerically equal to the ratio of k_1 and k_2 (i.e., $K_{eq} = k_1/k_2$). The parallel in the general order of compound retention in the tissues with trends in k_1 and K_{eq} values is illustrated in Figure 8. Retention in the kidney and blood are included as specific examples of tissue retention. The values of k_1, K_{eq}, and the tissue distribution data have been normalized relative to those for *cis*-$[Pt(NH_3)_2Cl_2]$. Comparison of these data shows closer agreement between trends in compound retention and rate constants than for equilibrium constants. If compound retention paralleled K_{eq} values more closely than k_1 values, substantially lower levels of retention in the blood and

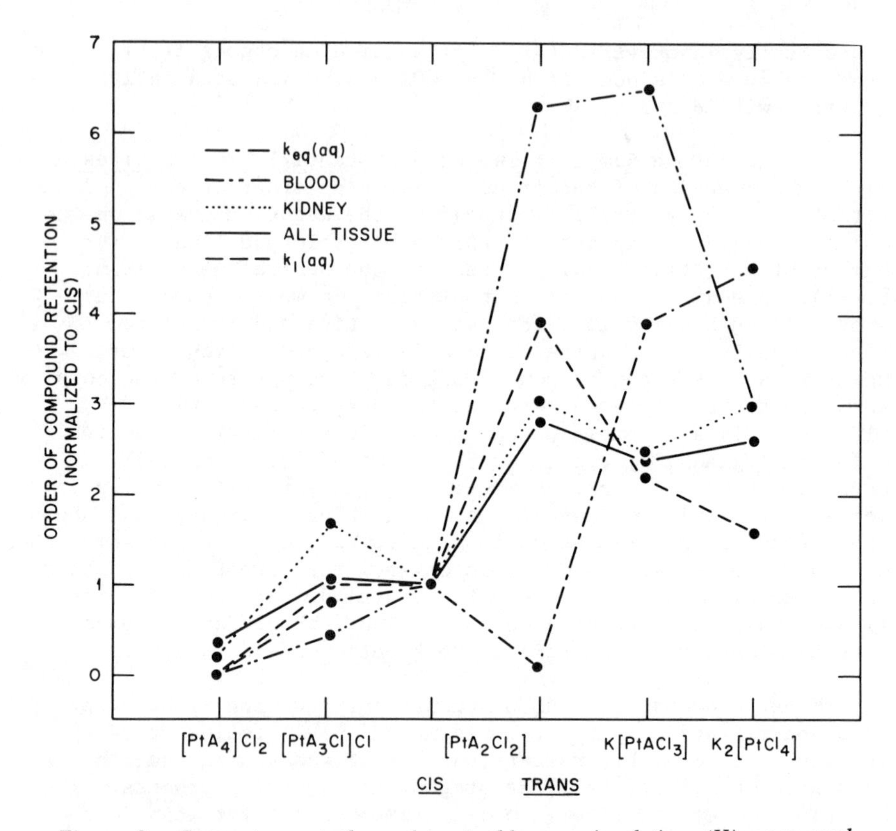

Figure 8. Comparison of the order of chloroammineplatinum(II) compound retention in all tissues with rate (k_1) *and equilibrium* (K_{eq}) *constants for aquation. Equilibrium and rate data (Ref.* 16*) are normalized relative to* cis-[$Pt(NH_3)_2Cl_2$].

kidney would have been anticipated for *trans*-[$Pt(NH_3)_2Cl_2$]. These results indicate that kinetic factors, as might have been predicted *a priori*, play a dominant role in the binding/retention and potential activity/toxicity of these complexes *in vivo*. Knowledge that kinetic factors appear to be important in determining the relative tissue concentrations of the chloroammines suggests the possibility that the relative tissue concentrations of similar substitution-inert complexes, particularly second-generation Pt(II) antitumor drugs, might be predictable, at least qualitatively, from basic kinetic data.

The trend in liver retention (not illustrated) does not show the pronounced uptake for *trans*-[$Pt(NH_3)_2Cl_2$] as in the kidney. Thus, the chemical-biogical processes leading to tissue retention (and perhaps organ toxicity) appear quite different for the liver *vs* kidney. In this context, it is worthwhile to note that while uptake of *cis*-[$Pt(NH_3)_2Cl_2$] in the kidney can lead to nephrotoxicity, reports of hepatotoxicity associated with *cis*-[$Pt(NH_3)_2Cl_2$] chemotherapy are rare.

Antitumor Activity. The order of compound retention in normal tissues does not parallel trends in the antitumor potential of these compounds. As illustrated in Figure 9, complexes V and VI exhibit the highest tissue levels but elicit virtually no antitumor activity, and complex IV, which shows only marginal antitumor activity, is retained to a greater extent than *cis*-$[Pt(NH_3)_2Cl_2]$, the most potent antitumor Pt(II) complex. Although there is no parallel between normal tissue retention and antitumor activity for these complexes, the amount of Pt bound per DNA base nucleotide (Rb) *in vivo* might indeed provide a useful, if not the ultimate, correlation. Preliminary Rb values (*in vivo*) determined for ^{195m}Pt-labeled *cis*-and *trans*-$[Pt(NH_3)_2Cl_2]$ (24,25) appear to parallel the antitumor biological activity of these complexes. Deoxyribonucleic acid isolated from liver and tumor (Reuber H-35 hepatoma) of ACI rats administered these complexes showed Rb values of $(5.3 \pm 2.0) \times 10^{-5}$ for *cis* and $(4 \pm 3) \times 10^{-6}$ for *trans*. Taking into account the statistical variability in these data, approximately 2–28 times more *cis* than *trans* is bound per DNA-base nucleotide. These data clearly indicate that *cis* binds more effectively than *trans* to DNA *in vivo*. It is of particular interest to know whether the amount of Pt bound, and/or, as expected, the nature of the binding correlates with the antitumor activity of these complexes.

Charged Pt(II) complexes, in general, have not exhibited significant antitumor activity (5) and, with the exception of *cis*- and *trans*-$(Pt(NH_3)_2Cl_2]$, the chloroammine complexes are charged species. The lack of antitumor activity for charged complexes has been attributed to biophysical factors such as transport and pharmacokinetic phenomena. This notion is consistent with the relatively low toxicity (i.e., high tolerance) observed for charged chloroammineplatinum(II) species and the relatively low tissue retention found for the cationic species, but is inconsistent with the high tissue retention observed for the anionic species, $[PtCl_4]^{2-}$ and $[Pt(NH_3)Cl_3]^-$. Thus, the inactivity of the charged complexes cannot be explained by insufficient tissue uptake. Tissue uptake is certainly a necessary, but not a sufficient criterion of antitumor activity; and other criteria such as the intrinsic lability (reactivity), capability and nature (specificity) of binding to the target site appear to be more critical than uptake *per se* in eliciting antitumor activity.

It is noteworthy that *cis*-$[Pt(NH_3)_2Cl_2]$ and $K[Pt(NH_3)Cl_3]$, the two antitumor active complexes (% T/C < 50), are also the two most mutagenic species as determined by the Ames and CHO/HGPRT test systems (27). The complexes are similar in that (a) through aquation the $[Pt(NH_3)Cl_3]^-$ ion can become a neutral molecule, *cis*-$[Pt(NH_3)(H_2O)Cl_2]^\circ$, and (b) both possess *cis*-reactive groups. Two important criteria for antitumor activity to be exhibited are that complexes should be neutral and possess *cis*-reactive groups.

Toxicity. Acute toxicity and antitumor activity data for the chloroammineplatinum(II) complexes is tabulated in Table VIII. The order of increasing toxicity is:

Table VIII. Antitumor Activity and Acute Toxicity Data for Chloroammineplatinum(II)

Complex	Antitumor Activity[a]		Toxic Threshold[b]	
	Single Acute	Daily[1-9c]	mg/kg	mmoles/kg
$K_2[PtCl_4]$	127^{25}	$72(5)^{10}$	45±5	0.11
$K[Pt(NH_3)Cl_3]$	95^{20}	$20^{(10-30)}$	>40	0.11
$[Pt(NH_3)_4][Pt(NH_3)Cl_3]_2$	35^{40}	-	∿50	0.11[d]
cis-$[Pt(NH_3)_2Cl_2]$	1^{8}	2^{1-2}	9-10	0.033
trans-$[Pt(NH_3)_2Cl_2]$	$110^{(10-40)}$	$96^{(20-40)}$	50±10[c,e]	0.17
$[Pt(NH_3)_3Cl]Cl$	$[71^{(10-100)}]$[g]	No Data Available	[>100][g]	(0.27)
$[Pt(NH_3)_4]Cl_2$	$100^{(5-200)}$	96^{200}	∿800[f]	2.39

[a] Form of entered data, $T(N)^D$: T = T/C; N = # survivors, D = dose, mg/kg; T/C = avg. wt. of tumor (treated)/avg. wt. of tumor (control). T/C values < 50 are considered active according to NCI protocol.

[b] Single (acute) dose for which % deaths ≥ 16% (1/6), Sarcoma 180 (solid) in Swiss mice.

[c] Best response for 9 consecutive daily injections; unpublished data of Van Comp and Rosenberg; remainder of data, ref. (5).

[d] Pro-rated for mmoles $[Pt(NH_3)Cl_3]^-$

[e] LD_{30} ∿ 60 mg/kg

[f] Estimate, LD_{50} ≳ 1000 mg/kg

[g] Estimate; data for close analog, [Pt(diethylenetriamine)Cl]Cl

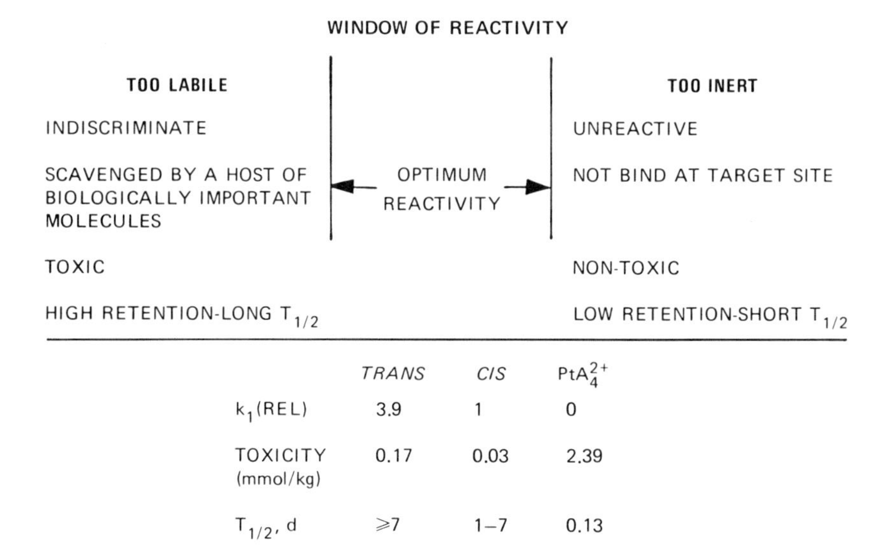

Figure 9. Illustration of the window of reactivity concept using data for cis- *and* trans-$[Pt(NH_3)_2Cl_2]$ *and* $[Pt(NH_3)_4]Cl_2$

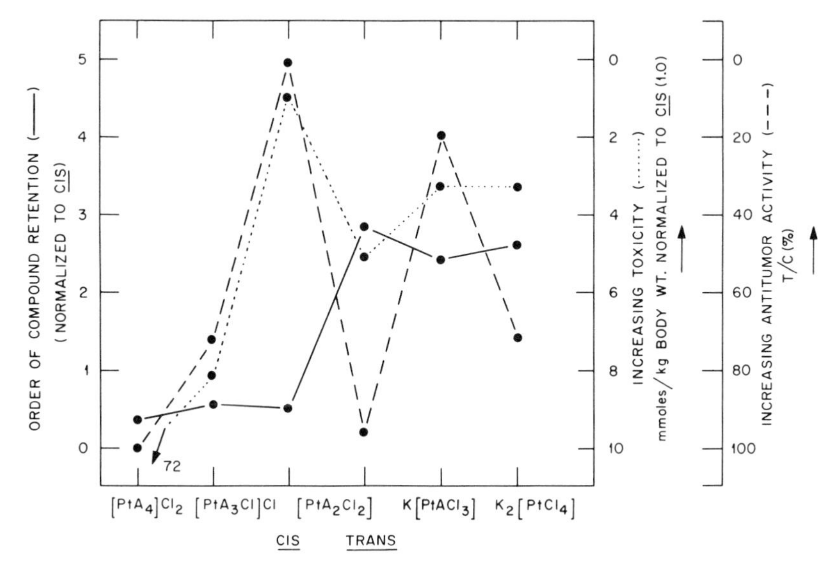

Figure 10. Trends in antitumor activity, relative acute toxicity, and order of compound retention in the tissues for chloroammineplatinum(II) compounds. Antitumor data (– – –): the best response (lowest % T/C) on administering 9 daily optimal doses to Swiss mice bearing the Sarcoma 180 (——) tumor. Response is measured in terms of the % T/C, the ratio of the average weight of the tumor for the treated group to that for the control group expressed as a %. Relative acute toxicity data (· · ·) are approximate LD_{16-30} values (mmol/kg mouse body weight) normalized to the value for cis-$[Pt(NH_3)_2Cl_2]$. *Order of compound retention (——) is the relative order of increasing total retention.* See *Tables VII and IX.*

$[PtCl_4]^{2-}$ > $[PtA_2Cl_2]°$ *trans* > $[PtACl_3]^-$ > $[PtA_3Cl]^+$ > $[PtA_2Cl_2]°$ *cis* >> $[PtA_4]^{2+}$

[A = NH_3; normalized LD_{16-30} data (relative to *cis*) are in parentheses.] The least toxic complex, $[Pt(NH_3)_4]^{2+}$, is 72 times less toxic than *cis*-$[Pt(NH_3)_2Cl_2]$, the most toxic complex. A comparison of toxicity, antitumor activity, and order of compound retention (Figure 10) indicates that the order of increasing toxicity (as mmol/kg mouse body weight) (7), (a) parallels the order of compound retention, except that the relative positions of *cis*- and *trans*-$[Pt(NH_3)_2Cl_2]$ are interchanged and (b) roughly parallels trends in antitumor activity, with the exception of the K_2PtCl_4 analog. It is clear that *cis*-$[Pt(NH_3)_2Cl_2]$ is both the most potent antitumor agent (% T/C ∿ 1) and the most toxic chloroammine complex (LD_{16} ∿ 0.033 mmol/kg). The relatively higher toxicity but lower tissue retention for *cis* negates any notion that the antitumor potency or whole-animal toxicity are directly related to the amount of Pt taken up into the tissues. The origin of toxicity, as well as antitumor activity, probably depends on the chemical nature and/or biological expression of critical target lesion(s) rather than to a general heavy metal burden *per se*.

The approximately four-fold greater reactivity of *trans* compared to *cis*-$[Pt(NH_3)_2Cl_2]$ (a factor which is comparable to the ratio of their molar toxicities) could indeed be a mediating factor in the lower toxicity and, to a certain extent, to the extreme difference in antitumor response.

As regards the kidney, it would be interesting to determine if kidney toxicity for these complexes is proportional to the amount of Pt retained [irrespective of the nature of the chloroammineplatinum(II) complex used] or whether the structure-activity criteria for kidney toxicity parallel those for antitumor activity (5). Using uptake at 24 h post-injection as a measure of the relative kidney toxicity, the order of decreasing toxicity would be: (A = NH_3)

$[PtA_4]^{2+}$ (72) << $[PtA_3Cl]^+$ (8) < $[PtA_2Cl_2]°$ (5), *trans* < $[PtACl_3]^-$ (3) ∿ $[PtCl_4]^{2-}$ (3) < $[PtA_2Cl_2]°$ (1), *cis*

This order suggests that the anionic complexes would be more toxic to the kidney than the cationic complexes and that *trans*-$[Pt(NH_3)_2Cl_2]$ would be more toxic than *cis*-$[Pt(NH_3)_2Cl_2]$. The *trans/cis* order of kidney retention (toxicity) is exactly the reverse of that found for whole animal toxicity. It would be interesting to see if kidney toxicity, as measured by BUN (blood urea nitrogen) and creatinine clearance tests, would indeed parallel this predicted order based on gross organ uptake.

Window of Reactivity (Lability). The relationship between tissue distribution and kinetic reactivity of the chloroammineplatinum(II) complexes gives credence to the "Concept of the Window of Reactivity." Although ideally applied to isostructural complexes of the same net charge, this concept is of sufficient generality to be applicable to this series of complexes. According to this concept (5), complexes which are chemically too inert will be too unreactive and will either not reach the target site or will not react with the target site at a sufficient rate or to a sufficient extent to elicit the critical biological response. Such inert compounds are anticipated to exhibit a short $T_{1/2}(B)$, have relatively low retention in the body as well as be relatively non-toxic. On the other extreme, complexes which are chemically too reactive will be indiscriminate in their binding and will be scavenged by a host of biologically important molecules. These complexes are expected to be less active biologically, exhibit a relatively long $T_{1/2}(B)$, have a high tissue retention, and be relatively toxic. Between the extremes of too inert and too reactive lies a region of optimal reactivity (lability) within which one might expect to find complexes of optimal biological activity. The data for *trans*-$[Pt(NH_3)_2Cl_2]$ compared to $[Pt(NH_3)_4]Cl_2$ (Figure 9) illustrate that this concept can be validly applied to the chloroammineplatinum(II) complexes since these two complexes lie at the extremes of reactivity and toxicity for this series of complexes and are virtually devoid of antitumor activity.

Summary and Conclusions

Pt-195*m*-labeled chloroammineplatinum(II) complexes have been synthesized on a microscale and used to study the distribution of these complexes in normal rats. Qualitative correlations of the distribution data, principally at 24 h post-injection, with available physico-chemical, biological, and structure-antitumor activity data were carried out to gain insight into the nature, fate, and potential utility of these and related agents in biological systems. Highlights of the results are as follows.

(1) Tissue distribution patterns are well-defined and reflect the unique chemistry of these substitution-inert complexes.
(2) The order of tissue retention at 24 h post-injection was: Kidney > liver > lung > genitals > spleen > bladder > adrenals > colon > heart > pancreas > small intestine > skin > stomach > brain.
(3) Although the antitumor drug, *cis*-$[Pt(NH_3)_2Cl_2]$, shows relatively high uptake in the rat genitals and is clinically effective against tumors of genitourinary origin, gross tissue uptake in normal rat tissues does not necessarily predict chemotherapeutic effectiveness for this drug.
(4) The tetraammine species, $[Pt(NH_3)_4]^{2+}$, a dipositive cation, exhibits the highest brain uptake which suggests the possible utility of multi-charged cationic species/metal ions as brain imaging agents.
(5) Kidney retention (toxicity) of Pt for *cis*-$[Pt(NH_3)_2Cl_2]$ is unaltered on varying basic pharmacological parameters such as dose and concentration of the administered *cis*-$[Pt(NH_3)_2Cl_2]$.

(6) The order of increasing compound retention reflects the kinetic influence of the *trans*-effect and suggests that kinetic factors, to a first approximation, play a dominant role in tissue uptake/binding and the potential activity/toxicity of these complexes *in vivo*. The data suggest further that chemical kinetic data (at least for aquation) might be useful in predicting the relative (gross) tissue concentrations of similar complexes, particularly for potential second-generation antitumor Pt(II) drugs.

(7) Membrane transport, as modeled by partition coefficient data, and tissue uptake do not appear to be limiting factors in the sequence of event(s) responsible for antitumor activity.

(8) Antitumor activity does not parallel the order of compound retention but does approximately parallel trends in toxicity (LD_{16-30} data). The number of Pt atoms bound/DNA base nucleotide (Rb values) appears to parallel the antitumor activity of *cis* vs. *trans*.

(9) The high tissue uptake but low antitumor activity for charged species indicates that uptake *per se* is necessary but not a sufficient criterion of activity and that other factors such as the intrinsic reactivity and the nature of binding to the target site are more critical in eliciting antitumor activity.

(10) Comparative data for *trans*-$[Pt(NH_3)_2Cl_2]$ and $[Pt(NH_3)_4]Cl_2$ validate the "Concept of the Window of Reactivity."

Abstract

Pt-195*m*-labeled chloroammineplatinum(II) complexes have been synthesized on a microscale and used to study the distribution of these complexes in normal Fischer 344 rats. Correlations of the distribution data with physico-chemical, biological, and structure-antitumor activity data are being carried out in order to gain insight into the nature, fate, and potential utility of these and related agents in biological systems. Results 24 h post-injection indicate that compound retention generally increases in the order: (A=NH_3; *c*=*cis*; *t*=*trans*)

$$[PtA_4]^{2+} < [PtA_3Cl]^{+} < c\text{-}[PtA_2Cl_2] < [PtACl_3]^{-} < [PtCl_4]^{2-} < t\text{-}[PtA_2Cl_2]$$

In general, this order parallels trends in kinetic constants more closely than equilibrium constants for aquation, charge on the complex, number of labile coordination sites (Pt-Cl bonds), biological half-times, and, to a limited extent, partition coefficients. Compound retention does not parallel trends in antitumor activity. The concept of the window of reactivity (lability) appears to be validated by comparative data for *t*-$[PtA_2Cl_2]$ and $[PtA_4]^{2+}$. Tissue distribution (% dose/g tissue) is independent of both dose and concentration of *c*-$[PtA_2Cl_2]$. General details of the microscale syntheses are presented. (Research sponsored by the Office of Health and Environmental Research, U.S. Department of Energy, under contract W-7405-eng-26 with the Union Carbide Corporation.)

Acknowledgments

Research sponsored by the Office of Health and Environmental Research, U.S. Department of Energy under contract W-7405-eng-26 with the Union Carbide Corporation. The authors thank W. D. Gude for audioradiographic studies, and K. R. Ambrose, L. A. Ferren, and C. E. Guyer for technical assistance.

Literature Cited

1. Einhorn, L. H.; Donahue, J., *cis*-Diamminedichloroplatinum, Vinblastine, and Bleomycin Combination Chemotherapy in Disseminated Testicular Cancer, Ann. Intern. Med., 1977, 7, 293-298.
2. Holland, J., in "Cisplatin — Current Status and New Developments," Academic Press: New York, 1980 (in press).
3. Harder, H. C.; Rosenberg, B., Inhibitory Effects of Antitumor Platinum Compounds on DNA, RNA, and Protein Synthesis in Mammalian Cells *In Vitro*, Int. J. Cancer, 1970, 6, 207-216.
4. Roberts, J. J.; Thomson, A. J., The Mechanism of Action of Antitumor Platinum Compounds, Prog. Nucleic Acid Res. Mol. Biol., 1978, 22, 71.
5. Cleare, M. J.; Hoeschele, J. D., Studies on the Antitumor Activity of Group VIII Transition Metal Complexes. Part I. Platinum(II) Complexes, Bioinorg. Chem., 1973, 2, 187-210.
6. Cleare, M. J., Transition Metal Complexes in Cancer Chemotherapy, Coord. Chem. Rev., 1974, 12, 367.
7. Van Camp, L. (personal communication) and reference 5, p. 203.
8. Fernelius, W. C., Ed., "Inorganic Syntheses;" McGraw-Hill: New York, 1946, vol. 2, p. 250.
9. Tschugaev, L. A., A New Method of Preparing Chloro- and Bromotriammino-platinous Haloids, Trans Chem. Soc. (London), 1915, 107, 1247-1250.
10. Dhara, S. C., Rapid Method for the Synthesis of *cis*-$[Pt(NH_3)_2Cl_2]$, Indian J. Chem., 1970, 8, 193-194.
11. Kleinberg, J., Ed., "Inorganic Syntheses;" McGraw-Hill: New York, 1963, vol. 7, p. 239.
12. Elleman, T. S.; Reishus, J. W.; Martin, D. S., Jr., The Acid Hydrolysis (Aquation) of the Trichloroammineplatinate(II) Ion, J. Am. Chem. Soc., 1958, 80, 536-541.
13. Scandola, F.; Traverso, O.; Carassitti, V., Photochemistry of the Tetrachloroplatinate(II) Ion, Mol. Photochem., 1969, 1(1), 11-21.
14. Chatt, J.; Gamlen, G. A.; Orgel, L. E., The Visible and Ultraviolet Spectra of Some Platinous Ammines, J. Chem. Soc., 1958, 486-496.
15. Aprile, F.; Martin, D. S., Jr., Chlorotriammineplatinum(II) Ion, Acid Hydrolysis and Isotopic Exchange of Chloride Ligand, Inorg. Chem., 1962, 1, 551-557.

16. Tucker, M. A.; Colvin, C. B.; Martin, D. S., Jr., Substitution Reactions of Trichloroammineplatinate(II) Ion and the *trans* Effect, Inorg. Chem., 1964, 3, 1373-1383.
17. Hoeschele, J. D.; Butler, T. A.; Roberts, J. A., Microscale Synthesis and Biodistribution of ^{195m}Pt-labeled *cis*-Dichlorodiammineplatinum(II), *cis*-DDP, Proc., 2nd International Symposium on Radiopharmaceuticals, 1979, March 19-22, Seattle, Washington, 173-183.
18. Tobe, M. L.; Khokhar, A. R., Structure, Activity, Reactivity and Solubility Relationships of Platinum Diamine Complexes, J. Clin. Hemat. Oncol., 1977, 7, 127.
19. Hoeschele, J. D.; Van Camp., L., Whole-Body Counting and Distribution of *cis*-^{195m}Pt-[Pt$(NH_3)_2Cl_2$] in the Major Organs of Swiss White Mice in "Advances in Antimicrobial and Antineoplastic Chemotherapy," University Park Press: Baltimore, Md., 1972, vol. II, p. 241-242.
20. Wagner, H. N., Ed., "Principles of Nuclear Medicine," W. B. Saunders Co., Philadelphia, 1968, p. 663.
21. Crapper, D. R., Aluminium Neurofibrillary Degeneration and Alzheimer's Disease, Brain, 1976, 99(1), 67-80.
22. Hoeschele, J. D.; Butler, T. A.; Ambrose, K. R.; Roberts, J. A. (unpublished results).
23. Ward, J. M.; Grabin, M. E.; LeRoy, A. F.; Young, D. M., Modification of the Renal Toxicity of *cis*-Dichlorodiammineplatinum(II) with Furosemide in Male F 344 Rats, Cancer Treatment Reports, 1977, 61, 375-379.
24. DeSimone, P. A.; Yancey, R. S.; Coupal, J. J.; Butts, J. D.; Hoeschele, J. D., The Effect of a Forced Diuresis on the Distribution and Excretion (*via* Urine and Bile) of ^{195m}Pt *Cis*-Dichlorodiammineplatinum(II), Cancer Treatment Reports, 1979, 76, 951-960.
25. Hoeschele, J. D.; Johnson, N. P.; Rahn, R. O., Comparative Distribution Studies of ^{195m}Pt-labeled Dichlorodiammineplatinum(II), DDP in the Rat with Emphasis on the Localization in DNA, Biochemie, 1978, 60(9), 1054.
26. Rahn, R. O.; Johnson, N. P.; Hsie, A. W.; Hoeschele, J. D.; Lemontt, J. F.; Brown, D. H.; Masker, W. E.; Regan, J. D.; Dunn, W. C., Jr., "The Interaction of Platinum Compounds with the Genome: Correlation Between Binding Data and Biological Endpoint," Symposium on the Scientific Basis of Toxicity Assessment, Elsevier (in press).
27. Johnson, N. P.; Hoeschele, J. D.; Rahn, R. O.; O'Neill, J. P.; Hsie, A. W., Mutagenicity, Cytotoxicity, and DNA Binding of Platinum(II) Chloroammines in Chinese Hamster Ovary Cells, Cancer Res., 1980, 40, 1463-1468.

RECEIVED June 11, 1980.

12

Structural Studies of the Hydrolysis Products of Platinum Anticancer Drugs, and Their Complexes with DNA Bases

COLIN JAMES LYNE LOCK

Institute for Materials Research, McMaster University, Hamilton, Ontario L8S 4M1, Canada

The action of the drug cis-dichlorodiammineplatinum(II) and related drugs against cancers has been interpreted in terms of a mechanism (Figure 1), whereby the drug is introduced into the bloodstream, is transported to the tumor cell, where it undergoes hydrolysis because of the low chloride concentration, and the hydrolysed species then reacts with the pyrimidine or purine bases on the DNA chain, in some way interfering with replication, possibly because of cross-linkage (1-3). The hydrolysis step was invoked because of kinetic studies in the early 1960's (4-7) which suggested that in dilute aqueous media displacement reactions of platinum(II) always go through a hydrolysis step.

The interaction of platinum with DNA has been clearly established both in vitro (8-11) and in vivo (12-14), and most modelling studies indicate that, although platinum can interact with the phosphate group (15), the prime interaction is with the purine and pyrimidine bases. Ultra-violet-visible spectroscopy (16,17) nuclear magnetic resonance (18,19) and x-ray photoelectron spectroscopy (20) studies have been interpreted to suggest that sites of attack could be

guanosine - N7, N1, N7-06 chelate
adenosine - N7, N1, N1-6NH_2 N7-6NH_2 chelate
cytidine - N3, 02, 4NH_2
thymidine - only very slow reaction

At the start of this work, x-ray diffraction studies had established that the only platinum coordination sites which existed in isolable compounds prepared by reaction of chloroplatinum complexes, including $PtC\ell_2(NH_3)_2$, with purine and pyrimidine bases were N7 for guanosine (21), N1 and N7 for adenosine (22,23), N3 for cytidine (24) and no reaction with thymidine.

Since "aquo-complexes" of platinum(II), prepared by treating the chloro-complexes with aqueous silver nitrate solution, did interact with both thymine and uracil to give the platinum-pyrimidine blues (25), it was evident that the interaction of the chloro-complexes with DNA bases could be different from that

0-8412-0588-4/80/47-140-209$05.00/0

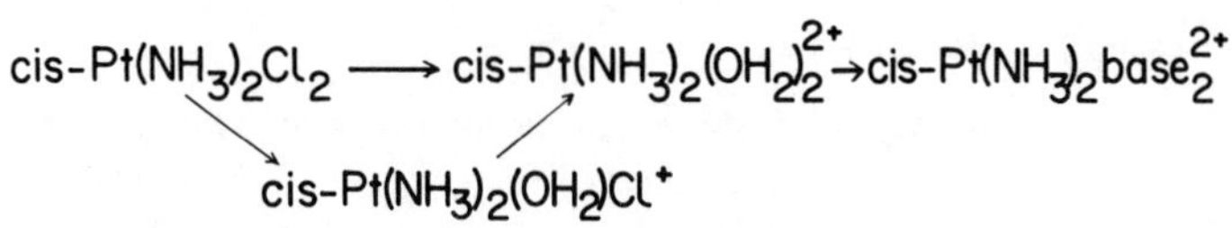

$$\text{cis-Pt(NH}_3)_2\text{Cl}_2 \longrightarrow \text{cis-Pt(NH}_3)_2(\text{OH}_2)_2^{2+} \rightarrow \text{cis-Pt(NH}_3)_2\text{base}_2^{2+}$$

$$\text{cis-Pt(NH}_3)_2\text{Cl}_2 \searrow \text{cis-Pt(NH}_3)_2(\text{OH}_2)\text{Cl}^{+} \nearrow \text{cis-Pt(NH}_3)_2(\text{OH}_2)_2^{2+}$$

Figure 1. The proposed mechanism for the interaction of cis-$PtCl_2(NH_3)_2$ with DNA bases

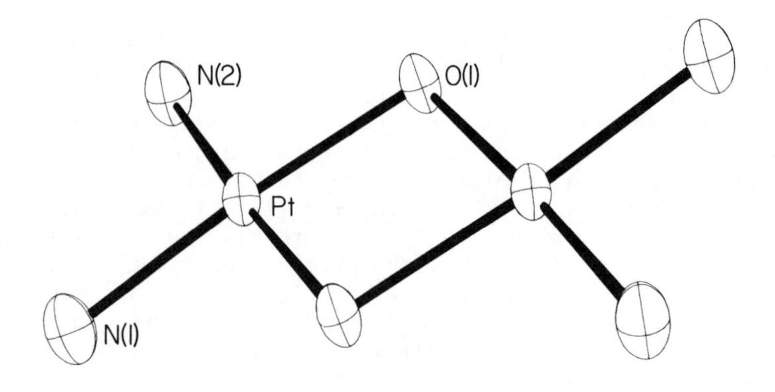

Figure 2. The structure of the $[(NH_3)_2Pt(OH)_2Pt(NH_3)_2]^{2+}$ cation (hydrogen atoms not shown in this, or any of the structures given below)

of the proposed intermediates formed in the cell.

A fairly comprehensive study has been undertaken of the so-called "aquo-species" formed by the silver nitrate reaction and we now know that the system is very complex. A number of hydroxo-bridged species have been isolated from solution including dimeric (26,27) (Figure 2) and trimeric (28,29) cations. The trimeric cations, which can exist in two forms crystallographically, roughly of C_2(28) (Figures 3,4) and C_{3v}(29) (Figure 5) symmetry, must interconvert rapidly in solution, since only one ^{195}Pt n.m.r. signal is obtained (30). In addition to the species known, at least one other species, probably with a single hydroxide bridge, has been isolated, but not characterized by x-ray crystallography. Surprisingly, no species involving coordinated water or terminal hydroxide groups could be isolated as solids. Attempts to break the hydroxide bridges by protonation at low pH led to the isolation of cis-dinitratodiammineplatinum(II) (31) (Figure 6). At the suggestion of Harry B. Gray, we investigated platinum(II) complexes of diethylenetriamine (dien) since these have only one replaceable coordination site and could not form oligomeric species involving double hydroxide bridges. So far, we have been unable to isolate either an aquo- or hydroxo-species, the material obtained from recrystallization in neutral water being nitratodiethylenetriamineplatinum(II) nitrate (32) (Figure 7).

We are faced, therefore, with the puzzling problem of why it is not possible to isolate any aquo-species and why all hydroxo-species are polymerized through hydroxide bridges. It is accepted that aquo-species exist in solution; Raman spectroscopy (Figure 8) shows that the coordinated nitrate groups are replaced immediately on solution in water (32), and n.m.r. studies have shown that these aquo-species are more slowly converted into the polymerised dimer and trimer hydroxo-bridged species (30).

A plausible explanation for this is based on I. D. Brown's acid-base model (33), which is based on a valence model of Brown and Shannon (34), which in turn is an extension of Pauling's valence concepts (35). Briefly, one can use bond order-bond length relationships to give an order to any chemical bond. Brown and Shannon's model adjusted these relationships so that the bond valence (equivalent to bond order) around an atom was always integral (e.g. 2 for oxygen, 1 for hydrogen). In particular, they considered that all neighbouring atoms interacted with the atom under consideration and had to be used in calculating the valence. Thus in addition to the normal bonding interactions, one has to consider weak interactions such as van der Waals interactions. This was clearly reasonable and the model was particularly successful in explaining hydrogen bonding (36,37).

Brown (33) extended the bond valence concept to individual ions and ligands and defined the residual bond valence as a measure of the acid or base strength depending on whether the

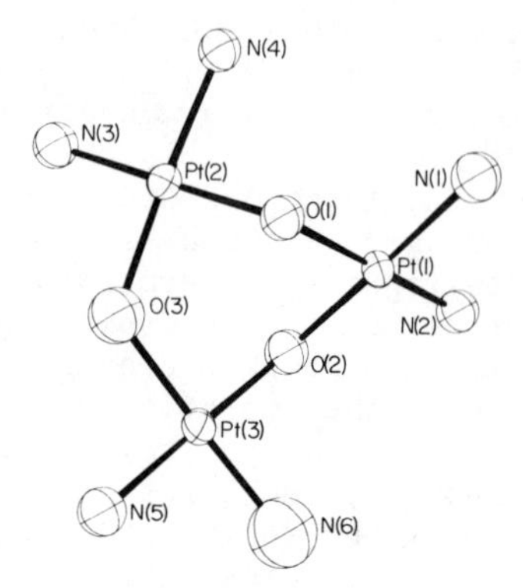

Figure 3. The structure of the $[\{(NH_3)_2Pt(OH)\}_3]^{3+}$ cation found in the nitrate salt, with Pt atoms in the plane of the page

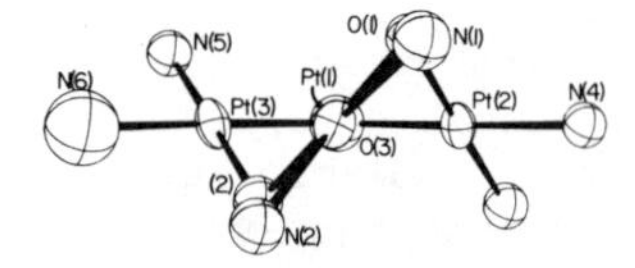

Figure 4. The same cation viewed down the C_2 axis

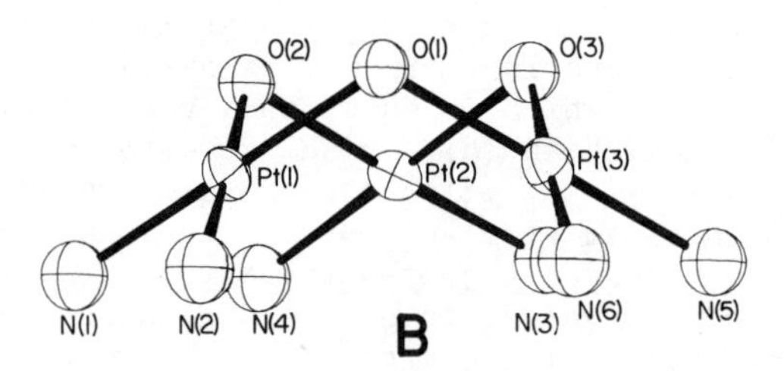

Figure 5. The structure of the $[\{(NH_3)_2Pt(OH)\}_3]^{3+}$ cation found in the sulphate salt

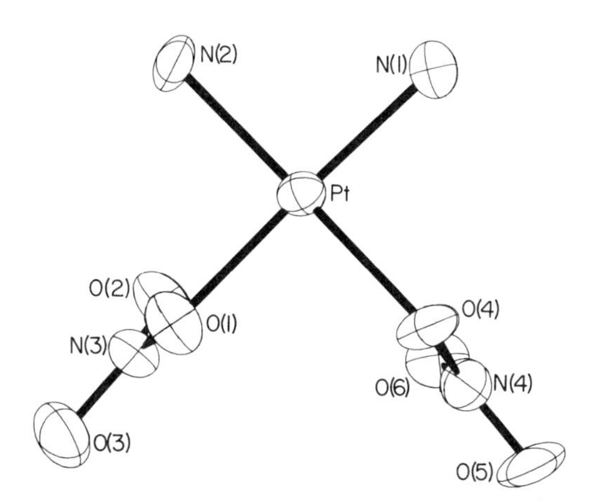

Figure 6. cis-*Dinitratodiammineplatinum(II)*

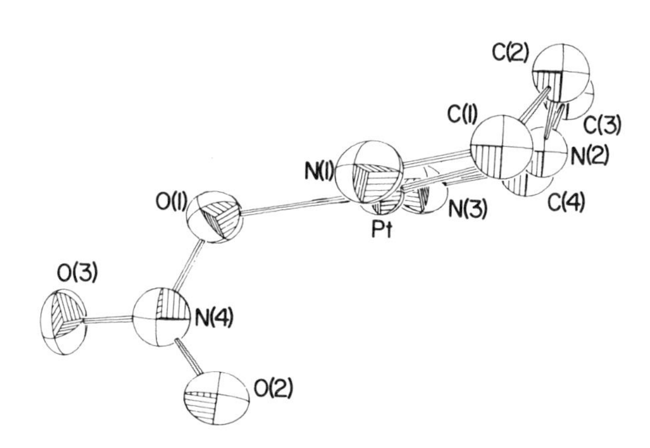

Figure 7. The nitrato(diethylenetriamine)platinum(II) cation

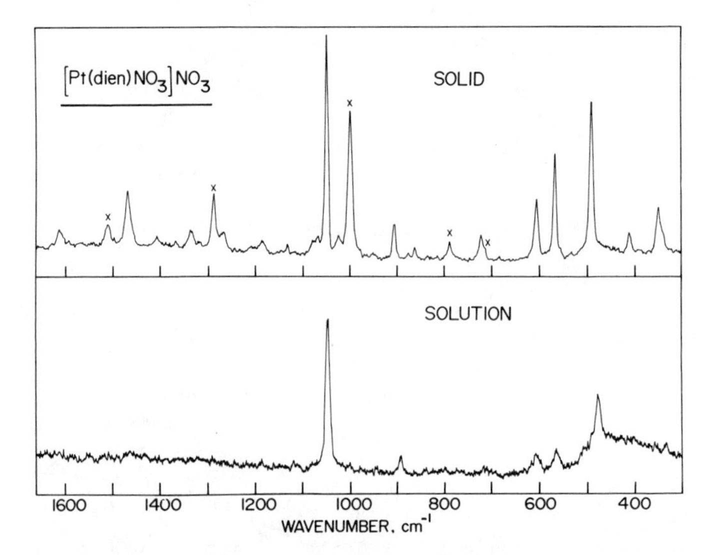

Figure 8. The Raman spectra of: a, solid nitrato(diethylenetriamine)platinum(II) dinitrate; b, an aqueous solution of the same salt. The disappearance of the bands marked x, characteristic of coordinated nitrate ion, show the coordinated nitrate ion is removed in solution.

group is an electron acceptor or donor. Consider the free nitrate ion. The nitrogen atom is formally N(V), thus the total bond valence about nitrogen must add up to five. Since the ion is symmetrical, the nitrogen-oxygen bond order (equal to the bond valence) is 1.67. Further since the valence at oxygen must add to two, the residual valence is 0.33. Thus we have

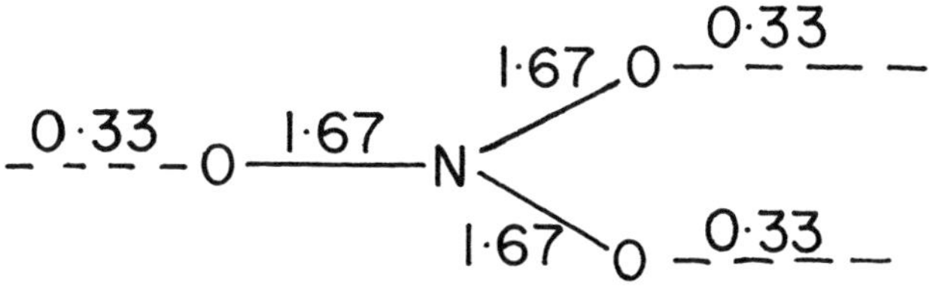

0·33 O 1·67 N 1·67 O 0·33 1·67 O 0·33

Since the nitrate ion is normally an electron donor, Brown defines the base strength of the nitrate ion oxygen atoms as 0.33. Similar consideration of water shows a base strength of 0.30 and an acid strength at each proton of 0.15.

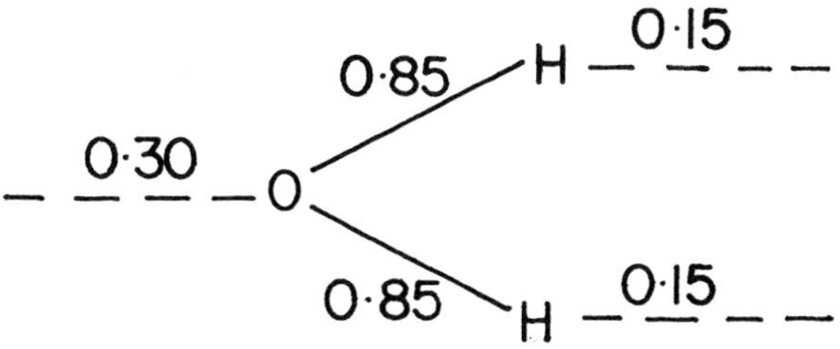

0·30 O 0·85 H 0·15 0·85 H 0·15

In the crystal, these residual bond valences will be satisfied by interactions with other ions or molecules. If the strength of Pt-N and Pt-O bonds is considered to be equal the bond order in each bond in square planar Pt(II) is 2/4 = 0.5. Thus for the $Pt(dien)^{2+}$ moiety the residual bond valence (and thus the acid strength) is 0.5. Now because a bond can have only one bond order, we arrive at one of Brown's important postulates, namely that, to form a bond, the acid and base strengths of the reacting species must match exactly. Clearly the base strengths of nitrate and water do not match the acid strength of $Pt(dien)^{2+}$. Nevertheless, it is possible for them to do so by some rearrangement. Thus if the residual valence at one oxygen atom of the nitrate group is raised to 0.5, the O-N bond order must drop to 1.5 and the total bond order of the remaining N-O bonds must increase by 0.17 units. Thus we might have

0·50 — O — 1·50 — N; N — 1·75 — O — 0·25; N — 1·75 — O — 0·25

Brown's equation for N-O bonds gives bond lengths of 1.29, 1.25 and 1.23Å for bond orders of 1.50, 1.67 and 1.83 respectively. We can examine this experimentally and have done so for the nitrate groups in cis-$Pt(NH_3)_2(NO_3)_2$ (31) and $[Pt(dien)NO_3]NO_3$ (32). The errors are large but the mean values are close to Brown's predictions

Pt — O1 — N — O2; N — O3

	NO_3A	NO_3B	Coord. nitrate	Non-coord NO_3
Pt-O	1.99(1)	2.03(1)	2.02(1)	-
O1-N	1.30(2)	1.28(2)	1.34(1)	1.24(1)
N-O2	1.22(2)	1.19(2)	1.25(1)	1.25(2)
N-O3	1.24(2)	1.22(2)	1.22(2)	1.25(2)

We can do a similar thing to water giving

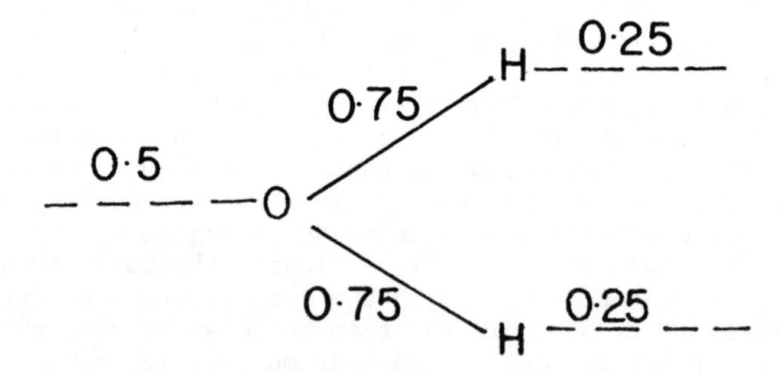

The net result is an increase in the acid strength of the protons; in other words water needs to form either more or substantially stronger hydrogen bonds to become coordinated to Pt(II). Apparently, nitrate in the crystal is not sufficiently good as a hydrogen bond acceptor and thus in competition, the nitrate ion is coordinated in preference to water. In aqueous solution, where water itself acts as an excellent hydrogen bond acceptor, water now coordinates in preference to nitrate.

The hydroxide ion provides a very different case. It is a very strong base

$$\overset{1.05}{\cdots\cdots}\quad O \overset{0.95}{\text{———}} H \quad \overset{0.05}{\cdots\cdots}$$

and it is not possible in any simple manner to reduce the residual valence to 0.5. Attempts to do so lead to

$$\overset{0.5}{\cdots\cdots}\quad O \overset{1.5}{\text{———}} H \quad \overset{-0.5}{\cdots\cdots}$$

and the <u>negative</u> bond valence at hydrogen is impossible. Formation of strong donor hydrogen bonds to oxygen might stabilize the hydroxide ion with a residual valence at oxygen of 0.5, but an alternative is

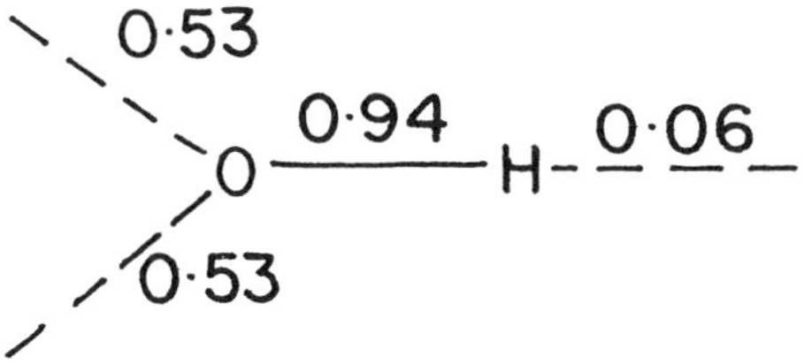

0·53
0·94
O———H- - 0·06 - -
0·53

The division into two residual valences of 0.53 almost exactly matches the residual valence of 0.5 for Pt(II) and thus it is hardly surprising the hydroxide ion forms bridges between two platinum(II) ions so readily.

Thus we can understand that, while aquo-complexes of platinum(II) exist in solution, it will be difficult to isolate them as solids unless a strong hydrogen-bonding species is used as a counter-ion. Further, monomeric hydroxo-species formed by the loss of a proton from the aquo-species will be thermodynamically unstable and will tend to oligomerize to hydroxo-bridged species.

We must now consider how the various "aquo species" interact with the DNA bases. So far, we have examined interactions of thymine, uracil and cytosine in detail and to a lesser extent with guanine. Reaction products from the aqueous solution produced by silver nitrate treatment of the chloride-complexes are complex mixtures, which is hardly surprising based on the known complexity of the aqueous solution. It is possible to prepare monomeric $(NH_3)_2\,Pt(base)_2^{2+}$ species similar to those obtained from the direct reaction of <u>cis</u>-$PtCl_2(NH_3)_2$ and the bases; we assume they come from monomeric aquo-complexes.

Dimeric platinum complexes are obtained by the reaction of pure $[(NH_3)_2Pt(OH)_2Pt(NH_3)_2](NO_3)_2$ with 1-methyl blocked thymine (38), uracil (39) and cytosine (40) (Figures 9-11). All involve platinum coordination at the N3 position with a second platinum attached to 04 or the deprotonated $4-NH_2$ group of the anions. Since for platinum(II) we suggest there is no Pt-Pt bond in the base complex, the formation of such a complex depends on the attacking species containing two platinum atoms. At the same time as these species are formed, one always obtains platinum-pyrimidine blues. The structural features of each of the dimeric-yellow species mentioned above is the head-to-tail arrangement of the pyrimidine ligands. We have commented before (38) that a head-to-head arrangement of the ligands allows further polymerization to form a four-membered chain, which on one electron oxidation will give a species analogous to Lippard's (41) α-pyridone blue. Lippert (42) has now isolated the head-to-head dimeric uracil complex. It is very unstable and very readily converts to a platinum-uracil blue.

The deprotonation of the $4-NH_2$ group on cytosine is interesting and unusual. The pK of this group (12.8) suggests the proton should only be removed in very basic conditions. The preparation of the compound at pH 7 suggests the pK is markedly shifted. We suggest this is caused by the initial coordination of a platinum-atom at N3. The n.m.r. shift of signals from protons on amine groups has been related to the proton pKs (43,44) and we have shown that a marked down-field shift occurs in the n.m.r. spectrum for the $4-NH_2$ group on platinum coordination (40) (Figure 12), consistent with a large pK change. One further interesting compound from this work contains a dimeric platinum cation with deprotonated cytosine, of almost the same structure as the platinum(II) cation but containing Pt(IIS) (Figure 13). The oxidation clearly removes an electron from a Pt-Pt anti-bonding orbital since the Pt-Pt distance is shortened by about 0.4Å. In addition, two N-bonded nitrite groups are added in the terminal positions.

Thus we see that the interactions of the hydroxide bridged dimer with DNA bases are markedly different from those of cis-$PtCl_2(NH_3)_2$ with the same bases.

This may be important since we know that while cis-$Pt(NH_3)_2Cl_2$ is an effective anti-tumor agent, trans-$Pt(NH_3)_2Cl_2$ is not. Further, while there is a marked difference in cross-linking of DNA by cis and trans-$Pt(NH_3)_2Cl_2$ in vivo systems, the cis being more effective by a ratio of 10:1, in vitro experiments show nearly equivalent behaviour of cis and trans compounds (45). It seems likely that the difference is caused by the $Pt(NH_3)_2Cl_2$ compounds being the reactive species in vitro whereas metabolites are active in vivo. The only clearly established difference in the chemistry of the cis and trans compounds which is structure dependent is the formation from cis-$Pt(NH_3)_2Cl_2$ of the cyclic hydroxo-bridge oligomers, which cannot be formed by trans-

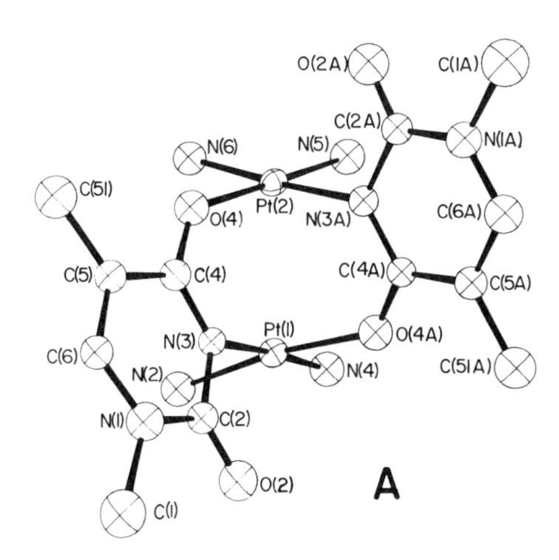

Figure 9. The structure of the cation $[(NH_3)_2Pt(1\text{-methylthyminate–N3,O4})_2Pt(NH_3)_2]^{2+}$

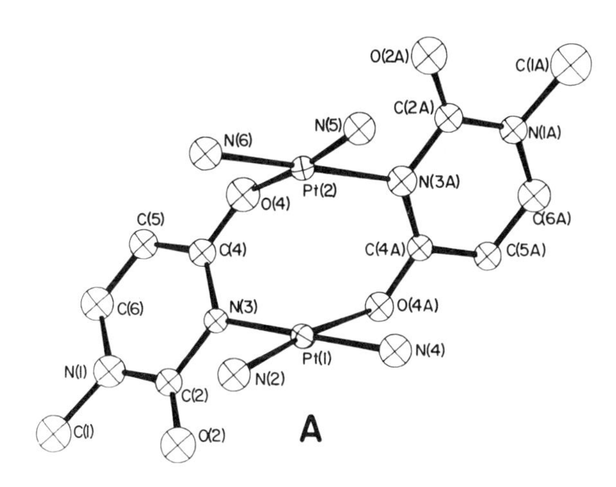

Figure 10. The structure of the cation $[(NH_3)_2Pt(1\text{-methyluracilate–N3,O4})_2Pt(NH_3)_2]^{2+}$

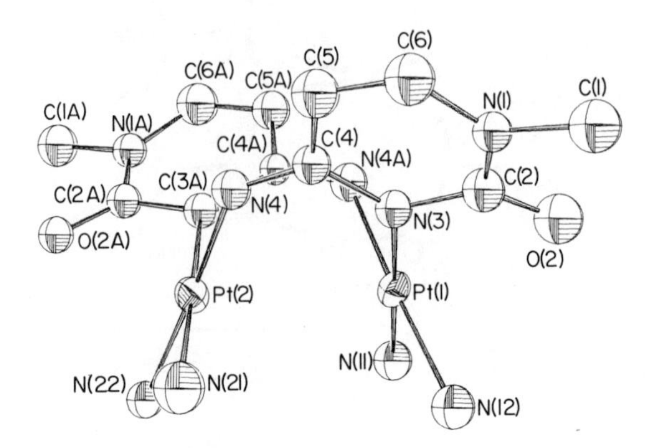

Figure 11. The structure of the cation $[(NH_3)_2Pt(1\text{-methylcytosinate–N3,N4})_2\text{-}Pt(NH_3)_2]^{2+}$

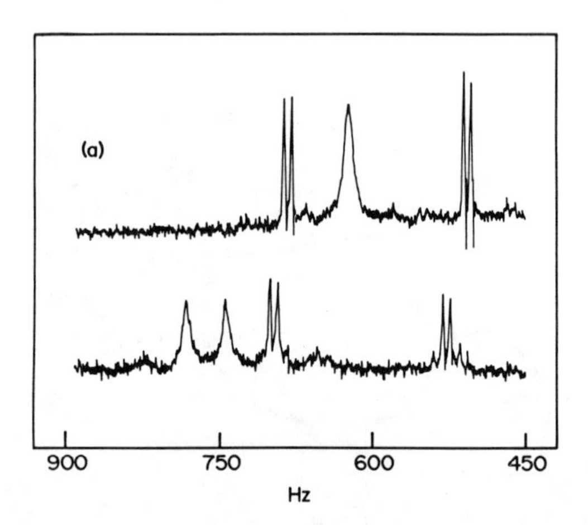

Figure 12. Portions of the NMR spectra of (upper) 1-methylcytosine in DMSO and (lower) 1-methylcytosine and K_2PtCl_4 *in DMSO*

$PtCl_2(NH_3)_2$. The hydroxo-bridged dimer cation acts in a different way towards pyrimidine bases compared to monomers (and may do so to purine bases) coordinating at positions involved in hydrogen bonding, and thus in chain cross-linking, rather than the preferred external N7 positions of the purine bases preferred by monomeric platinum compounds.

Unfortunately, these models are difficult to test *in vivo* and the pK studies listed above led to further work which suggests yet another model for how coordination of platinum(II) at N7 of guanosine may affect the replication process. Studies have suggested that the platinum(II) complexes attach preferentially to G-C base (46) pairs, and probably to G (47). We have made a study of 9-ethylguanine and 1-methylcytosine attached to the $(NH_3)_2Pt^{2+}$ moiety. We have examined structurally three compounds crystallized at different pH's (48). All three have the same basic molecular structure with platinum attached to N7 of 9-ethylguanine and N3 of 1-methylcytosine with the planes of the bases at roughly right angles to each other and to the square plane. There are interesting differences in intermolecular hydrogen bonding depending on the pH of crystallization. At low pH, one obtained crystals of $[(1\text{-MeC})(9\text{-EtG})Pt(NH_3)_2](ClO_4)_2$ and at high pH crystals of $[(1\text{-MeC})(9\text{-EtG-H})Pt(NH_3)_2](ClO_4)$, where 9-EtG-H indicates a 9-ethylguanine molecule deprotonated at N1. No peculiarities in hydrogen bonding are observed. The bonds involve principally interactions of the bases and ammonia with the anion. However, coordination of platinum has shifted the pK of the N1 proton from 9.2 to 7.8 (49), and as a result, at pH7 there are significant quantities of both of the above cations present. The crystals obtained at pH7 are $[(NH_3)_2Pt(1\text{-MeC})(9\text{-EtG})][(NH_3)_2Pt(1\text{-MeC})(9\text{-EtG-H})](ClO_4)_3$. These are interesting in that the two cations are hydrogen-bonded together as shown in Figure 14 giving a G-G pair.

In light of recent theories that cancer cells are deficient in some DNA damage repair mechanisms, and that the reason they can be attacked preferentially by chemotherapy, is because one causes further DNA damage which cannot be repaired, there has been an increased interest in methods of inducing base mispairing. As can be seen in Figure 14, the G-G mispairing could replace a G-C pair; the steric differences are not great. The important point is that this mispairing was induced by coordination of platinum at N7 of 1-ethylguanine, *outside the normal hydrogen bonding region*. Clearly, we have not yet run out of models for effects of platinum complexes on DNA replication which must be tested *in vivo* before we shall have a clear understanding of the action of platinum anti-cancer drugs.

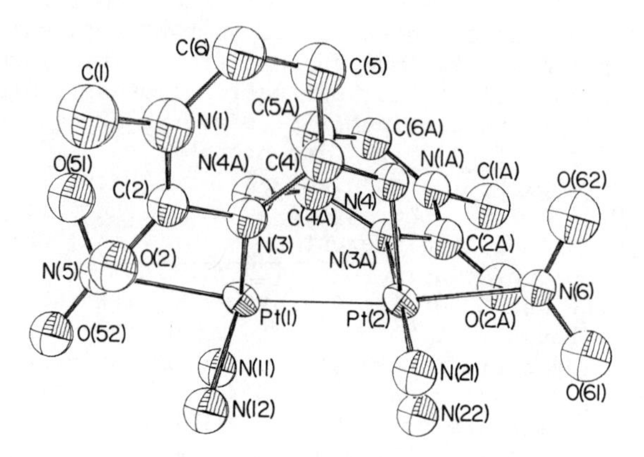

Figure 13. The structure of the cation $[NO_2(NH_3)_2Pt(1\text{-methylcytosinate–N3,-N4})_2Pt(NH_3)_2NO_2]^+$

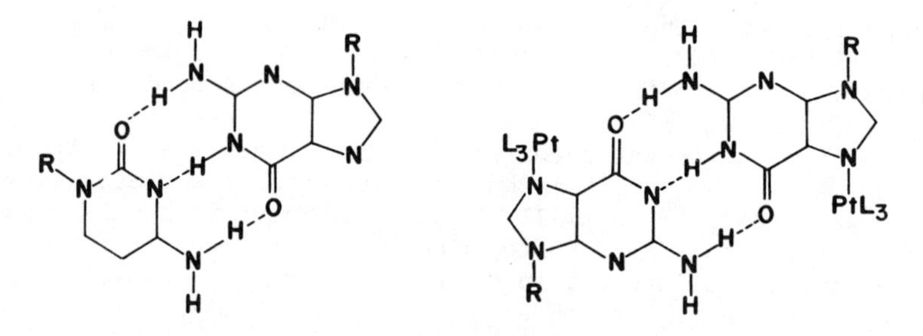

Figure 14. A comparison of the G–C *pair found in DNA with the* G–G *pair found in the hydrogen bonded dication* $[(NH_3)_2Pt(1\text{-methylcytosine})\ (1\text{-ethylguanine})]^{2+}[(NH_3)_2Pt(1\text{-methylcytosine})[1\text{-ethylguaninate}]^+$

Acknowledgements

Work of this nature proceeds most smoothly if a number of groups pool their respective expertises. We could not have proceeded in this field without the cooperation of the research groups of Professor Barnett Rosenberg (Michigan State University) and Dr. Bernhard Lippert (Technische Universität München) who supplied us with many of the crystals. We have had many detailed discussions with Professor I. D. Brown (McMaster University) concerning his bond valence model and he has been more than willing to help us. Members of my research group involved in various phases of this work have been J. Britten, R. Faggiani, P. Pilon, J. Pocé, R. A. Speranzini, Dr. G. Turner and M. Zvagulis.

Generous financial support has been received from the National Cancer Institute of Canada, Natural Science and Engineering Research Council of Canada, Medical Research Council of Canada, McMaster University Science and Engineering Research Board, Johnson Matthey and Mallory, and Hazell Coxall and Friends.

Literature Cited

1. Cleare, M. J. Coord. Chem. Rev. 1974, 12, 349 and references therein.
2. Thomson, A. J. "Platinum Coordination Complexes in Cancer Chemotherapy", 1974 Ed. Connor, T. A. and Roberts, J. J., Springer-Verlag, N.Y. 38 ff.
3. Cleare, M. J. J. Clinical Hemat. Oncol. 1977, 7, 1.
4. Banerjea, D.; Basolo, F.; Pearson, R. G. J. Amer. Chem. Soc. 1957, 79, 4055.
5. Basolo, F.; Gray, H. B.; Pearson, R. G. J. Amer. Chem. Soc. 1960, 82, 4200.
6. Basolo, F.; Pearson, R. G. "Mechanisms of Inorganic Reactions" (2nd ed.) 1967, Wiley, N.Y.
7. Belluco, U.; Ettore, R.; Basolo, F.; Pearson, R. G.; Turco, A. Inorg. Chem. 1966, 5, 591.
8. Wakelin, L. P. G. Biochem. Soc. Trans. 1974, 2, 866.
9. Macquet, J. P.; Theophanides, T. Biopolymers 1975, 14, 781.
10. Howle, J. A.; Gale, G. R.; Smith, A. B. Biochem. Pharmacol. 1972, 21, 1465.
11. Roos, I. A. G.; Arnold, M. C. J. Clinical Hemat. Oncol. 1977, 7, 374.
12. Harder, H.; Rosenberg, B. Int. J. Cancer, 1970, 6, 207.
13. Howle, J.; Gale, G. Biochem. Pharmacol. 1970, 19, 2757.
14. Howle, J.; Gale, G. J. Bact., 1970, 103, 259.
15. Bau, R.; Gellert, R. W.; Lehovec, S. M.; Louie, S. J. Clinical Hemat. Oncol. 1977, 7, 51.
16. Mansy, S.; Rosenberg, B.; Thomson, A. J. J. Amer. Chem. Soc. 1973, 95, 1633.

17. Roos, I. A. G.;Thomson, A. J.; Mansy, S. J. Amer. Chem. Soc., 1974, 96, 6484.
18. Kong, P. C.; Theophanides, T. Inorg. Chem. 1974, 13, 1167
19. Macquet, J. P.; Theophanides, T. Bioinorg. Chem. 1975, 5, 59.
20. Millard, M. M.; Macquet, J. P.; Theophanides, T. Biochem. Biophys. Acta 1975, 402, 166.
21. Gellert, R. W.; Bau, R. J. Amer. Chem. Soc., 1975, 97, 7379.
22. Terzis, A.; Hadjiliadis, N.; Rivest, R.; Theophanides, T. Inorg. Chim. Acta 1975, 12, L5.
23. Lock, C. J. L.; Speranzini, R. A.; Turner, G.; Powell, J. J. Amer. Chem. Soc. 1976, 98, 7865.
24. Lock, C. J. L.; Speranzini, R. A.; Powell, J. Can. J. Chem. 1976, 54, 53.
25. Lippert, B. J. Clinical Hemat. Oncol. 1977, 7, 26.
26. Faggiani, R.; Lippert, B.; Lock, C. J. L.; Rosenberg, B. J. Amer. Chem. Soc. 1977, 99, 777.
27. Lippert, B.; Lock, C. J. L.; Rosenberg, B.; Zvagulis, M. Inorg. Chem. 1978, 17, 2971.
28. Faggiani, R.; Lippert, B.; Lock, C. J. L.; Rosenberg, B. Inorg. Chem. 1977, 16, 1192.
29. Faggiani, R.; Lippert, B.; Lock, C. J. L.; Rosenberg, B. Inorg. Chem. 1978, 17, 1941.
30. Chikuma, M.; Pollock, R. J.; Ott, K. C.; Gansow, O. A.; Rosenberg, B. Manuscript submitted to J. Am. Chem. Soc.
31. Lippert, B.; Lock, C. J. L.; Rosenberg, B.; Zvagulis, M. Inorg. Chem. 1977, 16, 1525.
32. Britten, J.; Lock, C. J. L. to be published.
33. Brown, I. D. J. Chem. Soc. Dalton, in press.
34. Brown, I. D.; Shannon, R. D. Acta Cryst. 1973, A29, 266.
35. Pauling, L. "Nature of the Chemical Bond" 3rd ed. 1960, Cornell University Press, Ithaca, N.Y., p. 239.
36. Brown, I. D. Acta Cryst. 1976, A32, 24.
37. Brown, I. D. Acta Cryst. 1976, A32, 786.
38. Lock, C. J. L.; Peresie, H. J.; Rosenberg, B.; Turner, G. J. Amer. Chem. Soc. 1978, 100, 3371.
39. Faggiani, R.; Lock, C. J. L.; Pollock, R. J.; Rosenberg, B.; Turner, G. manuscript in preparation.
40. Faggiani, R.; Lippert, B.; Lock, C. J. L.; Speranzini, R. A. manuscript submitted to J. Amer. Chem. Soc.
41. Barton, J. K.; Rabinowitz, H. N.; Szalda, D. J.; Lippard, S. J. J. Amer. Chem. Soc. 1977, 99, 2827.
42. Lippert, B. private communication 1979/09/26.
43. Kokko, J. P.; Mandell, L.; Goldstein, J. H. J. Am. Chem. Soc. 1962, 84, 1042.
44. Grunwald, E.; Loewenstein,, A.; Meiboom, S. J. Chem. Phys. 1957, 27, 641.
45. Roberts, J. J.; Pascoe, J. M. Nature 1972, 235, 282.
46. Stone, P. J.; Kelman, A. D.; Sinex, F. M. Nature 1974, 251, 736.

47. Jankowski, J. P.; Macquet, J. P.; Butour, J. L. Biochimie 1978, 60, 901.
48. Faggiani, R.; Lippert, B.; Lock, C. J. L.; Speranzini, R. manuscript in preparation.
49. Lippert, B. manuscript in preparation.

RECEIVED April 21, 1980.

13

The Genetic Toxicology of Substitutionally Inert Transition Metal Complexes

G. WARREN, S. J. ROGERS, and E. H. ABBOTT

Department of Chemistry, Montana State University, Bozeman, MT 59717

In recent years it has become apparent that some metal ions have important genetic effects and that these effects may have serious biological consequences for humans. For example, epidemiological studies have shown that chromium (1), arsenic (2, 3, 4), nickel (5, 6), and possibly cadmium (7) are carcinogenic in man. Platinum(II) compounds have also been shown to be carcinogenic in laboratory animals (8). Interestingly, these platinum(II) compounds can, on the other hand, be potent anti-tumor drugs (9). This property has been linked to their reaction with DNA, but the exact mechanism is yet unclear (10).

Recently, bacterial systems have been developed to provide rapid, efficient screening for genetically active compounds. A good correlation has been observed between mutagenicity of organic compounds in one such assay, the Ames Salmonella his^- reversion test, and carcinogenicity in mammals (11). Metal ions, including As(III), Be(II), Cr(VI), Cu(II), Fe(II⇄III), Mn(II⇄III), Mo(VI), Pt(II), Se(VI) (12), Rh(I), Ru(II) (13), and Te(IV) have been shown to cause mutations in one of several bacterial systems (12, 13, 16). Correlation between mutagenicity and carcinogenicity of metals in the Ames test has not been as good as for organic compounds. Most metal salts are bacteriocidal and lethality often interferes with demonstration of mutagenesis. Metals such as nickel, cadmium, lead, titanium and zinc, thought to be carcinogenic, are negative with the Ames test. Those such as Cu, Mo, Se, Te, Rh, and Ru that are positive in some mutagenesis assays are not known carcinogens (12). DNA repair assays are more successful in detecting most known metal carcinogens (13), but there are still many more elements genetically active in this test that are not known to be carcinogenic.

After an extensive survey of metal salts for mutagenic activity and DNA-damaging ability, a systematic investigation of the genetic toxicology of some substitutionally inert metal complexes was begun in this laboratory. Such complexes were chosen because they largely retain their structures over the time necessary to observe genetic effects in microbial systems, and one such complex, Pt(II), had been shown to be a consistently potent mutagen in the

0-8412-0588-4/80/47-140-227$05.00/0

Ames Salmonella assay (14, 15). With these compounds it is possible to avoid the complications of multiple equilibria which occur in the study of metal salts.

Procedures

Two basic types of assays were used. Lethality studies compared the lethal effect of compounds on wild type bacteria (*Escherichia coli*) K-12 with the lethality on several mutant strains constructed with known DNA repair deficiencies. Compounds which are more toxic to such a repair deficient mutant than to the wild type are considered to be genetically toxic. Table I lists the bacterial strains used, together with their genotypes.

TABLE I

REPAIR DEFICIENT BACTERIA STRAINS USED IN REPAIR ASSAYS

Bacteria Strain	Genotype	Phenotype
AB 1157	Wild Type	Repair Proficient
AB 1899	lon⁻	Filaments
AB 1886	uvr A6	Excision Minus
PAM 100	lex C	Recombinationless
RH 1	rec A56, uvr A6	Recombinationless and Excision Minus
GW 801	rec A56	Recombinationless
PAM 5717	lex A2	Recombinationless
AA 34	rec A56, lex A2	(Recombinationless)[2]
P 3478	pol A1	Polymerase 1 Minus

A second type of test is the Ames test (11). This test utilizes a series of specially constructed Salmonella bacterial strains which are unable to produce the natural amino acid histidine because of known mutations in the histidine biosynthetic genes. The bacteria are exposed to a potential mutagen on a medium without histidine, and the number of bacterial colonies able to grow is scored. These colonies arise from a cell which has undergone a specific back mutation which results in the bacteria being proto trophic for histidine. Table II lists the Salmonella strains utilized in this phase of the work.

TABLE II

HIS⁻ BACTERIA USED IN REVERSION (AMES TEST) STUDIES

Bacteria	Defect	Plasmid (pKm101)
TA 92*	Base pair substituted	+
TA 98$^+$	Frame shifted, uvr$^-$, rfa$^-$	+
TA 100*	Base pair substituted, uvr$^-$, rfa$^-$	+
TA 1535*	Base pair substituted, uvr$^-$, rfa$^-$	---
TA 1537**	Frame shifted, uvr$^-$, rfa$^-$	---
TA 1538$^+$	Frame shifted, uvr$^-$, rfa$^-$	---

*All carry the same mutation (his G46)

+Both carry his D3052 frameshift

**Carries his C3076 frameshift

The coordination compounds used in this work were synthesized by published procedures.

Results

Rhodium Complexes. Table III presents the results of our investigations on rhodium complexes. Preliminary results indicated that mutagenic activity was optimized in dihalides. Since the dichlorobis(1,10-phenanthroline) complex and its bipyridyl analog have the *cis* geometry, we studied the *cis* and *trans* isomers of the dichlorobis(ethylenediamine)rhodium(III) cation. Here the *cis* isomer is an active mutagen, while the *trans* isomer is inactive. This is reminiscent of the anti-tumor activity of the dichlorodiamineplatinum(II) isomers, wherein the *cis* isomer is quite active, while the *trans* isomer is inactive (9). Interestingly, many of the tri halo complexes were not active. For those that are active, relatively rapid ligand exchange prevails. It is not clear what is the dominant species present at the time of mutation. The common denominator for activity therefore appears to be that the complex should be a cationic *cis* substituted complex.

Chromium Complexes. Table IV contains the results on chromium compounds. Like rhodium, a *cis* dihalo structure seems to confer mutagenic properties on the complex. Unlike the rhodium complexes, no activity is observed in the *cis* dihalopolyamine complexes. Several other complexes are active too, but these undergo sufficiently rapid ligand exchange that we are unsure as to the

TABLE III

GENETIC TOXICITY TESTING OF RHODIUM(III) COMPLEXES

Abbreviations: Pyr=Pyridine, Phen=1,10 Phenanthroline, En=Ethylene-diamine, Pn=1,2 Propanediamine, Ox=Oxalate, Trien=Triethylenetetramine, Bipy=Bipyridine, 3Pic=Picoline

COMPOUND	REPAIR STRAINS*	REVERSION STRAINS**
[Rh $(Pyr_3)Cl_3$]	Not Active	
[Rh $(Phen)_3$]Cl_3	Not Active	
[Rh $(En)_3$]Cl_3	Not Active	
trans-[Rh $(En)_2Cl_2$]Cl	Not Active	
[Rh $(NH_3)_5Cl$]Cl_2	Not Active	
[Rh $(Pyr)_3$ $(SCN)_3$]	Not Active	
[Rh $(En)_3$]Cl_3	RH 1 (2)	Not Active
[K_3Rh $(Ox)_3$]	Not Active	
cis-[Rh$(En_2)Cl_2$]Cl	RH 1 (16)	TA 92(0.04), TA 98(0.033), TA 100(0.067)
[Rh $(NH_3)_4Cl_2$]Cl	RH 1 (16)	TA 92(0.02), TA 100(0.079)
[Rh $(Pyr)_4Br_2$]Br	RH 1 (512), 1886 (16), 1889 (4), AA 34 (8), 801 (32)	TA 92(1.10), TA 98(114), TA 100(9.6)
[Rh $(Pyr)_4Cl_2$]Cl	RH 1 (256), 1886 (8)	TA 92(0.02), TA 98(9.8), TA 100(2.0)
[Rh (Trien)Cl_2]Cl	RH 1 (64), 1899 (2)	TA 92(0.02), TA 98(0.072), TA 100(0.01)
[Rh $(CH_3CN)_3Cl_3$]	RH 1 (64), AA 34 (4), 801 (4)	TA 92(0.36), TA 98(0.112), TA 100(0.126)
[Rh $(Bipy)_2Cl_2$]Cl	RH 1 (16)	TA 92(0.01), TA 98(.025), TA 100(0.097)
[Rh $(Phen)_2Cl_2$]Cl	RH 1 (64)	TA 92(0.08), TA 98(0.82), TA 100(0.30)
Rh $Cl_3(H_2O)_3$	RH 1 (8)	TA 92(0.06), TA 98(0.162), TA 100(0.240)
[Rh $(3Pic)_4Cl_2$]Cl	RH 1 (128), 1886 (16), 1899 (16), 801 (2)	TA 92(0.59), TA 98(125), TA 100 toxic

*numbers in parentheses refer to the fold increase in inhibition in microtiter repair in comparison with wild type AB1157.

**number in parentheses refers to his+ revertants per nanomole of compound in the plates. Calculated on a linear portion of the dose response curve for each compound.

TABLE IV

MUTAGENESIS TESTING OF CHROMIUM COMPOUNDS

Abbreviations: Ox=Oxalate, En=Ethylenediamine, Pyr=Pyridine, Bipy=Bipyridine, Phen=1,10 Phenanthroline

COMPOUND	REPAIR STRAINS*	REVERSION STRAINS**
$K_3[Cr(Ox)_3]$	Not Active	
$[Cr(En)_3]Cl$	Not Active	
$[Cr(NH_3)_4H_2OCl]Cl_2$	Not Active	
$[Cr\ Ox(NH_3)_4]Cl$	Not Active	
$[Cr(NH_3)_5H_2)]Cl_3$	Not Active	
cis-$[Cr(NH_3)_4Cl_2]Cl$	Not Active	
$[Cr(En)_2Cl_2]Cl$	Not Active	
cis-$[Cr(Pyr)_4F_2]Br$	Not Active	
$K_3[Cr(CN)_6]$	Not Active	
$[Cr(Bipy)_2Ox]I$	RH 1 (512), 1886 (8), 1899 (128), P 100 (4), AA 34 (8), 801 (2), 5717 (4)	TA 92(0.89), TA 98 (1.2)
$[Cr(Bipy)_2Cl_2]Cl$	RH 1 (512), 1886 (16), 1899 (64), P 100 (2), AA 34 (64), 801 (8), 5717 (2)	TA 92(0.56), TA 98 (0.48), TA 100(1.11)
$[Cr(Phen)_2Cl_2]Cl$	RH 1 (256), 1886 (16), 1899 (32), P 100 (2), AA 34 (32)	TA 92(0.49), TA 100 (2.8), TA 98(0.06)
$[Cr(Urea)_6]Cl_3$	RH 1 (16), AA 34 (8)	TA 98(0.03), TA 100 (0.3)
$[Cr(NH_3)_5Cl]Cl_2$	RH 1 (4), 1899 (2), P 100 (2)	
$[Cr(NH_3)_5H_2O]Cl_2$	RH 1 (4), 1899 (2)	

*numbers in parentheses refer to the fold increase in inhibition in microtiter repair in comparison with wild type AB1157.
**number in parentheses refers to <u>his</u>+ revertants per nanomole of compound in the plates. Calculated on a linear portion of the dose response curve for each compound.

dominant structure of the complex when it reaches the genetic material.

Other Metals. Complexes of Pt(II), Pt(IV), Ni(II), and Cobalt(III) have been surveyed. Results appear in Table V. In the cases of Pt and Co compounds, examples of differential lethality are observed. In the case of the two nickel compounds, no activity is observed despite the fact that both are relatively inert to ligand substitution. In the case of the cobalt(III) complexes, the absolute configuration of the complex affected the differential lethality, with the (-)tris(ethylenediamine)cobalt(III) cations considerably less active than the (+) complex.

Discussion

The results presented above suggest several points which merit further discussion. It has been recognized for some time that chromium workers suffer a high incidence of cancer (1). In attempting to identify the active form of chromium, various groups have searched for mutagenic effects of chromium species. Chromate and dichromate have been found to be mutagenic, but it has been recognized that these high oxidation states are short-lived in the cellular medium. A more plausible candidate for the mutagenic effects of chromium would be chromium(III); however, up to now the chromium(III) complexes that had been tested have been inactive (16). Therefore, it has been suggested that the carcinogenic agent may not be the metal, but rather an oxidation product formed when chromate is reduced in the cell. Our work clearly demonstrates the mutagenicity of chromium(III) compounds. The structures which are active are analogous to complexes which can be formed upon the reduction of the chromate ion in the cellular medium. Thus, chromium(III) complexes are potentially hazardous. They should be seriously considered as the active carcinogens in chromate exposure, and due care should be exercised by persons working with chromium(III) itself.

It is interesting and important to inquire into the mechanisms by which these substitutionally inert complexes cause mutagenesis. First of all, a striking range of activities are observed, depending on the nature of the ligands. This indicates that the biologically active species are the starting complexes or species closely related to them. The fact that different activity is observed for the two different optical isomers of the tris(ethylenediamine)cobalt(III) cations is further evidence for this.

A considerable number of genetically toxic compounds have been examined by both reversion and lethality studies. Since the mechanism by which DNA is damaged is well understood for a number of organic compounds, it is interesting to compare our results to those for mutagens of known mechanisms. A summary by Green is particularly useful for this purpose (16). Our repair assay results for the rhodium compounds bear a strong resemblance to those obtained with crosslinking agents like mitomycin C. In lethality

TABLE V

GENETIC TOXICITY TESTING OF MISCELLANEOUS METAL IONS

COMPOUND	REPAIR STRAINS*	REVERSION STRAINS**
Cd^{+2} ($CdAc_2$)	RH 1 (4), 1886 (4)	
Hg^{+2} ($HgBr_2$)	2494 (2), RH 1 (8), 1886 (4)	
Mn^{+2} ($MnCl_2$)	801 (2)	
Mo^{+5} (Polymolybdate)	AA 34 (2)	
Ni^{+2} ($NiCl_2$)	P 100 (2)	
Sb^{+3} ($SbCl_3$)	RH 1 (2)	
As^{+5} ($NaAsO_2$)	RH 1 (4), 1886 (4)	
Zn^{+2} ($ZnCl_2$)	RH 1 (2), AA 34 (2)	
$[Co(NH_3)_6]Cl_3$	RH 1 (16), P 100 (8), 1886 (2), 1899 (4)	
$[Co(Pn)_3]Cl_3$	Not Active	
$+[Co(En)_3]I_3$	RH 1 (8), 1886 (2)	
$-[Co(En)_3]I_3$	RH 1 (16), 1886 (4)	
d-cis-$[Co(En)_2(NO_2)_2]Br$	RH 1 (2)	
l-cis-$[Co(En)_2(NO_2)_2]Br$	RH 1 (2)	
$[Pt(En)_3]Cl_4$	RH 1 (16), 1886 (4), P 100 (4)	
cis-$[Pt(NH_3)_2Cl_2]$	RH 1 (16), 1886 (2), 1899 (2)	TA 92(29), TA 98 (0.61), TA 100(47)
trans-$[Pt(NH_3)_2Cl_2]$	RH 1 (2)	TA 92(st), TA 100 (st)***

*fold increase in minimal inhibitory concentration in microtiter repair assay.

**number in parentheses refers to <u>his</u>+ revertants per nanomole of compound in the plates. Calculated on a linear portion of the dose response curve for each compound.

***spot test.

studies, we find sensitivity in 1886, a uvr⁻ strain. Greater sensitivity is observed for the double mutant Rh1, which is rec⁻, uvr⁻. In the reversion studies TA 92 is affected much more than TA 100. The TA 100 strain is uvr⁻ and uvr excision is the first step in crosslink repair. Evidently, the absence of uvr in TA 100 prevents the initiation of repair which could revert TA 100, although the repair mechanism of pKm 101, the ampicellin plasmid of both TA 92 and TA 100 is required for expression of mutation in uvr⁻ strains. On the basis of the results at hand, we propose that the activity of the rhodium compounds is due to interstrand crosslinks in the DNA caused by rhodium bridging of the strands. The chromium compounds appear to have a different pattern of activity in the lethality studies and may be acting by a different mechanism. It is also interesting to note that all the complexes cause reversion only in the plasmid containing strains TA 92, TA 100, and TA 98 consistent with the Pt(II) results. The complexes are extremely lethal for uvr⁻ *Salmonella* strains TA 1535 and TA 1538 which do not contain the plasmid pKm 101.

Table V shows that some activity is observed for metal salts in lethality studies. These metal ions are not as active as the inert complexes in the lethality assay, and none of these ions has been found to be active in reversion studies. Perhaps the difference is that the metal ions rapidly come to equilibrium in the cell and form kinetically labile and non-toxic complexes with DNA. As strand separation occurs, the labile ions that are on the DNA can be rapidly displaced and thus are not able to interfere with replication. On the other hand, the inert ions undergo substitution at rates which are slow compared to cell division. When they undergo ligand substitutions and become attached to DNA, they remain fixed to the DNA long enough to cause errors during replication.

We should note the fact that in the bacterial systems, the substitutionally inert complexes which are mutagenic and active in the repair assay all have effects similar to those of the platinum complexes known to be anti-tumor drugs. Since repair effects are closely correlated with activity of the platinum compounds (17), then the anti-tumor activity could be related to the substitutional inertness of our complexes. We have noted that many of the complexes we have studied are far less bacteriocidal than the platinum compounds , and yet comparable in repair activity. This suggests that further study of substitutionally inert metal complexes may yield anti-tumor drugs which are as effective as the platinum compounds, and yet lack their undesirable toxic side effects.

Finally, it should be pointed out that our work substantiates the health hazard of substitutionally inert metal ions. Environmental pollution by these ions is potentially a serious problem. It should be considered as these metals are widely used in the catalytic converters of automobiles, and because more extensive combustion of coal will introduce more such metal ions into the environment.

Conclusions

1. Certain complexes of a variety of substitutionally inert metal ions are highly mutagenic, depending on the structure and the charge of the complex. Mutagenicity appears to be more closely related to the rate of ligand substitution than to the nature of the metal.
2. The mutagenicity of chromium(III) complexes suggests that they could be responsible for the high incidence of lung cancer noted for chromium workers.
3. The complexes of substitutionally inert metal ions should be carefully investigated for their anti-tumor activity because they may present superior alternatives to the toxic platinum(II) compounds.
4. Environmental pollution by substitutionally inert metal ions is a matter of concern and may result in serious health problems.

Abstract

A variety of compounds of rhodium(III), chromium(III), cobalt (III), and platinum(IV) have been tested for their genetic toxicology using lethality assays on selected repair deficient *Escherichia coli* bacterial strains, and by reverse mutation assays in Salmonella. The nature of the ligands, charge of the complex, and stereochemistry of the complex profoundly affect the mutagenicity of these compounds. For rhodium(III), unicationic complexes with adjacent leaving groups are most active. The pattern of differential lethality indicates that it involves interstrand DNA crosslinking. A different pattern of activity is demonstrated for chromium(III). The strong mutagenic effects of chromium(III) compounds suggest that this valence state may be responsible for the carcinogenic activity of chromium. The general activity of substitutionally inert metal ions suggests that this class of compounds should be systematically examined for anti tumor activity. It is apparent that the toxic effects of heavy metals as environmental pollutants can be influenced by the oxidation state of the metal and the available (ambient) ligands, and thus rather complicate the prediction of toxicological patterns.

Literature Cited

1. Enterline, P. E., J. Occup. Med., 1974, 16, p. 523.
2. Ott, M. G., Holder, B. B., and Gordon, H. L., Arch. Environ. Health, 1974, 29, p. 250.
3. Tokudome, S. and Kuratsune, M., Int. J. Cancer, 1976, 17, p. 310.
4. Lee, A. M. and Fraumeni, J. F., J. Nat'l Cancer Inst., 1969, 42, p. 1045.
5. Pedersen, E., Hogetveit, A. C., and Anderson, A., Int. J. Cancer, 1973, 12, p. 32.

6. Doll, R., Morgan, L. G., and Speizer, F. E., Br. J. Cancer, 1970, 24, p. 623.
7. Lemen, R. A., Lee, J. S., Wagener, J. K., and Blejer, H. P., Ann. N. Y. Acad. Sci., 1976, 271, p. 273.
8. Leopold, W. R., Miller, E. C., and Miller, J. A., Cancer Research, 1979, 39, p. 913.
9. Rosenberg, B., Die Naturwissenschaften, 1973, 60, p. 399.
10. Rosenberg, B., Cancer Chemotherapy Reports, 1975, 59, p. 589.
11. McCann, J., Choi, E., Yamasaki, E., and Ames, B. N., Proc. Nat'l Acad. Sci., U.S., 1975, 72, p. 5135.
12. Sigel, H., Ed., Metal Ions in Biological Systems, Vol. 10, Chapter 2, Flessel, C. P., Furst, A., and Radding, S. B., 1979, Marcel Dekker, New York (in press).
13. Nishioka, H., Mutat. Res., 1975, 31, p. 185.
14. Beck, D. J., and Brubaker, R. R., Mutat Res., 1975, 27, p. 281.
15. Monti-Bragadin, C., Tamaro, M., Banfi, E., and Zassinovich, G., Abstr. IUPKC Intern. Symp. Chin. Chem. Chem. Toxicol. of Metals, Monaco, 1977.
16. Newbold, R. F., Amos, J., Connell, J. R., Mutat Res., 1979, 67, p. 55.
17. Green, M. H. L., and Muriel, W. J., Mutat Res., 1976, 38, p. 3.
18. Tamaro, M., Venturini, S., Eftimiadi, C., and Monti-Bragdin, C., Experentia, 1977, 33, p. 317.

RECEIVED May 22, 1980.

14

The Iron Bleomycins

JAMES C. DABROWIAK, FREDERICK T. GREENAWAY, and FRANK S. SANTILLO
Department of Chemistry, Syracuse University, Syracuse, NY 13210

STANLEY T. CROOKE and JOHN M. ESSERY
Bristol Laboratories, Thompson Road, Syracuse, NY 13206

The bleomycins (BLM) are a group of Streptomyces-produced glycopeptides which show antitumor activity (1). Clinically employed bleomycin, marketed under the trade name Blenoxane, is a mixture of eleven antitumor-active bleomycins. The drug is widely used for the treatment of squamous cell carcinomas, lymphomas and testicular carcinomas (2).

Current evidence suggests that the antibiotic functions by a unique mechanism of action. It has long been known that bleomycin is capable of causing single and double strand breaks in DNA (1), and it is now believed that DNA breakage by bleomycin occurs via a metal-mediated redox mechanism involving iron. Horwitz and coworkers (3,4,5) were the first to show that Fe(II)BLM is an air sensitive compound and that it readily oxidizes in the atmosphere to Fe(III)BLM. They postulated that this oxidation process occurs in the tumor cell while the iron-bleomycin is bound to DNA, and that DNA damage is the result of radical production in the vicinity of the biopolymer by iron-bleomycin.

Recent results (6-9) have shown bleomycin to be bifunctional. One part of the drug is capable of binding to DNA (the bithiazole moiety) while a second region (involving the amino-pyrimidine-imidazole region and possibly the carbamoyl group) is capable of binding metal ions. Although the ligating properties of a number of transition metal ions, especially Cu(II) and Zn(II), towards bleomycin have been studied (9), the physical and chemical properties of the biologically important iron bleomycins are only now being investigated. This report discloses our latest findings on the iron bleomycins.

Experimental

Bleomycin A_2•HCℓ 1, the most abundant component of Blenoxane, was used as obtained from Bristol Laboratories. The synthesis and characterization of the new bleomycin analogues 2-4 will be reported elsewhere (10). The structures of 1-4 are shown in Figure 1.

0-8412-0588-4/80/47-140-237$05.00/0

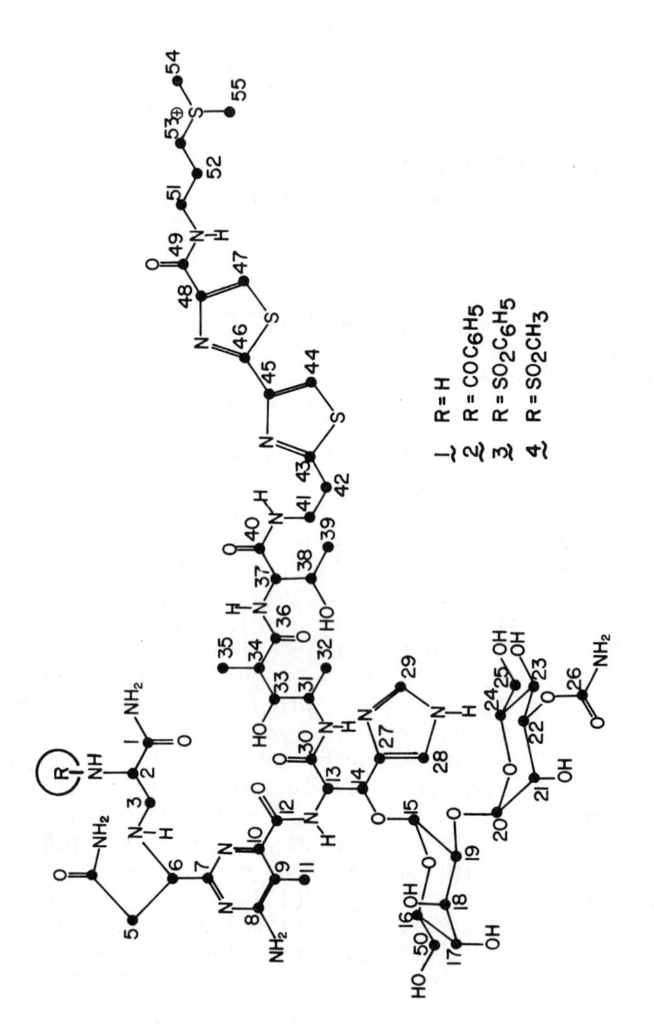

Figure 1. The structures of BLM-A_2, 1, and the new bleomycin analogues, 2–4

The preparation and handling of Fe(II)BLM was done in an inert atmosphere box. Unless otherwise noted, complexes were prepared by the addition of stoichiometric amounts of either $Fe(II)(C\ell O_4)_2 \cdot 6H_2O$ or $Fe(III)(C\ell O_4)_3 \cdot 6H_2O$ to the drug, followed by the adjustment of the pH to a specific value. The pH_m (the pH meter reading without correction for the deuterium isotope effect) of the Fe(II)-drug complex in D_2O was adjusted to 6.5 prior to obtaining the ^{13}C nmr spectrum. For nmr purposes, the Fe(III)BLM complex was prepared by oxidizing the divalent compound with molecular oxygen. The pH_m of the oxygenated solution was adjusted to 6.5 prior to obtaining the spectrum.

The variation in the amounts of the high and low spin forms of Fe(III)BLM as a function of pH was determined using esr. Since the linewidth of the low spin esr signal was independent of pH, the peak height of the signal was used as a measure of the amount of low spin Fe(III)BLM present. Although the linewidth of the high spin species varied as a function of the pH, especially in the presence of buffers, the peak height of the signal was taken as rough measure of the amount of high spin Fe(III)BLM present in the mixture. The esr as well as the nmr and absorption data were collected in the earlier described manner (11,12,13).

The Fe(II) complexes of the new bleomycin derivatives, 2-4, were prepared in the presence of the atmosphere by the addition of $Fe(II)(C\ell O_4)_2$ to a 10^{-3}M solution of the bleomycin. The pH was adjusted by addition of small amounts of 1M HCℓ or 1M NaOH to the solution containing the complex. No oxidation to Fe(III) occurred.

The dc polarographic studies were carried out in buffered and in unbuffered aqueous solutions. Potentials were referenced against a Ag/AgCℓ saturated NaCℓ electrode. The electrochemical assignments and the apparatus used to collect the polarographic data have been previously described (14,15).

The Iron Binding Site of Bleomycin

The location of the iron binding site on bleomycin is critical to understanding the role that iron plays in the DNA-degrading process. The ^{13}C nmr spectra shown in Figure 2 and the structure shown in Figure 3 clearly show that the pyrimidine-imidazole-sugar regions of both Fe(II)BLM and Fe(III)BLM are involved in metal binding. ^{13}C nmr resonances attributed to carbon atoms in that area of the drug are either missing from or are broadened in the spectra of the iron complexes. Only the narrow unshifted resonances were assigned. The following carbon atoms of Fe(II)BLM were observed as narrow resonance lines (Figure 2b): C, 41-49 and C, 51-55. Broadened and unassigned resonances for Fe(II)BLM occurred at 23.2, 57.5, 93.0, 168.5 and 192.5 ppm. A number of broad, overlapping resonances also occurred in the sugar region of the nmr spectrum (60-75 ppm).

Fe(III)BLM exhibited narrow resonance lines for carbon atoms

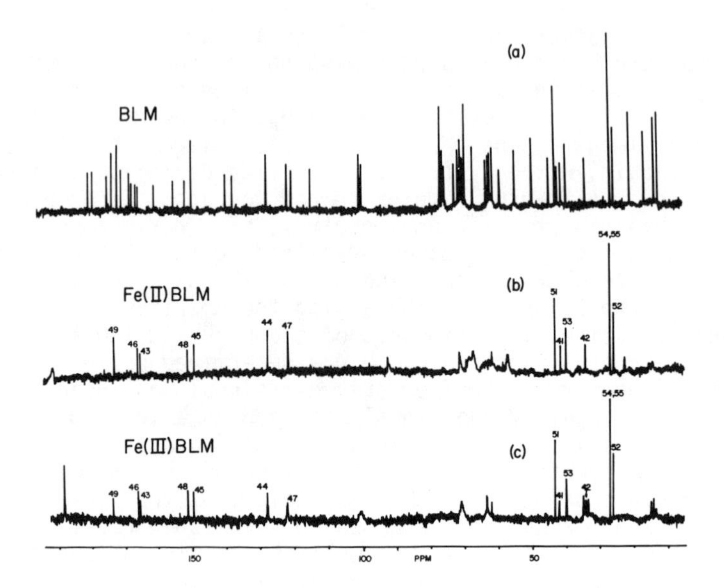

Figure 2. The ^{13}C NMR spectra of: a, BLM-A_2, b, Fe(II)BLM-A_2, and c, Fe(III)BLM-A_2

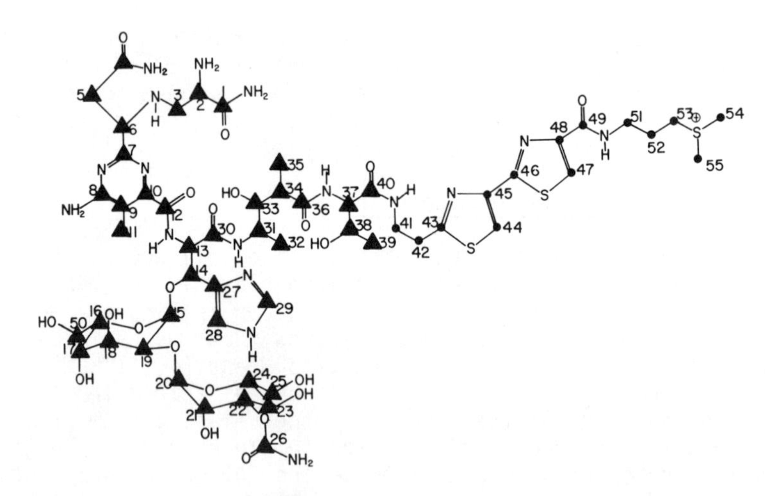

Figure 3. The structure of BLM-A_2 showing which ^{13}C NMR resonance lines are missing from the spectra (▲) of Fe(II)BLM-A_2 and Fe(III)BLM-A_2

C, 41-49 and C, 51-55 (Figure 2c). Broad unassignable resonances were observed at 64 and 71 ppm but the sugar region was simpler than was observed for Fe(II)BLM. In addition, broad and apparently shifted resonances were found at 14, 34, and 101 ppm. The ^{13}C nmr spectrum of Fe(III)BLM also contained a number of weak but narrow resonances at 14.2, 15.2, 33.9, 34.2 and 188.5 ppm. These resonances have not been previously observed in the spectrum of the drug or any of its metalloderivatives and they may be due to bleomycin fragments which occur upon oxidation of Fe(II)BLM (16).

Thus, the ^{13}C nmr results of Fe(II)BLM and Fe(III)BLM are similar to those earlier obtained for Cu(II) and Zn(II)BLM. Although the bithiazole is an important group for DNA binding (17-19), the ^{13}C nmr data for the iron complexes graphically show that the nitrogen-sulfur heterocycle of BLM is not involved in iron binding. Its resonance lines are narrow and unshifted and are not affected by the presence of the paramagnetic cations.

Specific bleomycin donor atoms of Fe(II) and Fe(III) have been identified. Potentiometric titration studies by us (6) and by Sugiura *et al*. (20) have shown that the primary amine function and the imidazole moiety are bound to Fe(II) in Fe(II)BLM. Sugiura *et al*. (20) have also pointed out that the secondary amine function is bound to Fe(II). The assignment of the fourth and final proton loss accompanying the binding Fe(II) to the antibiotic has been attributed to the removal of an amide proton.

UV visible absorption and electrochemical studies have identified an additional ligating group of both Fe(II)BLM and Fe(III)BLM. Polarographic measurements show that the reduction at -1.22V, assigned to the two electron irreversible reduction of the 4-amino pyrimidine moiety of the antibiotic (15) is sensitive to iron binding. Figure 4 shows that this wave is absent from the dc polarogram of the Fe(II,III)-drug complexes. The binding of the 4-amino pyrimidine moiety to Fe(II) and Fe(III) dramatically shifts the reduction potential of the heterocycle out of the polarographic window of observation. Similar effects have observed for other metallobleomycins and metallotallysomycins (15).

Further evidence that the pyrimidine is bound in the iron derivatives can be seen in the difference spectra shown in Figure 5. Both of the complexes exhibited a pyrimidine π-π^* electronic transition which was shifted from its original position in bleomycin. This band occurs at ~230 nm in the free drugs and shifts to ~250 nm in their metalloderivatives. In addition, the iron complexes exhibit a second strong envelope at ~300 nm. Since transitions due to two chromophores present in the antibiotics, the $n \rightarrow \pi^*$ of the pyrimidine and the π-π^* of the bithiazole moiety, overlap at 290 nm, the origin of the band at ~300 nm in the difference spectra of the metalloderivatives is difficult to determine.

Fe(II)BLM exhibits two weak absorptions at 475 nm (ε_M 360) and 375 nm (ε_M ~600). The positions and intensities of these two

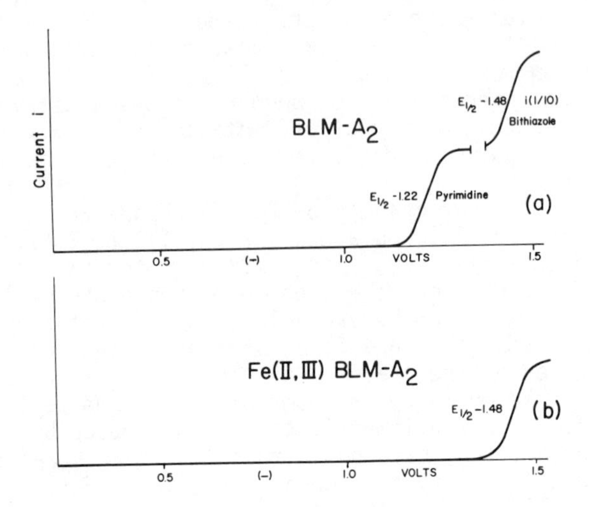

Figure 4. Filtered DC polarograms of: a, BLM-A_2 and b, Fe(II, III)BLM-A_2 in unbuffered aqueous solution, pH, 7.0. The wave of catalytic origin at −1.48 V is associated with the bithiazole function of BLM.

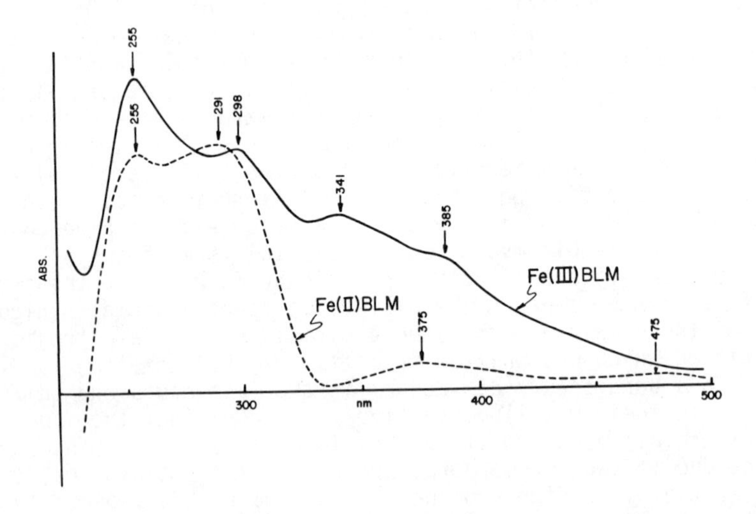

Figure 5. Difference spectra: (– – –) Fe(II)BLM vs. BLM; (——) Fe(III)BLM vs. BLM

absorptions are in good agreement with those earlier reported by Sausville et al. (3). Although the bands appear to be of d-d origin the assignment of the transitions must await the determination of the ground state. Fe(III)BLM has two strong (εM 1,000-2,000) but poorly defined absorptions at 341 and 365 nm. The origin of these bands is unknown. The difference spectra of Fe(III)-BLM produced by air oxidation of Fe(II)BLM and the complex formed by combining $Fe(III)(ClO_4)_3$ and the antibiotic at pH, 7.0 were identical.

The spectroscopic evidence taken together with the titration results for Fe(II)BLM definitively establish the amino-pyrimidine-imidazole-sugar region of bleomycin as important for binding Fe(II) and Fe(III). The detailed geometry and stereochemistry of the iron binding site remains to be elucidated.

The Spin State of the Complexes

The spin state of the divalent complex, Fe(II)BLM, has not yet been determined. However, the fact that the ^{13}C nmr spectrum of the Fe(II) antibiotic complex has many carbon resonance lines missing and others that are broadened (Figure 2b), shows that Fe(II)BLM is paramagnetic. Diamagnetic complexes such as Zn(II)-BLM yield shifted but narrow ^{13}C nmr lines (11). Thus, the only spin states possible for Fe(II)BLM are S = 1 and S = 2. Magnetic, and Mössbauer measurements are under way to resolve this issue.

The spin state of the Fe(III) complex was easily determined by esr. Preparing Fe(III)BLM by combining molar equivalents of $Fe(III)(ClO_4)_3$ and the antibiotic in water in the absence of a buffer yielded at least two esr-active forms of Fe(III). One form, with a g value of 4.3 has been assigned to a high spin (S = 5/2) Fe(III) ion in a completely rhombic crystal field (6,7,20). Since the peak at g, 4.3 occasionally shows some splitting, particularly in the presence of buffers, the signal may be due to the presence of more than one form of high spin Fe(III). The fact that $Fe(III)(ClO_4)_3$ dissolved in water at pH, 7.0 gives only a very weak signal at g, 4.3 while the signal observed in Fe(III)-drug solutions is strong suggests that the high spin signal is due to Fe(III) which is in some manner bound to the drug. The esr signals having g values of 2.412, 2.182 and 1.898 emanate from a low spin (S = 1/2) Fe(III) ion in a rhombic crystal field (21). No other esr active species were observed between pH 3.0 and 10.5. As has been previously reported (6,7), producing Fe(III)BLM by air oxidation of Fe(II)BLM yields a transient low spin Fe(III) species having g values of 2.254, 2.171 and 1.939. The esr signatures of that species and those of the previously mentioned compounds are shown in Figure 6.

Stabilities of the Iron Complexes

Esr studies have shown that the relative amounts of low and high spin Fe(III) are dependent on pH, and on the presence of buffer ions. The pH dependency of the spin states of Fe(III)BLM is shown in Figure 7. Within the pH range 7-9 the low spin form of Fe(III) dominates and the amounts of the two spin forms of Fe(III) are relatively insensitive to pH. At pH values outside this range, increasing amounts of high spin Fe(III) form, at the expense of the low spin species, until at pH ~3 and ~11 high spin Fe(II) predominates. These processes are reversible but at low pH the rate of conversion of the low spin species to the high spin species upon lowering the pH is much greater than the reverse process. These results suggest that Fe(III)BLM is a low spin complex and that it exists in equilibrium with high spin forms which predominate at high and low pH. Electrochemistry shows that at least one metal ligating group, the pyrimidine moiety, is metal bound in the pH range 4-9.

The low spin-high spin equilibrium is also affected by the presence of buffers. At pH, 7.0 and 8.0 in either a phosphate or a borate buffer the high spin esr signal of the metal complexes dominates. A pH study using phosphate buffers has shown that the loss of the low spin signal and the increase in the high spin signal for Fe(III)BLM can be correlated with the presence of the buffer species $H_2PO_4^-$. It is possible that this buffer species is binding to the metal ion thereby reducing the crystal field and yielding a high spin Fe(III) complex. At pH values >10, where $H_2PO_4^-$ is in low concentration and HPO_4^{-2} predominates, the esr spectra of Fe(III)BLM are nearly identical to those obtained in water at pH, 10.0. Similar counter ion effects were observed for borate buffers.

The visible absorption spectra of both Fe(II)BLM and Fe(III)BLM were found to be insensitive to pH within the pH range pH 4-9. However, outside of this range the drug appeared to demetallate as is evidenced by the loss of the absorption spectrum of the metalloderivative. The absorption spectra of both Fe(II) and Fe(III) complexes were affected by the presence of buffer ions. In general, the spectra taken in tris buffer solutions, as opposed to those obtained in phosphate and borate media, most closely approximate spectra obtained in unbuffered solutions.

Since DNA degrading by BLM has been shown to be sensitive to the nature and concentration of the buffer ions which are present in solution (5), we examined the effects of various buffers of the esr spectra and the electrochemical properties of the Fe(III) complexes. Electrochemistry allows a clear view of how one of the metal ligating sites of BLM, the 4-amino pyrimidine moiety, behaves in the presence of buffer ions. If Fe(III)BLM is synthesized from the free drug and $Fe(III)(ClO_4)_3$ in the absence of buffers, electrochemistry shows that the 4-amino pyrimidine remains bound to the metal ion within the pH range 4-9 where esr shows that low

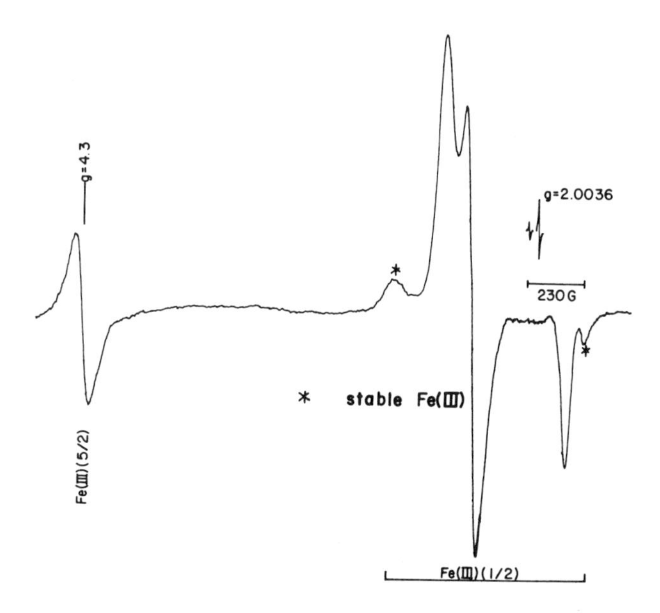

Figure 6. The ESR spectrum of Fe(III)BLM taken immediately after oxidation of Fe(II)BLM with molecular oxygen

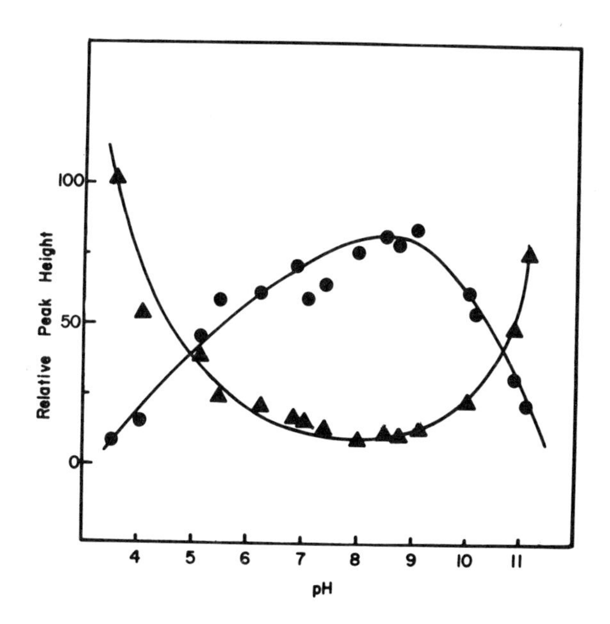

Figure 7. The relative amounts of (▲) high and (●) low spin Fe(III)BLM as a function of pH as determined by ESR

spin Fe(III) predominates. In a pH 7 or 8 phosphate buffer (0.1M) electrochemistry shows that the pyrimidine moiety is not bound to the metal ion. Esr confirms that these solutions contain only high spin (S = 5/2) Fe(III). Synthesizing Fe(III)BLM by anaerobically combining $Fe(II)(ClO_4)_2$ and the drug in a pH 7 or 8 phosphate buffer, followed by air oxidation of the resulting Fe(II)-drug complex yields a low spin Fe(III)BLM complex which contains a bound pyrimidine function. If the oxidized solutions are allowed to stand at room temperature for a few hours, esr and visible absorption spectra show that a redistribution of the spin states occurs; high spin Fe(III) again forms at the expense of low spin Fe(III). Electrochemistry, however, shows that the pyrimidine is still bound to the metal ion. Thus, our results show that Fe(III)BLM is high spin when the pyrimidine moiety is not bound to the metal, but that some high spin forms also exist where the pyrimidine is bound to the cation. These observations, when viewed in the light that almost all of the DNA degrading experiments with the antibiotic have been carried out in buffered media, underscore the importance of carefully defining the metal ligating properties of the drug. It is interesting to note that the highest DNA degrading activity achieved by iron bleomycin is in a phosphate buffer— a medium which apparently significantly alters the primary coordination sphere of the metal ion (5).

The Bleomycin Derivatives

In an effort to more clearly understand the mechanism by which bleomycin degrades DNA, three new bleomycin analogues were synthesized, 2-4. All three analogues were formed by the condensation of the primary amine function of $BLM-A_2$ with either benzoyl chloride (2) or a sulfonyl chloride (3,4) using Schotten-Baumann conditions (10). The compounds were purified via chromatographic means were characterized using ^1H and ^{13}C nmr and potentiometric titration methods. Particularly noteworthy in the titration curves of the derivatives is the absence of inflections at pH values of 2.9 and 7.4. For $BLM-A_2$, inflections at these pH values correspond to the loss of a proton from the protonated, but hydrogen bonded, secondary amine function (Eq. 1) and a loss of a proton from the protonated primary amine function respectively (Eq. 2).

$$-NH_2^+ \longrightarrow -NH + H^+ \quad pK_a, 2.9 \tag{1}$$

$$-NH_3^+ \longrightarrow -NH_2 + H^+ \quad pK_a, 7.4 \tag{2}$$

In contrast to $BLM-A_2$, neither of the new derivatives facilitated the oxidation of Fe(II) to Fe(III) in the atmosphere (10). Esr studies of aqueous solutions containing equimolar amounts $Fe(II)(ClO_4)_2$ and the analogues, 2-4 at pH, 7.0 showed the lack of formation of Fe(III). Similar observations have been reported for dep BLM (22). However, more importantly, incubation of closed circular PM2 DNA with any of the analogues, 2-4 in the presence of Fe(II) salts and reducing agents did not lead to DNA breakage. Identical reaction conditions for BLM-A_2, on the other hand, lead to DNA breakage. These recent observations substantially strengthen the proposed metal-mediated mechanism of action of the antibiotic and underscore in importance of the primary amine function of bleomycin in the DNA degrading process.

Conclusions

Spectroscopic evidence shows that for the Fe(II,III) complexes of BLM the metal ions are bound to the pyrimidine-imidazole-sugar region of the drug leaving the bithiazole moiety free to bind to DNA. The paramagnetic Fe(II) complex can be oxidized to to a low spin (S = 1/2) Fe(III) complex which is rapidly and irreversibly transformed to a second low-spin Fe(III) complex. The low spin species is in a pH- and buffer-dependent equilibrium with high spin (S = 5/2) Fe(III). Electrochemical studies show that the binding of the 4-amino pyrimidine moiety of bleomycin to Fe(III) is sensitive to both pH and the presence of certain buffer species. UV visible absorption spectra show similar equilibria for the Fe(II) complexes. Because *in vitro* DNA-cleavage occurs in a wide range of buffer solutions it appears that the displacement of some of the metal ligating atoms does not inhibit the action of the iron-drug complex. Finally, studies with three new bleomycin analogues have sharply defined the importance of the primary amine function of bleomycin in the drugs ability to degrade DNA by a metal mediated mechanism.

Abstract

Using electrochemistry, uv-visible absorption, esr, and ^{13}C nmr spectroscopy, and with the help of three new bleomycin analogues, the Fe(II) and Fe(III) binding sites of the anti-tumor antibiotic bleomycin have been located. The drug appears to utilize the amine-pyrimidine-imidazole region for iron binding. The physical properties of the metal complexes and how those properties relate to the proposed mechanism of action of the anticancer agent is also discussed.

Acknowledgement

We acknowledge the National Institute of Health for support of this work (Grant, CA-25112-01). F.T.G. is also grateful for support through an N.I.H. Institutional Award.

Literature Cited

1. Umezawa, H. Biomedicine, 1973, 18, 459.
2. Blum, R. H.; Carter, S. K.; Agree, K. A. Cancer, 1973, 31, 903.
3. Sausville, E. A.; Peisach, J.; Horwitz, S. B. Biochem. Biophys. Res. Commun., 1976, 73, 814.
4. Sausville, E. A.; Peisach, J.; Horwitz, S. B. Biochemistry, 1978, 17, 2740.
5. Sausville, E. A.; Stein, R. W.; Peisach, J.; Horwitz, S. B. Biochemistry, 1978, 17, 2746.
6. Dabrowiak, J. C.; Greenaway, F. T.; Santillo, F. S. in "Bleomycin: Chemical, Biochemical and Biological Aspects" (S. Hecht ed.), Springer-Verlag, New York, New York, 1979, p. 137.
7. Sugiura, Y.; Kikuchi, T. J. Antibiot., 1978, 31, 1310.
8. Oberley, L. W.; Buettner, G. R. FEBS Lett., 1979, 97, 47.
9. Dabrowiak, J. C. in "Metal Ions in Biological Systems" (H. Sigel ed.), Vol. 11, Marcell Dekker, Inc., New York, 1980, p. 306.
10. Crooke, S. T.; Dabrowiak, J. C. unpublished results.
11. Dabrowiak, J. C.; Greenaway, F. T.; Grulich, R. Biochemistry, 1978, 17, 4090.
12. Dabrowiak, J. C.; Greenaway, F. T.; Longo, W. E.; Van Husen, M.; Crooke, S. T. Biochim. Biophys. Acta, 1978, 517, 517.
13. Greenaway, F. T.; Dabrowiak, J. C.; Van Husen, M.; Grulich, R.; Crooke, S. T. Biochem. Biophys. Res. Commun., 1978, 85 1407.
14. Macero, D. J.; Burgess, Jr., L. W.; Banks, T. M.; McElroy, F. C. Proceedings, Symposium on Microcomputer Based Instrumentation, National Bureau of Standards, Guithersburg, Maryland, p. 45, June, 1978.
15. Dabrowiak, J. C.; Santillo, F. S. J. Electrochem. Soc., 1979, 126, 2091.
16. Takita, T.; Muraoka, Y.; Nakatani, T.; Fujii, A.; Iltaka, Y.; Umezawa, H. J. Antibiot., 1978, 31, 1073.
17. Chien, M.; Grollman, A. P.; Horwitz, S. B. Biochemistry, 1977, 16, 3641.
18. Povirk, L. F.; Hogan, M.; Dattagupta, N. Biochemistry, 1979, 18, 96.
19. Hiroshi, K.; Naganawa, H.; Takita, T.; Umezawa, H. J. Antibiotics, 1978, 31, 1316.
20. Sugiura, Y.; Ishizu, K.; Miyoshi, K. J. Antibiot., 1979, 32, 453.
21. Peisach, J.; Blumberg, W. E. Adv. Chem. Ser. 1971, 100, 271.
22. Sugiura, Y. Biochem. Biophys. Res. Commun., 1979, 88, 913.

RECEIVED May 21, 1980.

CHELATION THERAPY FOR REMOVAL OF IRON OVERLOAD IN COOLEY'S ANEMIA

15

Iron Chelation in the Treatment of Cooley's Anemia

W. FRENCH ANDERSON

Laboratory of Molecular Hematology, National Heart, Lung, and Blood Institute, National Institutes of Health, Bethesda, MD 20205

I will discuss iron chelation in the treatment of the human genetic disease beta-thalassemia, otherwise known as Cooley's anemia. Cooley's anemia, named after Dr. Thomas Cooley who first described the disease in detail in 1925 (1), is a lethal hereditary anemia in which the infants cannot make their own blood (2). They must be transfused every three to four weeks starting at six months of age for the rest of their lives. Failure to transfuse results in death from profound anemia. According to the World Health Organization, the group of diseases called the thalassemias, of which Cooley's anemia is the most severe, is the largest health problem in the world for single locus genetic diseases.

The major problems these patients have, however, are caused not directly from their disease, but from the treatment. Every unit of transfused blood contains several hundred milligrams of iron. The body is not equipped to excrete such large amounts of iron and so it is deposited in the heart, liver, endocrine glands and other organs producing severe medical problems ultimately leading to death in early teenage. Iron chelation therapy was attempted by a few physicians back in the late 1960's, but in general such therapy was considered useless or even harmful (3). A major effort has been made over the past 5 years, however, to find means whereby the presently available iron chelator, Desferal , could be effectively administered to remove large quantities of iron from iron-overloaded patients. This effort has proven successful, and Desferal is now used world-wide for iron chelation therapy (4). Life expectancy and quality of life of treated patients now appear greatly improved.

From the beginning it was realized, however, that Desferal[R] is not an ideal drug (3). It can only be given by injection and these can sometimes be very painful, it becomes effective only when the body's iron load is already ten-times normal, and it is very expensive to produce. In 1973, under the sponsorship of the National Institute of Arthritis, Metabolism and Digestive Diseases, a new program was instituted designed to develop new iron chelators for clinical use (5). Several of the chemists here today have been a part of that very successful program. Over 100 compounds have been tested and several are either in or being prepared for human clinical trials (6).

I am very pleased to have been asked to summarize this field for you prior to the major papers to follow. I will review the disease Cooley's anemia, outline the medical problems and complications, and then will give you a brief summary of the concepts behind iron chelator development.

Description of Cooley's Anemia

Plates 1A and B* are photographs of a patient with Cooley's anemia. At first glance, he appears perfectly normal, but there are several characteristics of his disease that are apparent. These patients cannot make useful blood because of their genetic block, but their bone marrows try to make blood and in so doing become markedly overactive. Therefore, certain bones become grossly enlarged because of the greatly expanded bone marrow inside them. The upper jaw enlarges, disproportionately, producing prominent front teeth, a large space between the nose and upper lip, and high prominent cheeks. The forehead swells or "bosses" out. Because of the large amount of total body metabolism going into ineffective blood production, body growth is abnormal and greatly delayed. The arms and legs are spindly and the abdomen is large and protuberant because of a greatly enlarged liver. The spleen becomes so large early in life that it fills the whole abdomen and has to be removed.

The x-ray shown in Figure 1A demonstrates the large amount of bone marrow in the skull. A normal skull shows a single thin line of bony tissue. As can be seen here, there is a considerable quantity of fuzz, and that fuzz represents the enormously enlarged bone marrow. This characteristic is called "hair-on-end" because the x-ray looks like hair standing up out of the skull. Figure 1B shows an x-ray of the hand. The insides of the bones are dark on the x-ray, resulting from the hyperactive bone marrow which has hollowed out the normally solid bones. This is true of all the long bones: the fingers, arms, and legs. At times the bones become so thin that they fracture easily.

A blood smear of normal blood showing healthy red corpuscles in Plate 2A*. In contrast, Plate 2B* shows the blood smear of a patient with Cooley's anemia. Note how pale the blood cells are; they are of all different sizes and shapes, with many of them grossly abnormal in shape. A scanning electron microscope picture of these cells is shown in Figure 2. Most of these cells have, not surprisingly, a very short half-life.

A photograph of Dr. Max Perutz's two angstrom-per-centimeter x-ray diffraction model of hemoglobin is shown in Plate 3*. Hemoglobin, the blood pigment that carries oxygen and makes the blood red, is composed of four chains, two alpha globin chains

* Color plates are located in the Appendix.

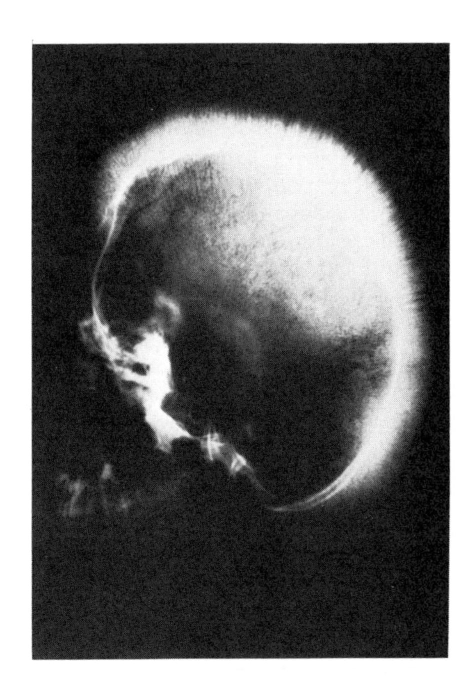

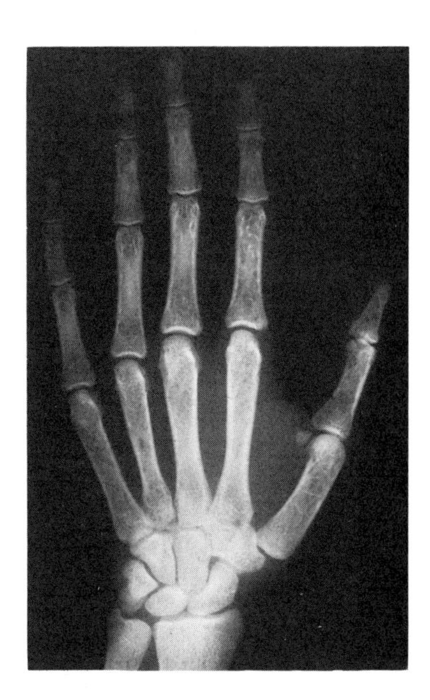

Figure 1. X-rays of the skull and hand of a patient with Cooley's anemia

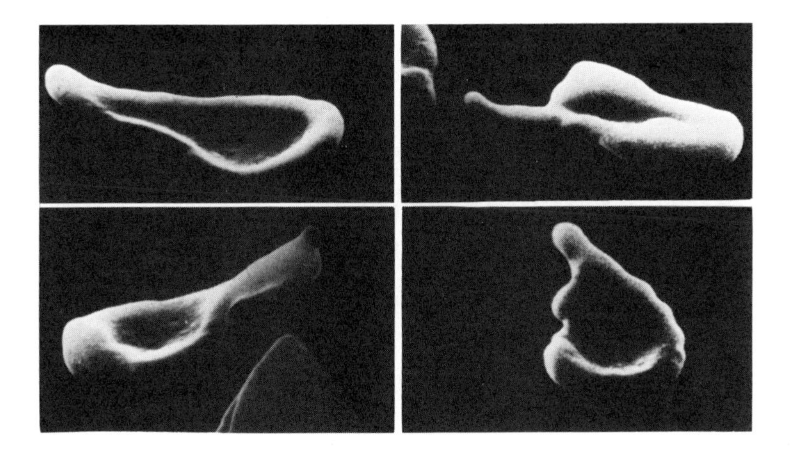

Springer–Verlag

Figure 2. Scanning electron microscope photographs of red blood cells in Cooley's anemia (12)

(in black) and two beta globin chains (in white). The red disc represents heme which contains one molecule of iron. In Cooley's anemia, the bone marrow cells cannot make the beta (or white) chains.

Chromatograms of the globin chains synthesized by the immature red blood cells of patients are shown in Figure 3 (7). The chains are separated on carboxymethyl-cellulose using a urea-phosphate buffer. The top panel shows a chromatograph of normal human alpha and beta globin chains. There is a roughly equal amount of alpha globin (on the right) and beta globin (on the left). In contrast, the middle panel shows the globin chains of a patient with homozygous beta thalassemia (i.e., Cooley's anemia). There is a normal amount of alpha globin but a grossly decreased amount of beta globin and a slight compensatory increase in gamma (or fetal) globin. The small number of beta and gamma globin chains complex with alpha globin chains to produce a limited number of hemoglobin molecules. This produces hypochromic red blood cells (RBC). What is even worse is that the extra uncomplexed alpha globin chains, which are insoluble, precipitate within the cell, cause membrane damage, and as a result the cells break down. The abnormality here is the inability of the small number of non-alpha globin chains to complex with all the alpha globin chains. The iron in these broken down RBCs is released and is deposited in body organs. The bottom panel shows the globin chains from a newborn infant. The baby has fetal blood normally and is only beginning to switch to adult hemoglobin. Here, however, the number of beta plus gamma globin chains exactly equals the number of alpha globin chains so that no globin chain imbalance exists.

This inability to make the beta globin chains in Cooley's anemia, which consequently means an inability to make hemoglobin, together with the precipitation of uncomplexed alpha globin chains causing all the red cells made to be lysed, accounts for the signs and symptoms seen in the disease Cooley's anemia.

Medical Complications from Iron Overload

The only treatment for this disease is blood transfusion, but with every unit of blood, as we said before, several hundred mg of iron are deposited in the body and this deposition of iron results in severe symptoms and ultimately in death. The body is normally equipped to handle about one mg of iron a day in the diet. What happens to the organs of patients with this massive deposition of excess iron?

Plate 4* is a photograph of the heart from a patient who died from transfusional hemosiderosis - i.e., iron overload from transfusions; it shows large quantities of iron. Note the dark brown or rust coloration; that is iron. One cannot slice an organ with this much iron - one has to saw it. It is hard to imagine how a heart like this could have kept functioning for as

* Color plates are located in the Appendix.

long as it did. This patient was 23 years old when he died.

A section of cardiac muscle from the heart shown in Plate 4*is shown in Plate 5*. The iron deposition in each cell is shown by the blue stain. As can be seen, every cell has large quantities of iron.

Plate 6* shows a photograph of the liver from the same patient. The iron deposits can be seen as brown patches all over the surface. This liver is very heavy and, obviously, functioned very poorly.

An iron stained section of the pituitary gland is shown in Plate 7*. It is no wonder that these patients grow slowly and have delayed and abnormal sexual development.

A bar graph which summarizes the amount of iron in each of the body tissues in this patient is shown in Figure 4. Most tissues had large amounts of hemosiderin and ferritin and many organs are grossly loaded. Most normal tissues have no detectable hemosiderin or ferritin.

Now, what can be done? Obviously, the ideal solution would be a treatment which would allow the bone marrow to make normal amounts of alpha and beta globin chains, thereby correcting the abnormality. Until this becomes possible, however, the next alternative is to improve techniques for removing excess iron from the body tissues or even to chelate and excrete excess iron before it is ever deposited in the body.

Development of New Iron Chelators

Let me now spend a few minutes summarizing the general field of iron chelator development. The best organic chemists for producing effective iron chelators are the microbes (8). They have been doing it for millions of years, and have developed extremely effective molecules for chelating iron. Dr. Neilands, the expert on bacterial iron chelators, will be the next speaker.

Microbes utilize primarily two types of chemical groups for chelation: hydroxamic acids [-N(OH)-CO] (Figure 5) and catechols which are two adjacent OH groups on a phenyl ring (Figure 6).

The simplest iron chelator in existence appears to be hadacidin (Figure 5). This little molecule has one hydroxamic acid group and, therefore, has two bonds available to chelate iron. It is not particularly effective since it can only form two bonds, while iron utilizes six. At the other extreme is ferrachrome, an absolutely beautiful molecule, which is like a Venus Fly Trap with three hydroxamic acid groups - one in the back of each of three arms. Once iron goes in, it is held in place very tightly by six bonds. Ferrachrome is one of the most effective iron chelators in nature.

But neither ferrachrome, which is too complex, nor hadacidin, which is too simple, are effective in chelating iron from the tissues of animals. The type of molecule that one wants is one which is fairly simple and which has three hydroxamic acids

* Color plates are located in the Appendix.

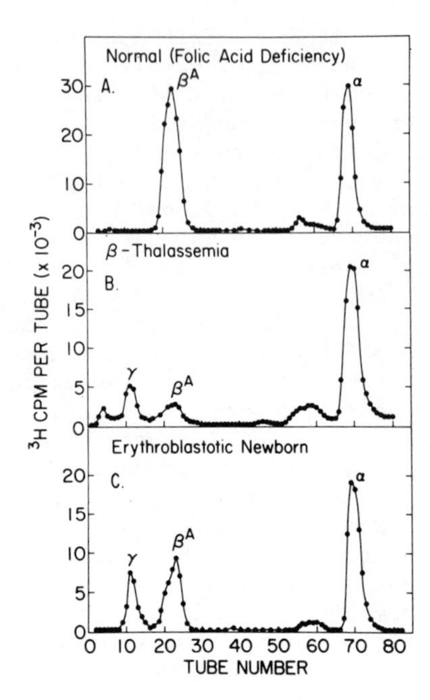

Figure 3. Globin chain synthesis in the immature blood cells of an adult with no globin abnormality (top), a patient with Cooley's anemia (middle), and a newborn with no globin abnormality (bottom); for experimental details see Ref. 7

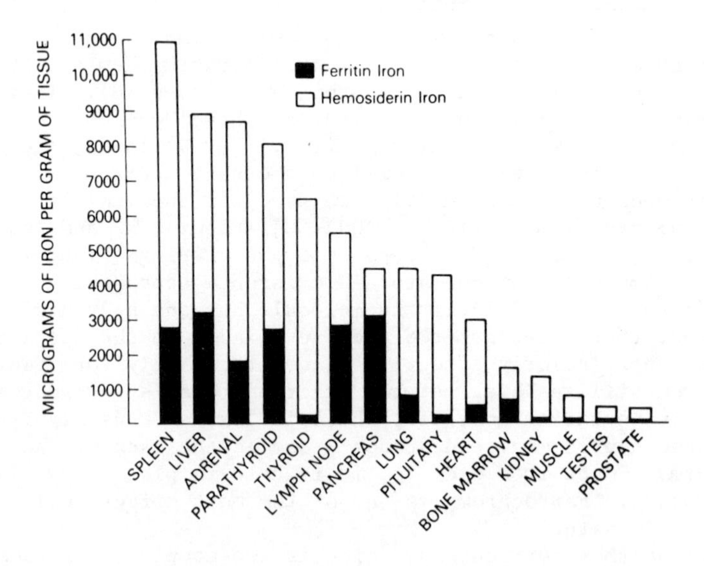

Figure 4. Iron content of body organs from the same patient

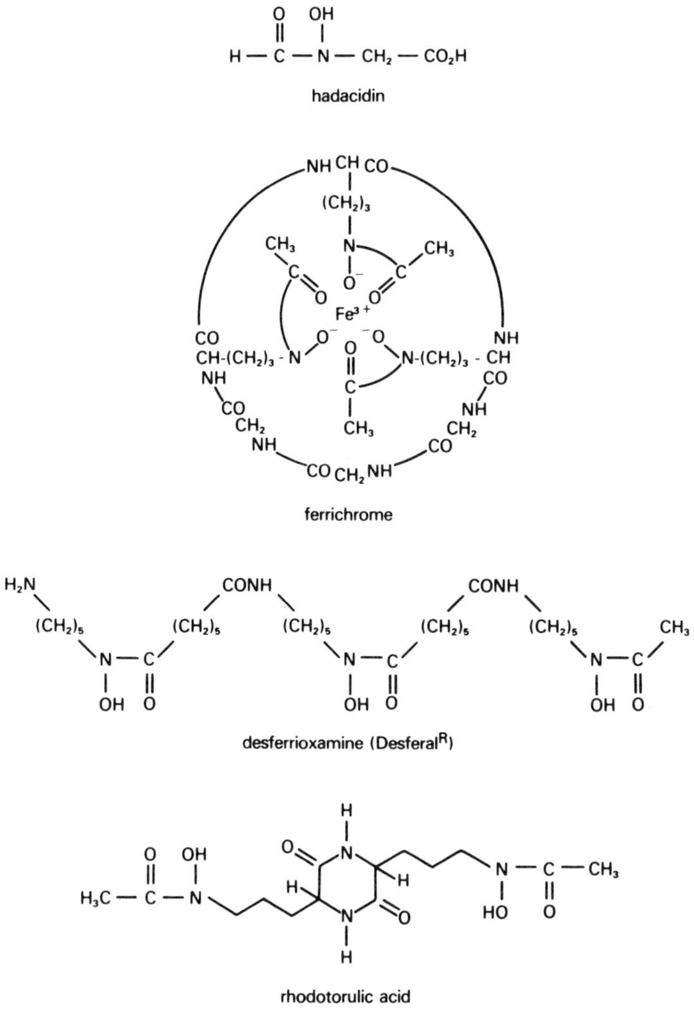

hadacidin

ferrichrome

desferrioxamine (Desferal[R])

rhodotorulic acid

Figure 5. Microbial iron chelators containing hydroxamic acids

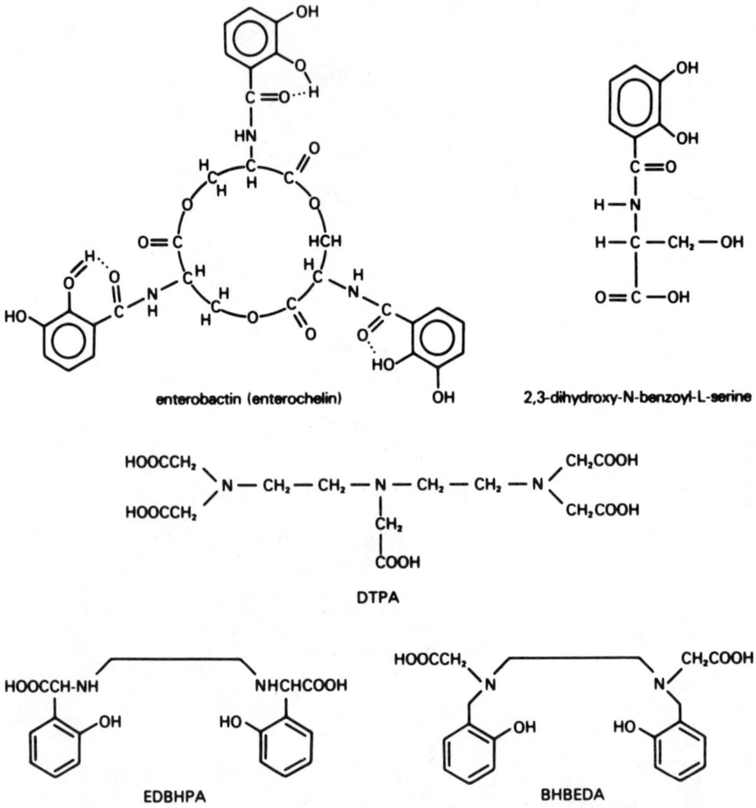

enterobactin (enterochelin)

2,3-dihydroxy-N-benzoyl-L-serine

DTPA

EDBHPA

BHBEDA

Figure 6. Microbial and synthetic iron chelators

in positions so that six stable bonds can be formed to chelate iron. Such a molecule is desferrioxamine or Desferal^R: three hydroxamic acids are separated by several carbons to provide adequate spacing for the iron binding groups along with several additional oxygens and nitrogens to make the molecule more water soluble. This iron chelator is effective in man.

A chelator does not have to have three hydroxamic acids to be effective, however: rhodotorulic acid only has two. Drs. Anthony Cerami, Robert Grady and their colleagues at Rockefeller University have shown that this molecule can remove iron reasonably effectively in iron-overloaded animals (9). There are many other types of molecules which possess hydroxamic acid residues, but these examples serve to illustrate the basic principles.

The second group of compounds, which are the catechols, are represented by enterobactin, which has three catechol groups (and, therefore, six bonds) (Figure 6). This iron chelator, which will be discussed in detail by both Dr. Neilands and Dr. Raymond and appears to have the highest iron binding constant yet measured, is isolated from *E. coli* and is composed of three identical subunits of 2,3-dihydroxybenzoic acid attached to a serine. If the serine is removed, the simple molecule, 2,3-dihydroxybenzoic acid, is obtained. The Rockefeller group have shown that 2,3-dihydroxybenzoic acid is a moderately effective oral iron chelator (10).

Can we gain sufficient knowledge from the microbes in order to synthesize chemically even more effective iron chelators? The answer is yes. EDTA (ethylenediaminetetraacetic acid) is a molecule probably everyone here has used for chelating heavy metal ions. If one adds to it another nitrogen group and another acetic acid, we obtain DTPA (diethylenetriaminepentaacetic acid). DTPA and Desferal are the two molecules that have been used clinically in man. Desferal has an iron binding constant of $10^{30.6}$. DTPA has a constant of 10^{27}. Both are very strong. Transferrin the iron carrying protein in the blood, has, of course, a very high iron binding constant: 10^{36}.

By substituting two phenyl groups for acetic acids in EDTA, Dr. Arthur Martell of Texas A & M University has synthesized one of the strongest synthetic iron chelators ever made: BHBEDA (N,N'-Bis((2-hydroxybenzyl))ethylenediamine-N,N'-diacetic acid) (11). Its iron binding constant is 10^{40}! And yet it is not a particularly effective iron chelator in animal studies.

There are many reasons why a molecule can be very effective *in vitro* (i.e., have a high iron binding constant) while being ineffective *in vivo*. First, the effective iron binding capacity under physiologic conditions is very different from the iron binding capacity under optimal *in vitro* conditions due to a number of factors including concentrations of the metal, the chelator and other ions, particularly $[H^+]$, $[OH^-]$, and $[Ca^{++}]$. Second, the iron binding constant is an equilibrium value, not a kinetic one. Third, bioavailability and biostability are major factors. Many compounds are rapidly broken down in the blood stream or are very quickly cleared by the liver or kidneys. Oral administration adds another series of problems - the acid pH of the stomach, absorption barriers in the intestine, etc. Thus, a compound's iron binding constant is only a rough guide for indicating potential effectiveness.

An isomer of BHBEDA, EDBHPA (ethylenediamine-N,N'-bis((2-hydroxyphenylacetic acid)), although 10^6 lower in iron binding capacity *in vitro* (10^{34}), appears to be a very effective iron chelator in animals (6). This molecule, and modifications of it, will be mentioned by Dr. Pitt shortly. It might be the hoped-for "magic bullet" which could effectively and simply remove the iron overload from patients receiving multiple blood transfusions.

Thus, using a knowledge of bacterial iron chelators and applying modern organic chemistry techniques, it does appear possible to synthesize molecules that are more effective in removing iron from animal tissues than the natural chelators like Desferal .

Literature Cited

1. Cooley, T.B.; and Lee, P. Series of cases of splenomegaly in children with anemia and peculiar bone changes. *Tr. Am. Pediat. Soc*. 1925, 37, 29.

2. Weatherall, D.J.; and Clegg, J.B. "The Thalassaemia Syndromes"; Blackwell Scientific Publications, Oxford, 1972, 374 pp.

3. Waxman, H.S.; and Brown, E.B. Clinical usefulness of iron chelating agents. *Progress in Hematology*, 1969, 6, 338-373.

4. Anderson, W.F.; Bank, A.; and Zaino, E.C., Eds. "Fourth Conference on Cooley's Anemia"; *Ann. N.Y. Acad. Sci*., 1980, in press.

5. Anderson, W.F.; and Hiller, M.C., Eds. "Development of Iron Chelators for Clinical Use"; Proceedings of a Symposium. DHEW Publication No. (NIH) 77-994, 1977.

6. Pitt, C.G.; Gupta, G.; Estes, W.E.; Rosenkrantz, H.; Metterville, J.J.; Crumbliss, A.L.; Palmer, R.A.; Nordquest, K.W.; Sprinkle-Hardy, K.A.; Whitcomb, D.R.; Byers, B.R.; Arceneaux, J.E.L.; Gaines, C.G.; and Sciortino, C.V. The selection and evaluation of new chelating agents for the treatment of iron overload. *J. Pharmacol. and Exper. Therap*. 1979, 208, 12-18.

7. Nienhuis, A.W.; Canfield, P.H.; and Anderson, W.F. Hemoglobin messenger RNA from human bone marrow. J. Clin. Invest. 1973, 52, 1735-1745.

8. Neilands, J.B.; Iron and its role in microbial physiology. In: "Microbial Iron Metabolism. A Comprehensive Treatise." Neilands, J.B. Ed., Academic Press, New York, 1974, pp. 3-34.

9. Grady, R.W.; Graziano, J.H.; Akers, H.A.; and Cerami, A. The identification of rhodotorulic acid as a potentially useful iron-chelating drug. Blood 1974, 44, 911.

10. Peterson, C.M.; Graziano, J.H.; Grady, R.W.; Jones, R.L.; Vlassara, H.V.; Canale, V.C.; Miller, D.R.; and Cerami, A. Chelation studies with 2,3-dihydroxybenzoic acid in patients with beta-thalassemia major. Brit. J. Haematol. 1976, 33, 477-485.

11. L'Eplattenier, F; Murase, I.; and Martell, A.E. New Multidentate Ligands. VI. Chelating tendencies of N,N'-(2-hydroxybenzl) ethylenediamine-N-N'-diacetic acid. J. Amer. Chem. Soc. 1967, 89, 837-843.

12. Bessis, Marcel; "Living Blood Cells and their Ultrastructure"; Springer-Verlag, New York, 1972, p. 230.

RECEIVED June 20, 1980.

16

High Affinity Iron Transport in Microorganisms

Iron(III) Coordination Compounds of the Siderophores Agrobactin and Parabactin

J. B. NEILANDS, T. PETERSON, and S. A. LEONG

Department of Biochemistry, University of California, Berkeley, CA 94720

The insolubility of free ferric ion at physiological pH ($K_{sp.} = [Fe^{3+}][OH^-]^3 = 10^{-39}$) has required the evolution of special mechanisms in aerobic and facultative anaerobic microbial species for the assimilation of this critically important metal ion. Two pathways have been defined, namely, low affinity and high affinity. The former, still poorly understood, is comparatively inefficient and non-specific. The high affinity system, aspects of which will be the subject of this paper, is itself comprised of two parts: relatively low molecular weight (500-1500 daltons), virtually ferric ion specific ligands termed siderophores, and the matching, membrane-associated receptor complex which recognizes and transports the siderophore in its iron laden form. Recent reviews regarding the operation and biomedical significance of the high affinity system, in all its variant forms among the bacteria and fungi, are available (1,2,3).

Siderophores as chemical entities can generally be classed as either hydroxamates or catechols. Ferrichrome and enterobactin are prototypical members of the two classes, respectively. The tri-catechol siderophore, enterobactin, is of special interest since it has been demonstrated repeatedly that it can supply iron to bacteria in the presence of certain ferric tri-hydroxamate type siderophores ($K_f \cong 10^{30}$) not utilized by the organisms (4). In 1975 Tait (5) isolated a putative siderophore from cultures of *Micrococcus* (now *Paracoccus*) *denitrificans* derepressed for iron and proposed for its structure the compound shown in Figure 1 (R=H), which he called "Compound III". Recently (6) we obtained agrobactin from the phytopathogen *Agrobacterium tumefaciens* and characterized it as an analogue of Compound III containing three residues of 2,3-dihydroxybenzoic acid and the oxazoline form of the centrally attached residue of L-threonine (Figure 2, R=OH). The protonated oxazoline was shown to open slowly in aqueous media to afford agrobactin A (Figure 1, R=OH). These findings inspired a reinvestigation of "Compound III" and led to the discovery (7,8) of the oxazoline ring in the siderophore from *P. denitrificans*, now renamed parabactin (Figure 2, R=H). By

0-8412-0588-4/80/47-140-263$05.00/0

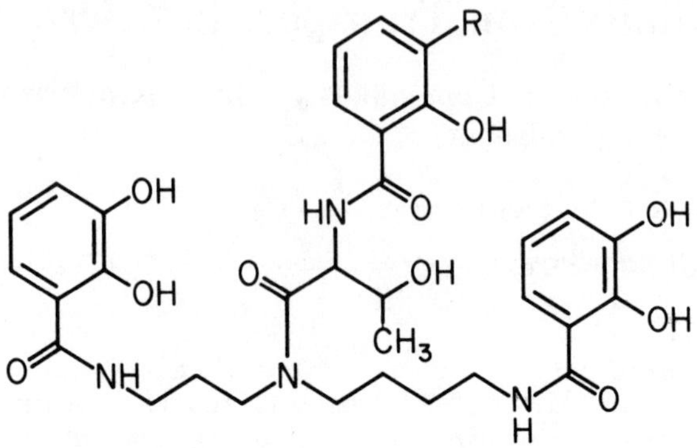

Figure 1. Parabactin A, R = *H; agrobactin A,* R = *OH*

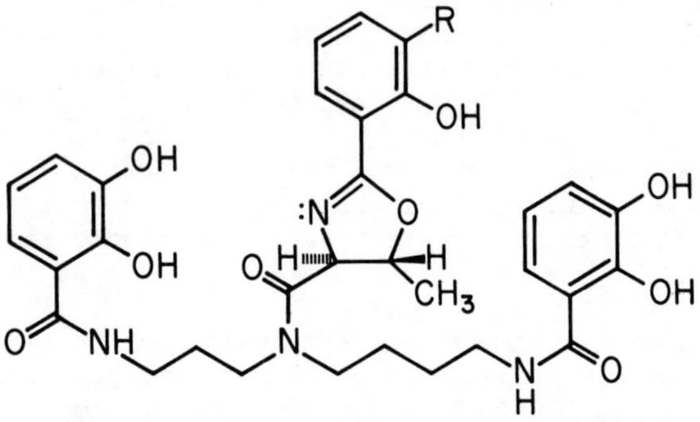

Figure 2. Parabactin, R = *H; agrobactin,* R = *OH*

analogy with the agrobactin-agrobactin A pair, the open form formula previously assigned to "Compound III" has now been designated parabactin A (Figure 1, R=H).

The tertiary N atom of the oxazoline ring in parabactin and agrobactin provides a binding site for the six-coordinate ferric ion. Inspection of molecular models of agrobactin suggests two possible modes of ferric ion complexation to the 2,3-dihydroxyphenyloxazoline ring system, viz., via the catechol groups or, alternatively, through the tertiary N and the orthohydroxyphenyl function. The data reported herein indicate that in the case of both siderophores binding is to the oxazoline nitrogen and that the conformation and configuration of the ferric complexes may be as depicted in Figure 3.

I. Structure and Properties

A. Preparation of Siderophores.

Published methods were used for the isolation of agrobactin (6), agrobactin A (6), parabactin (8), enterobactin (9), rhodotorulic acid (10), ferrichrome (11) and ferrichrome A (12). Parabactin A was obtained by a synthetic procedure, to be reported separately. The catechol type siderophores were checked for purity by thin layer chromatography on silica gel in 4:1 chloroform:methanol and by measurement of the absorption intensity of the band in the near ultraviolet. Although both parabactin and agrobactin can form hydrochlorides (pKa' = 2.3), the siderophores were obtained by extraction into ethyl acetate at neutral pH and were hence in the unprotonated form.

B. Ligand Deprotonation Curves and Proton Stoichiometry.

a. Direct Titration with Ferric Chloride.

In agrobactin the binding of ferric ion in the tri-catechol mode would be expected to release six protons. Alternatively, chelation to two catechols plus the *o*-hydroxyphenyloxazoline group would generate only five protons per mole. Exactly 2 μmoles of siderophore were weighed on a Cahn G-2 electrobalance and dissolved in 1-2 ml of ethanol in a 10 ml beaker. After the addition of 0.1 ml of methanol containing 2 μmoles of ferric chloride the solution was diluted with water to ca. 5 ml and standard 0.100 N NaOH introduced with the recording titration apparatus (13). The complexes, which were initially blue, became purple and then wine in color as the titration progressed into neutral pH. From the data shown in Figure 4, it is apparent that between 4 and 5 protons per mole were released at neutral pH. Also striking was the fact that the deprotonation curves for agrobactin and parabactin remained superimposable at all except very alkaline values of pH.

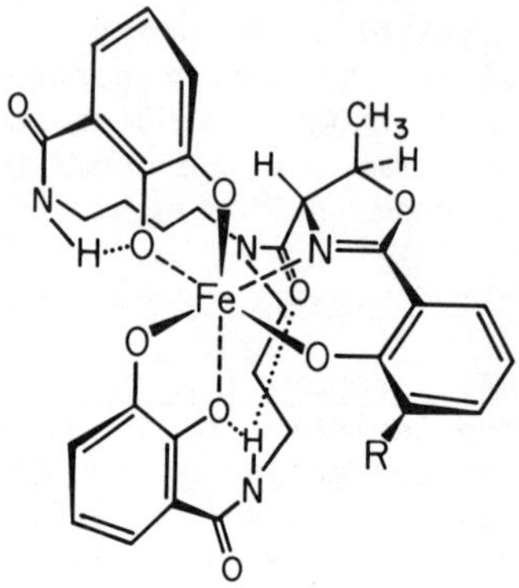

Figure 3. Ferric parabactin, R = *H; ferric agrobactin,* R = *OH*

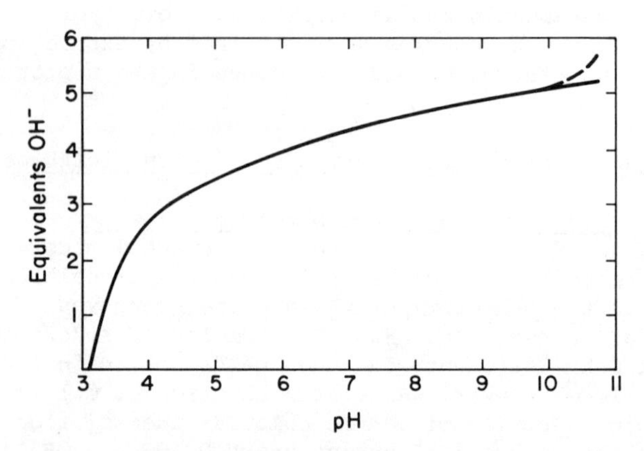

Figure 4. Titration of agrobactin and parabactin in the presence of an equimolar quantity of ferric chloride. The siderophores differed significantly only at very high pH, where (– – –) agrobactin displayed a new buffer zone not present in (——) parabactin.

b. Ligand Exchange. The well-known tendency of ferric catecholates to undergo internal oxidation-reduction at lower values of pH (14,15), concern for the stability of the oxazoline moieties, and the opportunity to diminish the proton discrepancy between tri-catechol *vs* bi-catechol-plus-oxazoline binding in agrobactin from 1 in 6 down to 1 in 3 prompted us to supply iron to the siderophores via a ligand exchange reaction of the type:

$$\text{Agrobactin + ferric rhodotorulate = ferric agrobactin + } 2 \text{ or } 3\ H^{+} + \text{rhodotorulic acid.}$$

Rhodotorulic acid, a hydroxamate type siderophore, binds 2/3 mole ferric ion to give an uncharged, orange colored chelate which is fully formed at neutral pH (10).

Titrations were performed as above except that 2 μmoles of ferric chloride and 3 μmoles of rhodotorulic acid were mixed and neutralized prior to introduction of 2 μmoles of catechol type siderophore dissolved in ethanol. The equivalents of standard alkali required to neutralize the solution were then noted.

Upon introduction of 2 μmoles of agrobactin the color of the neutral ferric hydroxamate solution changed from orange to blue and the pH fell to 3.6. Exactly 3.98 μmoles of alkali were required to raise the pH to the end point of pH 6.5 to 7, corresponding to 2 moles H^{+} per Fe^{3+}. In the case of parabactin, the pH went to just less than 4.0 upon addition of the catechol and the color of the solution was brown, thus indicating the presence of some ferric hydroxamate. Restoration of the pH to 6.5 to 7, which again appeared to be an end point, required 1.7 H^{+} per Fe^{3+}. For agrobactin A, the pH fell to 4.4 and the solution, although wine colored, upon back titration was still buffering strongly in the pH range 7-8.

From these data we conclude that agrobactin and parabactin form similar coordination compounds with ferric ion in which one of the atoms linked to the iron in the central bidentate portion of the complex is the tertiary N of the oxazoline ring. At neutral pH agrobactin A, in contrast to the tri-catechol enterobactin (16), appeared to be not fully coordinated to the ferric ion.

Similar experiments were performed by ligand exchange with the 1:1 ferric complex of nitrilotriacetate. The total proton yields per iron to the neutral pH zone were 4.7 and 4.6 for agrobactin and parabactin, respectively. Thus introduction of the iron (III), either directly or by ligand exchange, gave complexes which for both agrobactin and parabactin should result in divalent anions at neutral pH.

C. Net Electrical Charge. The iron complexes of agrobactin, agrobactin A, parabactin and enterobactin were prepared by neutralization of the ligands in the presence of ferric chloride and their electrophoretic mobilities compared with that of ferrichrome

A on paper at pH 6.6 in 0.1 μ phosphate buffer. The results are recorded in Table I.

Table I

Electrophoretic Mobility of Siderophore Iron (III) Complexes

Complex	MW	Net Negative Charge	Relative Mobility pH 6.6
Fe (III) agrobactin	687	2	0.85
Fe (III) agrobactin A	704	2-3	1.03
Fe (III) parabactin	671	2	0.87
Fe (III) enterobactin	719	3	1.20
Ferrichrome A	1052	3	1.00

While the mobility of substances on paper in an electric field is known to be influenced by a number of factors, charge and mass are certainly the two most significant parameters affecting the rate of migration. The data in Table I illustrate that ferric agrobactin and ferric parabactin are mononuclear and move at comparable rates; the slightly enhanced rate of ferric parabactin would be expected in view of its lower molecular weight, provided both complexes bore 2 negative charges. If ferric enterobactin were reduced from a net charge of 3^- to 2^-, its anticipated mobility relative to ferrichrome A would be ~0.8. The somewhat greater rates shown for ferric agrobactin and ferric parabactin, 0.85 and 0.87, respectively, are in inverse proportion to their lower molecular weights. Ferric agrobactin A could be speculated to carry between 2 and 3 charges/mole at this pH. All of the ferric catechols listed in Table I were wine colored at pH 6.5. In contrast, an iron complex of parabactin A was purple at both pH 6.5 and 7.4, and it displayed an unusually low mobility, thus indicating it to be incompletely formed and/or multi-nuclear.

In sum, the electrophoresis experiments support the concept of iron (III) binding to the oxazoline N in both agrobactin and parabactin to give, in each case, mononuclear di-anionic complexes.

D. Electronic Absorption Spectra. The absorption spectra of the complexes formed by exchange from ferric nitrilotriacetate were determined in 0.1 M phosphate, pH 7.4, with a Beckman Model 25 recording spectrophotometer over the range 400-700 nm. Agrobactin and parabactin gave wine colored iron complexes with a broad adsorption band centered at about 500 nm. At pH 7.4, the approximate a_{mM}^{max} for agrobactin was 4.1 at 505 nm; for parabactin the tentative figure was 3.3 at 512 nm. Tait (5) reported a value at pH 7.4 of 3.5 at 515 nm for his "Compound III", which we believe to have the structure shown in Figure 2, R=H. The absorption maximum for ferric parabactin A was shifted substantially to the red and lay at 535 nm.

Since the spectra of ferric agrobactin and ferric parabactin do show minor differences, a number of experiments were tried in which an attempt was made to mimic these differences via examination of the 1:3 ferric complexes of model salicyl and 2,3-dihydroxyphenyloxazolines and the 1:1:1 ferric complexes of the model oxazolines with Tait's "Compound II", N^1,N^8-bis-(2,3-dihydroxybenzamido)spermidine.

Ferric tris-salicyloxazoline precipitated upon its formation by the addition of three equivalents of alkali. The neutral complex was dissolved in ethanol to give an orange colored solution with a_{mM} = 4.5 at the maximum, 465 nm. A specimen of ferric tris-2,3-dihydroxyphenyloxazoline had a maximum at 520 nm with a_{mM} of 2.8 at pH 7.5; at pH 10, where the complex was fully formed, the a_{mM} was 4.6 and the maximum had shifted to 495 nm. These data, and those obtained by admixture of the oxazolines with Tait's Compound II, were not particularly illuminating as regards the mode of iron complexation in the natural products. Examination of the complexes in the ultraviolet might have proved instructive, but this aspect was not pursued.

E. Circular Dichroism. The siderophores studied herein contain one bidentate ligand mounted on an optically active substituent, namely, L-threonine or its oxazoline. Thus the particular type of coordination isomer considered by Corey and Bailar (17) could result in a chelate chromophore which is capable of rotation of plane polarized light.

The circular dichroism spectra of the ferric derivatives of the two siderophores, their "A" analogues, and enterobactin were determined in a suitably equipped Cary Model 60 spectropolarimeter. The data are recorded in Figure 5. It is apparent that the curve for ferric agrobactin A resembles that of enterobactin, which is known to yield a Δ,*cis* complex with ferric ion (18). In contrast, the curves for ferric agrobactin and ferric parabactin suggest clearly that in this case the configuration around the iron is predominantly a left-handed propeller, i.e., Λ,*cis*, as in ferrichrome (19). Steric constraints rule out the possible presence of geometrical isomers of the *trans* variety.

F. Formation Constant.

a. Equilibration with Ferrichrome. The data in Figure 6 illustrate the ability of agrobactin to remove iron from ferrichrome under the specific experimental conditions employed. Similar results were obtained with parabactin, enterobactin and a syn-

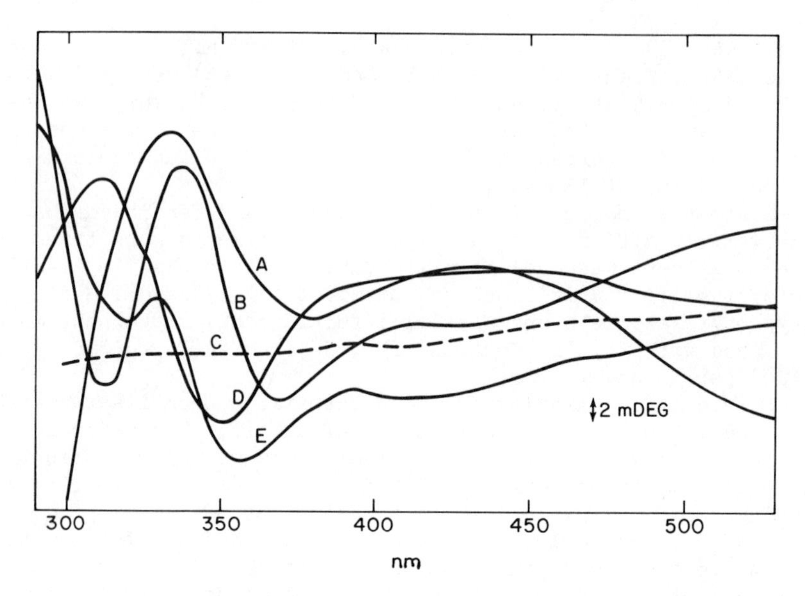

*Figure 5. Circular dichroism spectra of approximately 0.2m*M *solutions of A, ferric enterobactin; B, ferric agrobactin; C, blank; D, ferric agrobactin A; and E, ferric parabactin in 0.1*M *phosphate pH 7.4*

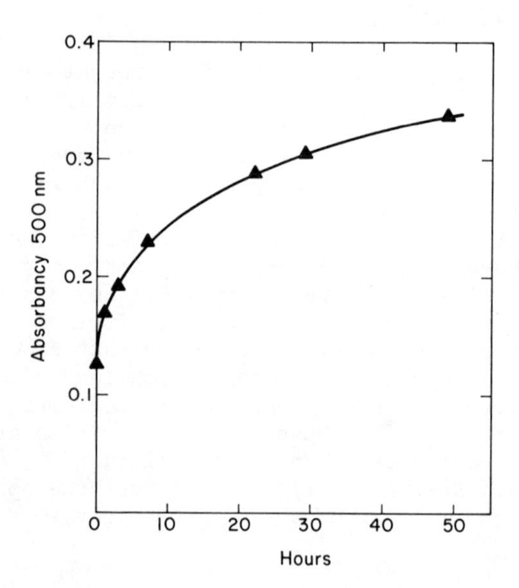

*Figure 6. Equilibration of 0.1m*M *ferrichrome and 0.1m*M *agrobactin in 50% ethanol–50% 10m*M *Na^+HEPES buffer pH 7.2 at 25°C*

thetic analog of enterobactin, cis, 1,5,9-tris(2,3-dihydroxybenzamido)cyclododecane (20). It is apparent that all of these catechol compounds are thermodynamically capable of capturing iron from ferrichrome, which is reported to have $K_f = 10^{29.1}$ (21). Since in each case the iron was completely transferred, we estimate that the K_f for the tri-catechols must be at least several orders of magnitude greater than those reported for the trihydroxamate type siderophore ligands.

b. Equilibration with Ferric Enterobactin. Having established that the catechol type siderophores based on spermidine are superior iron (III) complexing agents (in comparison with ferrichrome) we next sought to contrast them with enterobactin. The relative avidity of agrobactin, parabactin and enterobactin for iron (III) was estimated by observation of the progress of the following reactions from left to right:

Ferric enterobactin + agrobactin + H^+ = ferric agrobactin + enterobactin

Ferric enterobactin + parabactin + H^+ = ferric parabactin + enterobactin

Ferric agrobactin + enterobactin = ferric enterobactin + agrobactin + H^+

Ferric parabactin + enterobactin = ferric enterobactin + parabactin + H^+

A titration vial was charged with 0.5 ml ethanol, 2 μmoles of catechol type siderophore in 0.2 ml ethanol, 2 μmoles of $FeCl_3$ in 0.1 ml methanol and 0.5 ml of water. The pH was brought to about 7 with standard 0.1 N NaOH and 0.2 ml containing 2 μmoles of competing catechol type siderophore was then added. If necessary, the pH was readjusted to ~7.

Upon adding either agrobactin or parabactin to neutral solutions of ferric enterobactin there was little change in pH. However, mixing of enterobactin with either ferric agrobactin or ferric parabactin caused the pH to fall to less than 6 and ca. 1 to 2 μmoles of alkali were required to neutralize the solutions.

The neutral solutions were allowed to stand at room temperature for 1 hr and then subjected to paper electrophoresis for 30 minutes at pH 6.6. All four mixtures separated into about equal amounts of the two ferric complexes; a similar pattern was observed after incubation for 24 hours at room temperature. We conclude from these results that ligand exchange was rapid and that all three siderophores have about equal affinity for iron. Buffering by the _o_-hydroxyl aromatic functions, which may have pK_a' values <8 (16), probably precluded a pH rise upon transfer of iron from ferric enterobactin to the spermidine containing siderophores.

The surprisingly large affinity of the spermidine siderophores for iron was confirmed by re-examination of the equilibria between enterobactin, agrobactin and ferric ion on a more quantitative basis.

Stock 10 mM solutions of enterobactin and agrobactin in ethanol were prepared by dissolving, respectively, 3.736 mg in 0.56 ml and 3.421 mg in 0.54 ml. The ferric chloride was dissolved in methanol to give a 20 mM solution. A titration vial was loaded with 0.5 ml ethanol, 0.2 ml (2 μmoles) catechol solution, 0.1 ml (2 μmoles) ferric chloride solution and 0.5 ml water. The pH was raised to ca. 7 by addition of 0.1 N NaOH, at which point exactly 0.2 ml (2 μmoles) of solution of competing catechol was added and, when the latter was enterobactin, the pH readjusted to ~7. The neutral solutions were sealed and stored overnight at room temperature to assure equilibration, although the latter appeared to be achieved immediately.

Exactly 0.5 ml of each of the two solutions was applied to individual 20 cm wide sheets of Whatman No. 1 paper and the separation performed in phosphate buffer pH 6.8 for 30 min. @ 30 ma and 2.5KV. Colored bands occurred at 9 and 13 cm from the origin, representing ferric agrobactin and ferric enterobactin, respectively. The zones were eluted into 5 ml of pH 6.8 phosphate and the spectra recorded from 600 to 400 nm.

Starting with the iron in ferric enterobactin, the ratio of the maximum absorbancies of the ferric enterobactin and ferric agrobactin solutions prepared as above was 0.329/0.216 or 1.52, which would represent [ferric enterobactin]/[ferric agrobactin], assuming equal extinctions at the maxima. Taking an arbitrary value of 5.0 as the maximum millimolar extinction for both compounds, the yield of complexed iron would be ca. 80%. In a corresponding experiment starting with the iron in ferric agrobactin, the ratio of absorbancies for ferric enterobactin and ferric agrobactin was 0.277/0.192 or 1.44.

In an experiment in which equimolar amounts of enterobactin, agrobactin and ferric chloride were mixed and analyzed after titration to pH levels of 6, 7 and 8 the absorbancy ratios found, respectively, were 2.26, 1.36 and 1.55, all in favor of enterobactin, after electrophoretic separation at the same pH and elution of the bands into equivalent volumes of pH 6.8 phosphate buffer.

In view of the several assumptions, such as the lack of a precise number for the extinction of ferric agrobactin, an exactly quantitative value cannot be placed on the affinity of agrobactin for ferric ion. However it is clear that the binding strength is comparable to that of enterobactin, which at pH 7.4 is reported (18) to have a formation constant 10^6-larger than ferrichrome, a cyclohexapeptide trihydroxamate siderophore with a well established log K_f of 29.1 (21). Apart from the difference in charge, the two wine colored complexes are spectrally distinct. Ferric enterobactin, which has a salmon hue, has a maximum at 495 nm; ferric agrobactin has a more purple shade and the peak is at 505 nm.

II. Biological Activities

Agrobactin has already been shown to serve as a siderophore in *A. tumefaciens*, in which organism it has the capacity to reverse inhibition of growth invoked by the presence of ethylenediamine-di-(*o*-hydroxyphenylacetic acid)(EDDA). While Tait (5) referred to his "Compound III" as a "siderochrome", the previous name for siderophore, he did not demonstrate that the substance possessed this activity in *P. denitrificans*.

Table II affords the evidence that parabactin and agrobactin, but not their open-form analogues, can counteract growth retardation of *P. denitrificans* caused by EDDA. The simplest interpretation of these data is that the synthetic chelator, like deferri-ferrichrome A, forms a non-transportable or otherwise unavailable anionic complex with ferric ion which effectively denies iron to the cell in the absence of a strongly competitive ligand which is, in fact, utilized (22). Table II also records the activity of these compounds with *Escherichia coli* RW193, an organism defective in the synthesis of enterobactin. Thus it would appear that the Δ,*cis* complexes are active in *E. coli* while those with an enantiomorphic configuration around the iron, Λ,*cis*, are utilized by *P. denitrificans*.

Table II

Growth Response of *Paracoccus denitrificans* and *Escherichia coli* K-12 RW193 to Spermidine-Containing Siderophores

Siderophore	*P. denitrificans*[a]	*E. coli* RW193[b]
Agrobactin	+	-
Agrobactin A	-	+
Parabactin	+	-
Parabactin A	-	+

[a]Ligand tested by application of 0.1 and 0.5 nanomoles on nutrient agar plates containing a 1/7.5 x 10^5 dilution of minimal broth culture and 0.1 mg ethylenediamine-di-(*o*-hydroxyphenylacetic acid) per ml (22).

[b]Iron (III) complexes tested by application of 2.5, 25 and 250 picomoles in the assay system previously described (6).

III. Discussion

Our evidence suggests that both parabactin and agrobactin form analogous complexes with iron (III), as illustrated in Figure 3.

Although we have characterized the siderophore of P. denitrificans as parabactin (Figure 2, R=H) and not parabactin A (Figure 1, R=H), there is some question as to which form was isolated from the same organism by Tait (5). The relative stability of the oxazoline to acid hydrolysis, the spectral shifts observed by Tait in acidic media and not found in our parabactin A (8), and the properties he ascribes to the iron complex can only be reconciled with the structure in Figure 2, R=H. Although crude preparations of ferric enterobactin contain a number of colored species, the reddish form is the tris-catecholate; bluish tints are associated with oxidized/polymerized or otherwise coordinated forms of iron (23). Parabactin A yields a relatively inferior complex with ferric ion which fails to develop a red color even at quite alkaline pH.

The proton count upon iron complexation at neutral pH appears to be five/mole for both parabactin and agrobactin. The slight shortfall noted in Figure 4 could be attributed to various factors such as failure to achieve equilibrium, to destruction of the ligand, or to the presence of undetected impurities in the siderophores. Nonetheless it is apparent that at biological pH the oxazoline N atom in both ferric agrobactin and ferric parabactin is bonded to the iron. In a 2,3-dihydroxyphenyloxazoline the three possible binding sites are sterically constrained from simultaneous attachment to a central metal ion. Hence agrobactin can only be hexadentate.

In agrobactin O^- & N vs O^- & O^- coordination leads to a vastly more compact structure, greater stability of the iron (II) species, charge delocalization, reduced overall charge, the Λ configuration and, finally, an end to competition with H^+ for the very weakly acidic m-hydroxyl group. This mode of binding iron must be fundamental in the siderophore series since it occurs in agrobactin, parabactin and mycobactin; in the latter instance it has been confirmed by crystallography (24). Ferric ion, as a "hard" acid, normally prefers coordination to oxygen rather than to the "softer" base, nitrogen. However, in addition to the steric and other reasons just cited, binding to N in agrobactin places the metal ion in a six-membered ring containing two double bonds, an environment known to favor its stability. It has been shown (25) that mycobactin, with a single o-hydroxyphenyloxazoline group flanked by two hydroxamic acid functions, is able to remove iron from the tri-hydroxamate type siderophore ferrioxamine B ($K_f = 10^{30.6}$).

Possible roles for the m-hydroxyl in agrobactin are acidification of the neighboring o-hydroxyl group and enhanced water solubility of the ligand. A 0.1 mM solution of agrobactin can be achieved in an aqueous solution containing 20% methanol; parabactin is substantially less soluble.

The deprotonation curve for agrobactin with Ga (III) is essentially similar to that shown for Fe (III) in Figure 4. The spermidine containing siderophores undoubtedly bind Tb (III), an ion which we observed to perturb the absorption band of enterobactin in the near ultraviolet. Parabactin may form complexes with molybdate since the presence of molybdenum in the low iron medium imparts a bright orange color to the culture fluid of *P. denitrificans*, an organism known to possess the molybdenum containing nitrate reductase. Thus in this species parabactin may play a dual role in scavenging both iron and molybdenum. This idea has already been advanced for *Bacillus thuringiensis* (26).

Ferrichrome is a convenient siderophore against which to measure the relative capacity of catechol type ligands to complex iron (III). It is available as a crystalline solid of high purity, the stability constant ($10^{29.1}$) is known precisely and the maximum of the charge transfer band occurs some 75 nm lower than that of the ferric tris-catecholates. We found that enterobactin and the synthetic carbocyclic analogue derived from 1,5,9 tri-amino-cyclododecane (20) were able to remove iron from ferrichrome completely and at almost identical rates ($t_{1/2} \cong 6$ hrs in 10 mM HEPES pH 7.2, 10% methanol). Under comparable conditions an enterobactin analogue based on tri-aminomethyl benzene, 1,3,5-tris (N,N',N"-2,3-dihydroxybenzamido methyl)benzene (27), was unable completely to abstract ferrichrome iron and appeared to equilibrate about 50-50 with ferrichrome A ($K_f = 10^{29.6}$), although the reaction was faster. The deprotonation curve for enterobactin differs somewhat from that shown for agrobactin and parabactin in Figure 4 in that it is shifted to more acidic pH values throughout its entire course and results in displacement of exactly six H^+ at physiological pH (16). Hence the cyclic twelve-membered enterobactin backbone seems exquisitely designed to fold around the iron (III) ion, a process which has been observed to involve a major conformational change of the cyclic triester (28). Our results demonstrating almost equivalent distribution of iron among the three catechol type siderophores indicates that a similar comment applies to the specific linear array of atoms in the spermidine peptide moiety of agrobactin and parabactin. Tait (5) had already concluded that the stability of the *P. denitrificans* siderophore Fe^{3+} complex is "much higher" than that of Fe^{3+}·EDTA. The absence of ester bonds obviously enhances the resistance of the spermidine siderophores to hydrolytic cleavage.

The data given in Table II suggest opposite chirality requirements for the siderophore transport systems in the two Gram negative organisms *P. denitrificans* and *E. coli*. This has not been demonstrated previously among the catechol type siderophores although it has been reported (29) that the racemic ferric-*cis*-1,5,9-tris(2,3-dihydroxybenzamido)cyclododecane displays about half of the activity of ferric enterobactin in competing for the solubilized outer membrane receptor of *E. coli*. This would localize one site of optical specificity to the surface receptor. We have initiated a systematic study of the cross reactivities of the three catechol type siderophores in the organisms producing them. Recently Winkelmann (30) produced convincing evidence that Λ,*cis* ferrichrome is the preferred isomer in *Neurospora crassa*.

The structure given in Figure 3 represents our concept, based on the available evidence and the examination of CPK models, for the constitution of the iron (III) complexes of agrobactin and parabactin. Possible H bonds have been indicated but not yet demonstrated. The problem requires further investigation by additional techniques, such as X-ray diffraction and high resolution NMR; experiments to this end are in progress. The solution structures of agrobactin and parabactin, and those of their metal complexes, are of interest from the point of view of microbial iron assimilation. Moreover, parabactin (31) and agrobactin (A. Jacobs, personal communication) have been found capable of deferration of mammalian cell lines and are potentially useful drugs in chelation therapy.

Abstract

Agrobactin and parabactin form analogous complexes with iron (III) in which the metal ion is bound in apparent Λ,*cis* configuration to the distal catechols and to the central *o*-hydroxyphenyloxazoline function. The open chain derivatives, agrobactin A and parabactin A, appear to yield chelates of the opposite chirality, a conclusion supported by growth tests with *Paracoccus denitrificans* and *Escherichia coli*. The affinities of the tri-catechol siderophores agrobactin, parabactin and enterobactin for iron (III) are closely comparable in magnitude and exceed by several powers of ten the formation constant of the cyclohexapeptide trihydroxamate ferrichrome (log K_f = 29.1).

Acknowledgement

This research was supported in part by grants AI04156 and AM17146 from the NIH and by grant PCM 78-12198 from the NSF.

Literature Cited

1. Neilands, J. B., in "Iron in Biochemistry and Medicine", A. Jacobs and M. Worwood, eds., 2nd Ed., Academic Press, New York, in press.

2. Emery, T., in "Metal Ions in Biological Systems", H. Sigel, ed., Vol. 7, Marcel Dekker, New York, 1978, p. 77.

3. Neilands, J. B., in "Bioinorganic Chemistry-II", K. N. Raymond, ed., American Chemical Society, Washington, D.C., 1977, p. 3.

4. Wayne, R.; Frick, K.; Neilands, J. B. J. Bacteriol., 1976, 126, 7.

5. Tait, G. H. Biochem. J., 1975, 146, 191.

6. Ong, S. A.; Peterson, T.; Neilands, J. B. J. Biol. Chem., 1979, 254, 1860.

7. Neilands, J. B., Abst. 178th National Meeting, American Chemical Society, Washington, D.C., Sept. 9-14, 1979, Inorg., 36.

8. Peterson, T.; and Neilands, J. B. Tetrahedron Letters, 1979, No. 50, 4805.

9. Pollack, J. R.; and Neilands, J. B. Biochem. Biophys. Res. Commun., 1970, 38, 989.

10. Atkins, C. L.; and Neilands, J. B. Biochemistry, 1968, 7, 3734.

11. Neilands, J. B. J. Am. Chem. Soc., 1952, 74, 4846.

12. Garibaldi, J. A.; and Neilands, J. B. J. Am. Chem. Soc., 1955, 77, 2429.

13. Neilands, J. B.; and Cannon, M. D. Anal. Chem., 1955, 27, 29.

14. Mentasti, E.; Pelizzetti, E.; and Saini, G. J. Chem. Soc. Dalton, 1973, 2609.

15. Hider, R. C.; Silver, J.; Neilands, J. B.; Morrison, I.E.G.; and Rees, L. V. C. FEBS Letters, 1979, 102, 325.

16. Salama, S.; Stong, J. D.; Neilands, J. B.; and Spiro, T. G. Biochemistry, 1978, 17, 3781.

17. Corey, E. J.; and Bailar, J. C. J. Am. Chem. Soc., 1959, 81, 2620.

18. Raymond, K. N.; and Carrano, C. J. Accts. Chem. Res., 1979, 12, 183.

19. Zalkin, A.; Forrester, J. D.; and Templeton, D. H. J. Am. Chem. Soc., 1966, 88, 1810.

20. Corey, E. J.; and Hurt, S. D. Tetrahedron Letters, 1977, No. 45, 3923.

21. Anderegg, G.; L'Eplattenier, F.; and Schwarzenbach, G. Helv. Chim. Acta, 1963, 46, 1400.

22. Miles, A. A.; and Khimji, P. L. J. Med. Microbiol., 1975, 8, 477.

23. Wayne, R.; and Neilands, J. B. J. Bacteriol., 1975, 121, 497.

24. Hough, E.; and Rogers, D. Biochem. Biophys. Res. Commun., 1974, 57, 73.

25. Snow, G. A. Biochem. J., 1969, 115, 199.

26. Ketchum, P. A.; and Owens, M. J. Bacteriol., 1975, 122, 412.

27. Venuti, M. C.; Rastetter, W. H.; and Neilands, J. B. J. Med. Chem., 1979, 22, 123.

28. Llinas, M.; Wilson, D. M.; and Neilands, J. B. Biochemistry, 1973, 12, 3836.

29. Hollifield, W. C.; and Neilands, J. B. Biochemistry, 1978, 17, 1922.

30. Winkelmann, G. FEBS Letters, 1979, 97, 43.

31. Jacobs, A.; White, G. P.; and Tait, G. H. Biochem. Biophys. Res. Commun., 1977, 74, 1626.

RECEIVED April 7, 1980.

17

The Design of Chelating Agents for the Treatment of Iron Overload

COLIN G. PITT
Research Triangle Institute, P.O. Box 12194, Research Triangle Park, NC 27709

ARTHUR E. MARTELL
Department of Chemistry, Texas A&M University, College Station, TX 77843

Iron is an essential component of man's biochemistry but, in common with other elements, becomes toxic when in excess (1,2). This arises in part because of the tendency of iron(III) to separate in tissues as very insoluble hydroxide and phosphate salts at the physiological pH and higher unless bound to transferrin, the iron transport protein, or to ferritin, the iron storage protein. Iron absorption via the diet is physiologically controlled, but the body has no regulatory mechanisms for eliminating a toxic excess introduced by accidental overdose or by multiple transfusions.

Cooley's anemia and its treatment provide an example of the difficulties of correcting deficient iron metabolism (3,4). Cooley's anemia is a genetic disease originating from errant biosynthesis of the β-chain of hemoglobin which can only be treated by an interminable transfusion regimen. The increased iron input (20-25 mg/day) exceeds the capacity of transferrin and ferritin, resulting in separation of insoluble iron in critical tissues, e.g. the heart, liver, pancreas. In principle, this ultimately fatal condition can be treated by administration of an iron chelating agent which promotes remobilization and excretion of the deposited iron. In practice, neither desferrioxamine B (DFB), the most commonly used iron chelator, nor any other agent clinically evaluated to date has been able to do more than retard the condition (5).

It has been stated (5) that the ideal iron chelating agent should be inexpensive and orally active. It should have a high and selective ability to bind iron(III) rapidly under physiological conditions, relative to ferritin and transferrin, while not interfering with hemoglobin, myoglobin, the cytochromes, and normal iron biochemistry. It should be free of both acute and chronic side effects, and resistant to metabolic changes which impair its ability to bind iron. These requirements may be achieved with some degree of predictability and are now discussed together with the results of animal screens of potential iron(III) chelators which have been reported by different laboratories (6-9).

0-8412-0588-4/80/47-140-279$08.50/0

Coordination of Ferric Iron

The basic requirement of an iron chelating drug is that it have a high and selective affinity to bind iron avidly under physiological conditions. The tripositive Fe(III) ion is a hard acid and consequently is bound most strongly by hard bases, the most effective of which are oxyanions, such as hydroxide, phenoxide, carboxylate, hydroxamate and phosphonate (10). The coordination number is usually six, although some seven-coordinate complexes are known. The most favorable geometry is an octahedral arrangement of donor atoms, permitting the maximum possible distance between their formal or partial negative charges. Charge neutralization is an important factor, and is optimum when the total charge of the six donor atoms is -3, as for the bidentate hydroxamate and tropolonate ligands. Steric effects can be important and, in the case of the EDTA complex of Fe(III), prevent the formation of the optimal octahedral arrangement of the six donor atoms; here binding to a solvent molecule or an additional oxyanion and formation of a seven coordinate species is preferred (11).

Stability Constants as a Measure of Affinity for Iron

The affinity of a ligand for iron(III) may be defined quantitatively in terms of the thermodynamic constants of the equilibria involved between the aquo metal ion and the deprotonated ligand L:

$$FeL + L \underset{}{\overset{K_2}{\rightleftharpoons}} FeL_2 \qquad FeL_2 + L \overset{K_3}{\rightleftharpoons} FeL_3 \qquad FeL_{n-1} + L \overset{K_n}{\rightleftharpoons} FeL_n$$

The formation or stability constants K_n for this series of equilibria are defined by equation (1).

$$K_1 = \frac{[FeL]}{[Fe][L]} \qquad K_2 = \frac{[FeL_2]}{[FeL][L]} \qquad K_n = \frac{[FeL_n]}{[FeL_{n-1}][L]} \tag{1}$$

These in turn may be expressed as the stability products, β_i, where

$$\beta_1 = K_1 \qquad \beta_2 = K_1K_2 \qquad \beta_n = K_1K_2 \cdots\cdots K_n \tag{2}$$

and

$$\beta_n = \frac{[FeL_n]}{[Fe][L]^n} \tag{3}$$

Stability constants are generally applicable to experimental conditions which are most favorable for chelate formation, i.e., in the absence of competing ions and at the optimum pH. It has been estimated (12) that under these conditions the stability product β of transferrin, the body's iron transport protein, is approximately 10^{36}. This indicates a very high order of stability. Classes of synthetic chelating agents which have a comparable or greater affinity for iron(III) under optimum *in vitro* conditions are shown in Figure 1.

The efficacy of a chelating drug *in vivo* is usually reduced substantially by competing ions present in the biological milieu (13). Calcium ions (10^{-3} M), hydrogen ions (10^{-7} M), and hydroxide ions (10^{-7} M) are the most serious interferences which compete with either the drug for iron(III) (OH^-), or with the iron(III) for the chelating drug (Ca^{2+}, H^+). For example, the greater the basicity of the chelator, the greater its affinity for iron(III), but this effect is paralleled by a higher affinity for protons. When the pK of the chelator is substantially greater than the physiological pH, proton competition will greatly decrease the concentration of the basic form of chelator, thus reducing the metal ion binding.

Computer simulation of such metal-ligand equilibria in biofluids has been reported with the object of chelation therapy. In one paper (14) the distribution of Ca^{2+}, Mg^{2+}, Fe^{2+}, Cu^{2+}, Zn^{2+}, and Pb^{2+} amongst 5000 complexes formed with 40 ligands was computed. However, in considering ligand design it is often more helpful to estimate interferences by use of an interference term α (13,15,16). For example, in the case of proton interference, α_L represents the fraction of ligand in its completely deprotonated form. If T_L is the total concentration of the uncomplexed drug in the medium, then equation (4) may be derived.

$$[L^{m-}] = T_L \alpha_L \qquad (4)$$

where

$$\alpha_L = (1 + [H^+]\beta_1^H + [H^+]^2\beta_2^H + \ldots\ldots [H^+]^m\beta_m^H)^{-1}$$

and β_i^H is the appropriate stability product of LH_i formation, i.e., $[H_iL]/[H^+]^i[L]$.

The corresponding expressions for calcium ion and hydroxide ion interference are equations (5) and (6), respectively, and the effective binding constant, K_{eff}, of the iron(III)-drug complex is defined by equation (7)

$$\alpha_{CaL} = ([Ca]\beta_{CaL})^{-1} \qquad (5)$$

$$\alpha_{Fe} = (1 + [OH]\beta_1^{OH} + [OH]^2\beta_2^{OH} + \ldots\ldots [OH]^m\beta_m^{OH})^{-1} \qquad (6)$$

$$\log K_{eff} = \log \beta_{FeL} - n\log(\alpha_{CaL}^{-1} + \alpha_L^{-1}) - \log(\alpha_{Fe}^{-1}) \qquad (7)$$

A

Name	Structure	Formula of Chelate	n	Log β_n	Log K_{eff}	Log K_{Sol} T_L=0.001	Log K_{Sol} T_L=1.00
2,3-dihydroxynaphthalene-6-sulfonic acid, H_3L	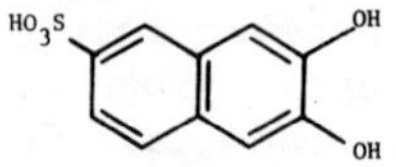	FeL_3^{6-}	3	44.2	18.20	0.89	6.89
Salicylic acid, H_2L	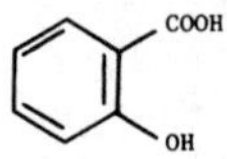	FeL_3^{3-}	3	35.3	8.01	-9.90	-3.90
8-Hydroxyquinoline, HL	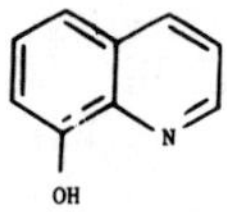	FeL_3	3	36.9	20.9	2.90	8.90
1,8-Dihydroxynaphthalene-3,6-sulfonic acid, H_4L	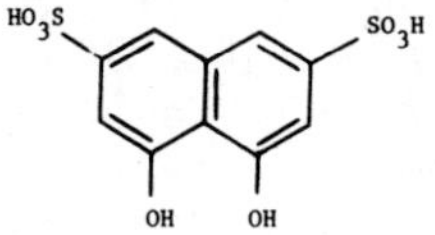	FeL_2	2	37.0	20.81	5.59	8.59

B

Name	Structure	Formula of Chelate	n	Log β_n	Log K_{eff}	Log K_{Sol} T_L=0.001	Log K_{Sol} T_L=1.00
3-Isopropyltropolone, HL	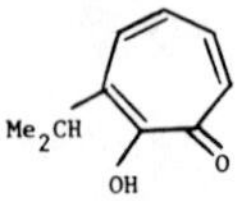	FeL_3	3	32.0	21.77	3.86	9.86
Acetohydroxamic acid, HL	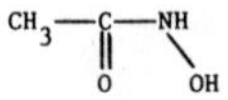 $CH_3—C(=O)—NH—OH$	FeL_3	3	28.3	13.10	-4.80	1.20
Nitrilotriacetic acid, H_3L	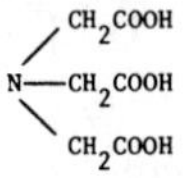 $N(CH_2COOH)_3$	FeL_2^{3-}	2	24.3	8.17	-6.74	-3.74
Ethylenediaminetetraacetic acid, H_4L	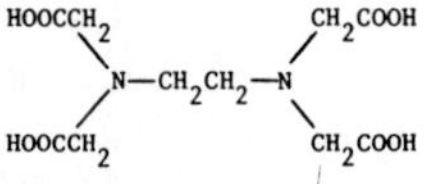 $(HOOCCH_2)_2N—CH_2CH_2—N(CH_2COOH)_2$	FeL^-	1	25.0	8.10	-3.81	-3.81
Diethylenetriaminepenta-acetic acid, H_5L	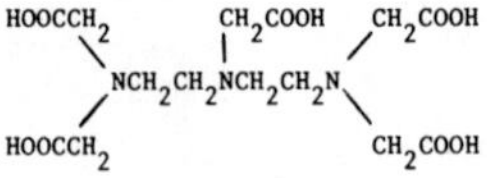 $(HOOCCH_2)_2NCH_2CH_2N(CH_2COOH)CH_2CH_2N(CH_2COOH)_2$	FeL^{2-}	1	28.0	10.96	-0.95	-0.95

Figure 1. Iron(III) affinities of chelating agents in terms of standard and effective stability constants

C

Name	Structure	Formula of Chelate	n	Log β_n	Log K_{eff}	Log K_{Sol} T_L=0.001	Log K_{Sol} T_L=1.00
Triethylenetetraminehexa-acetic acid, H_6L	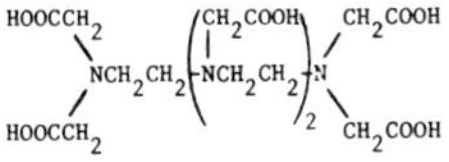 $HOOCCH_2$, $HOOCCH_2$ >NCH_2CH_2–(NCH_2CH_2 with CH_2COOH)$_2$–N< CH_2COOH, CH_2COOH	FeL^{3-}	1	26.8	10.61	-1.30	-1.30
N,N'-Bis(o-hydroxybenzyl) ethylenediamine-N,N'-diacetic acid, H_4L	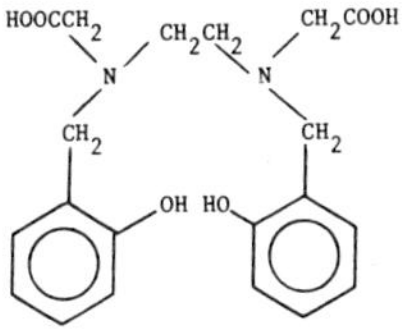 $HOOCCH_2$, CH_2CH_2, CH_2COOH, N, N, CH_2, CH_2, OH, HO	FeL^-	1	39.7	20.78	8.87	8.87
Ethylenebis-N,N'-(2-o-hydroxyphenyl)glycine, H_4L	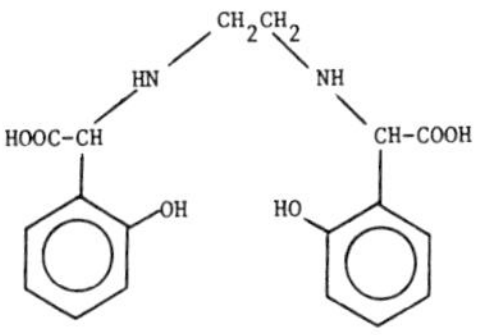 CH_2CH_2, HN, NH, HOOC-CH, CH-COOH, OH, HO	FeL^-	1	33.9	16.20	4.28	4.28

D

Name	Structure	Formula of Chelate	n	Log β_n	Log K_{eff}	Log K_{Sol} T_L=0.001	Log K_{Sol} T_L-1.00
Enterobactin, H_6L	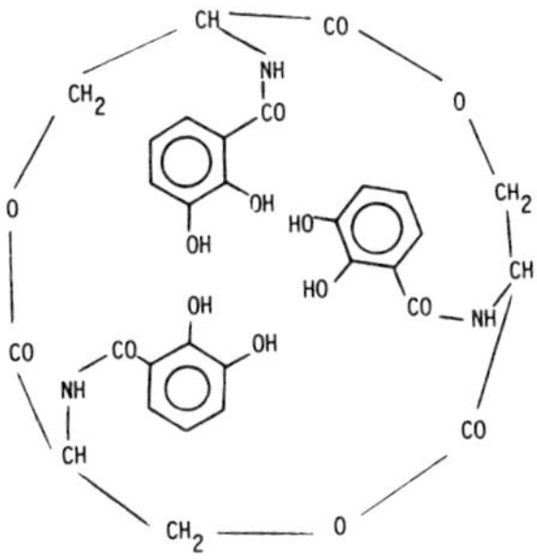 CH, CO, NH, CO, O, CH_2, O, OH, OH, CH_2, HO, HO, CH, CO, NH, OH, OH, CO, CO, NH, CH, CO, CH_2, O	FeL^{3-}	1	52	25	13	13
Desferrioxamine B, H_3L	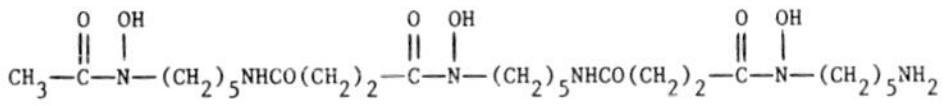 $CH_3-C(=O)-N(OH)-(CH_2)_5NHCO(CH_2)_2-C(=O)-N(OH)-(CH_2)_5NHCO(CH_2)_2-C(=O)-N(OH)-(CH_2)_5NH_2$	FeL	1	30.6	16.34	4.44	4.44

Standard data used in these calculations are : log β_1^{FeOH} = 11.09; log β_2^{FeOH} = 21.96; -log K_{sp} $Fe(OH)_3$ = 10^{41}; log β_4^{FeOH} = 24.4; - log K_w = 13.795; pH = 7.40.

Equation (6) only takes into consideration the formation of soluble mononuclear hydroxy complexes of iron. Since α_{Fe} has a fixed value of $10^{9.3}$ at pH 7.4, and is independent of other ligands, the relative efficacies of different chelating drugs may be compared by taking only calcium and hydrogen ion interferences into account.

Transferrin is believed (17) to bind iron(III) with three phenolate (tyrosine) residues and, being weakly acidic, proton interference is responsible for the significant difference between β (10^{36}) and K_{eff} (10^{16}). The phenolate group is also present in seven of the fourteen structures in Figure 1, which is testimony to the particularly strong and selective affinity of this group for iron(III). Proton interference increases with the number of phenolate groups and in the case of the tris complex of 2,3-dihydroxynaphthalene-6-sulfonic acid, proton interference reduces β_3 by a factor of 10^{25}.

Hydroxamic acids are stronger acids (pK_a 8-9) and, consequently, proton interference *in vivo* is less serious. This may be one reason why DFB, a hexadentate hydroxamic acid (Figure 2), has been of some utility in the treatment of iron overlaod. DFB is competitive with transferrin for iron(III) on the basis of their K_{eff} values and model *in vitro* studies, and the main drawbacks of DFB appear to be poor absorption when administered orally plus a susceptability to rapid metabolism and degradation in plasma (18).

Tropolones ought to be a most promising class of compounds for study. The pK_a of the pseudo-phenolic group is about 7 and, consequently, there is virtually no proton interference. Calcium ion interference is minor and in the case of 3-isopropyltropolone the value of log K_{eff} at pH 7.4 is 21.9 compared with a log β_3 value of 32. This value of log K_{eff} is the highest of the group of bidentate chelators in Figure 1, and 10^6 times greater than the K_{eff} of transferrin. A plasma concentration of 5×10^{-5} M (6 μg/ml) is calculated to be sufficient to sequester 10% of transferrin-bound iron, in spite of the fact that this tropolone is only bidentate. The major disadvantages of tropolones are expense, synthetic inaccessibility, some CNS toxicity, and difficulty in creating hexadentate forms.

Reliable equilibrium constants are not available for any phosphonic acids in the form of bidentate chelating agents, but comparison of the data for hexadentate agents shown in Table I demonstrates that the phosphonate group has a higher affinity for iron(III) than the carboxylate group.

Ferric Ion Hydrolysis and Ferric Hydroxide Solubilization

The most serious interference to Fe(III) binding and excretion is hydrolysis to produce aquo complexes and, in the most extreme case, to form a precipitate of ferric hydroxide. The first stage in the condition of iron overload *in vivo* occurs when the plasma iron level exceeds that of ferritin and transferrin.

TABLE I
Chelating Tendencies of N,N'-bis(o-hydroxybenzyl-N,N'-ethylenediaminedi(methylenephosphonic) acid (HBEDPO)

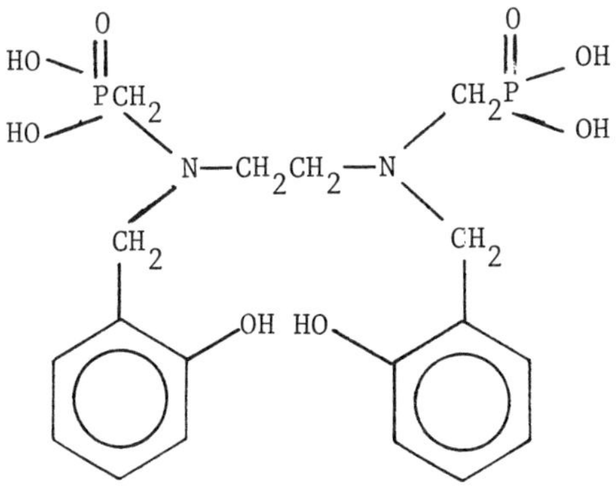

Log K_{ML}

M^{n+}	EDTA	HBED	HBEDPO
Cu^{2+}	18.70	21.38	24.00
Ni^{2+}	18.52	19.31	17.91
Co^{2+}	16.26	19.89	18.02
Ca^{2+}	10.61	9.29	8.36
Mg^{2+}	8.83	10.51	7.95
Fe^{3+}	25.0	39.68	>40

At this point, excess iron in the plasma is bound only weakly as non-specific protein complexes. As the iron concentration increases, the iron begins to separate in the form of insoluble phosphate or hydroxide complexes. An iron chelating drug must be capable of remobilizing this form of iron.

Assuming the worst case, where iron is separated as the less soluble (α) form of $Fe(OH)_3$, the iron concentration is given by equation (8) (19,20).

$$K_{sp} = 10^{-41} = [Fe^{3+}][OH^-]^3 = \frac{[Fe^{3+}]10^{-42}}{[H^+]^3} \qquad (8)$$

$$\text{i.e., } [Fe^{3+}] = 10[H^+]^3$$

One may then determine the effective solubilizing constant K_{Sol} of the chelating agent, which is the ratio of bound soluble iron (FeL_n) to the uncoordinated ligand in solution, T_L. Combining equations (4), (5), (8), and (9) and neglecting interference by other metal ions one may derive equation (10).

$$\frac{[FeL_n]}{[L]^n} = \beta_{FeL_n}[Fe^{3+}] \tag{9}$$

$$K_{Sol} = \frac{[FeL_n]}{T_L} = \frac{\beta_{FeL_n} 10[H^+]^3 T_L^{(n-1)}}{(\alpha_L^{-1} + \alpha_{Ca}^{-1})^n} \tag{10}$$

$$\text{i.e., } \log K_{Sol} = \log \beta_{FeL_n} + 1 - 3\ pH - n \log (\alpha_L^{-1} + \alpha_{Ca}^{-1}) + (n - 1)\log T_L$$

A value of log K_{Sol} equal to one or higher is indicative of a significant ability to dissolve $Fe(OH)_3$, the tendency increasing rapidly as the value of log K_{Sol} increases in magnitude. As an example, EDTA has a log β_{FeL} value of 25 and at pH 7.4 in the presence of 10^{-3} M calcium ion the interference term α_L is 10^{-12}. The calculated value of log K_{Sol} is -5.1 and ferric hydroxide is not expected to dissolve in this system. The values of log K_{Sol} of the bidentate chelating agents shown in Figure 1 indicate salicylic acid and related compounds will not dissolve $Fe(OH)_3$ while hydroxamic acids will be marginally effective. The natural sexadentate hydroxamic acid, desferrioxamine B, however, is predicted to be fairly effective. Sexadentate ligands containing two phenolic groups, EHPG and HBED, are seen to be highly effective in solubilizing ferric hydroxide, and the sexadentate triscatecholate ligand enterobactin is without doubt the most effective ligand listed in this paper.

The ability to bind and solubilize iron(III) in alkaline medium is not entirely predictable on the basis of the above considerations. For example, the three hydroxyethyl analogs of NTA in Figure 3 show an ability to sequester iron(III) in alkaline solution in the order TEA > DHG > HIMDA. As the number of hydroxyethyl groups increases, the resistance of the Fe(III) complex to disporportionation via ferric hydroxide precipitation extends increasingly into the alkaline region. In the case of the highest member of the series, triethanolamine, soluble, colorless Fe(III) chelates are formed at pH 14 and above, and even in solid alkali hydroxides. It has been demonstrated (21) that above pH 13, increasing $[OH^-]$ increases the effectiveness of triethanolamine in formation of stable complexes having the general formula $Fe_a(OH)_b(H_{-n}L)_c{}^{3a-b-nc}$, whereby b and the ratio a/c becomes much greater than unity at very high pH. While members of this series of compounds become more effective for Fe^{3+} as the pH increases, they become increasingly less effective at low pH, so that mix-

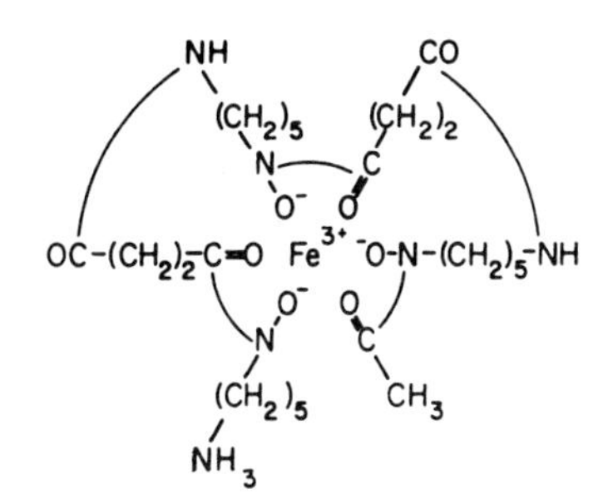

Figure 2. Iron(III) complex of desferrioxamine B

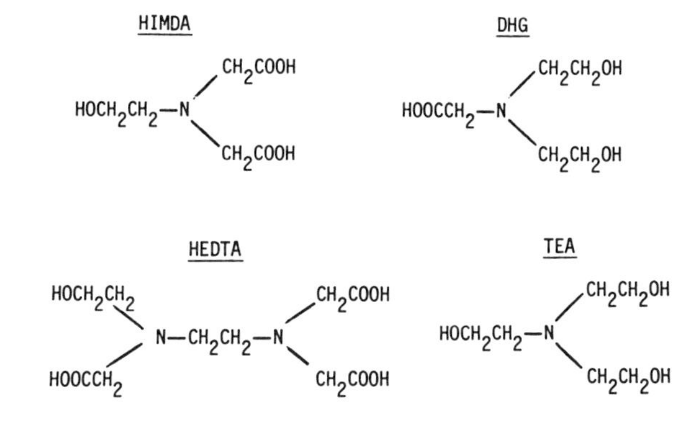

Figure 3. Hydroxyethyl derivatives of NTA and EDTA

tures of these ligands have been employed in industry to achieve effectiveness of Fe(III) complexing over a broad pH range. Frequently industrial sequestering agent preparations are composed of mixtures of these ligands with EDTA.

The replacement of one acetate function of EDTA by a hydroxyethyl group, to give HEDTA, also results in the formation of a compound which extends the useful Fe(III)-sequestering range of EDTA to higher pH (the EDTA-Fe(III) chelate system decomposes to precipitate $Fe(OH)_3$ around pH 8, depending on conditions and the concentration of excess EDTA). Since similar variation of the NTA structure provides "Fe(III)-specific" complexing agents through dissociation of the hydroxy group to negative alkoxide donors, it was first thought that HEDTA functions in a similar way, by the formation of chelates such as $FeH_{-1}L^-$ (where H_3L represents HEDTA). While such is certainly the case for Th^{4+}, which forms a unique polynuclear complex with HEDTA (22), evidence has been presented to show that the hydroxyethyl group remains intact in Fe(III) complexes, even at pH values high enough to produce hydroxo complexes and the corresponding μ-oxo dimer (23). This conclusion is further supported by the crystal structure of Fe(III)-HEDTA μ-oxo dimer (11). If one accepts structures for aqueous solution similar to those found in the solid state in which the hydroxyethyl group is not coordinated, one is still left with the problem of explaining the experimental fact that HEDTA-Fe(III) complexes are more stable than those of EDTA with respect to $Fe(OH)_3$ precipitation, even though the stability constant of the normal chelate (FeL^- for EDTA and FeL for HEDTA) is much greater in magnitude for EDTA. Under these circumstances, and in view of the x-ray data, the only reasonable explanation lies in the greater stabilities of the hydrolytic forms of the iron(III)-HEDTA chelate, for which equilibrium data are available (10), and which are illustrated schematically in Figure 4. Apparently the species $FeLOH^-$ and $FeL(OH)_2^{2-}$ are relatively resistant to disproportionation to ferric hydroxide, and the hydroxyethyl group must somehow be involved (e.g., through solvation) in the achievement of this effect.

Limitations of the Use of Stability Constants

While values of β, K_{eff}, and K_{sol} provide a quantitative assessment of the ability of a chelator to bind iron, it must be recognized that they are equilibrium constants and provide no information about the rate at which equilibria are established. This has been clearly shown by *in vitro* measurements of the rates of removal of iron(III) from transferrin by EDTA, citrate, and NTA (24). The percent iron transferred at equilibrium was found to follow the order EDTA > citrate ≈ NTA, which is the order predicted from the K_{eff} values of the chelates; however, equilibrium was established more rapidly with citrate and NTA (1 hr) than with EDTA (>12 hr under the specific experimental conditions).

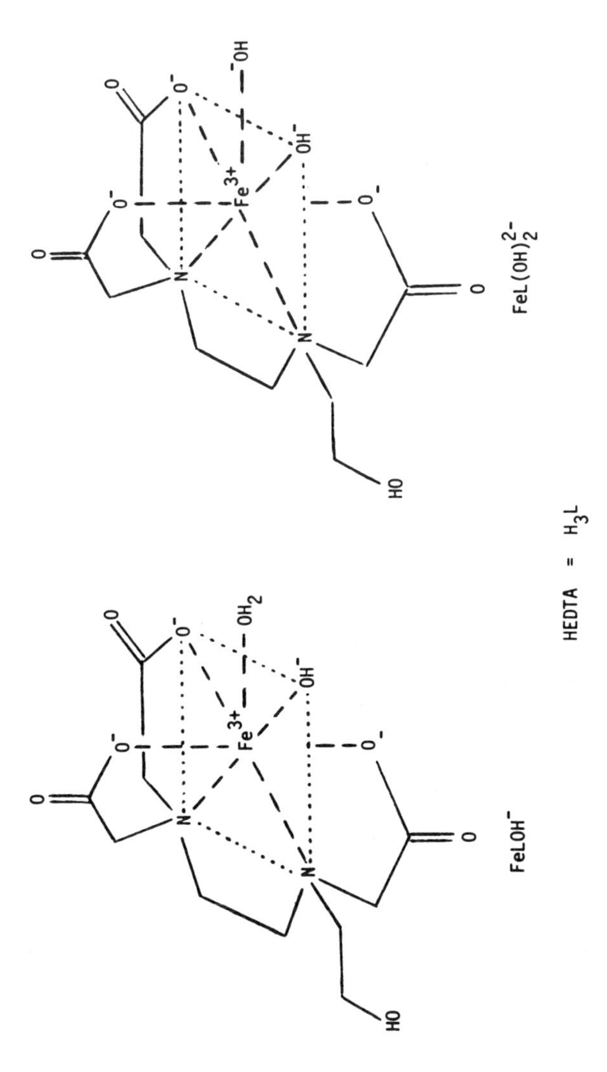

Figure 4. Hydrolyzed forms of HEDTA–iron(III) chelate

Obviously, the advantage of using a drug with a high K_{eff} value will be largely negated if the time required to establish equilibrium is substantially greater than the *in vivo* lifetime of the drug. This may be one reason why DFB is not a more effective iron chelating drug, for it is rapidly metabolized (18) by plasma enzymes yet is unable to remove iron from transferrin at a significant rate *in vitro* or *in vivo*. Recently, it has been shown that DFB will remove iron rapidly (hrs) from transferrin in the presence of a third, less effective ligand such as NTA, citrate, ATP, and 2,3-diphosphoglycerate (25,26,27). This confirms the kinetic nature of the problem, and suggests that an optimum iron chelating drug should incorporate some molecular feature which promotes exchange. Unfortunately, there is not a good understanding of what this functional group should be. A possible explanation for this effect is catalysis or transfer of the metal ion from one sexadentate ligand to another through the formation of mixed ligand intermediate ternary complexes, as described by Margerum (28) for analogous reactions in non-biochemical systems.

Equilibrium constants are possibly even less reliable when predicting the ability of chelators to mobilize ferritin-bound iron for it has been found that the percent iron removed in a given time for crystalline and non-crystalline ferritin by the above three chelators is NTA >> EDTA > citrate (29). This order shows no relationship to the calculated β, K_{eff}, and K_{Sol} values of the chelating agents and some other factor, e.g., the steric accessibility of the ferritin core (30) and associated kinetic factors must be involved. Redox mechanisms by which iron is mobilized from ferritin stores have been proposed and it has been shown that 2,2'-bipyridyl can mobilize several hundred iron atoms from ferritin in the absence of any other reducing agent with concomitant formation of 2,2'-bipyridyl-N-oxide (31). Thus, one of the mechanisms for mobilizing iron involves reduction of Fe(III) to Fe(II).

The importance of kinetic effects and specific biochemical mechanisms of iron transport do not negate the use of stability constants as guides to the design of new iron chelating drugs, but they do indicate the need to consider the bulk of the chelating agent and its possible effect on the kinetics of iron mobilization. It is quite conceivable that in some cases the advantages of a compact bidentate drug may offset the advantages of the chelate effect associated with the use of a hexadentate drug, or that the efficacy of a hexadentate drug may be enhanced by co-administration of a smaller, kinetically more labile iron chelator.

The Chelate Effect

Six coordinate iron(III) may be bound by either one molecule of a hexadentate drug, two molecules of a tridentate drug, three molecules of a bidentate drug, six molecules of a monodentate drug, or some combination thereof. The choice becomes important because of the extra stability associated with the chelate effect, which favors the use of compact multidentate ligands. The chelate effect arises because of (a) favorable stability constants associated with the formation of compact multidentate structures, and (b) a concentration factor, which becomes dominant at the concentrations likely to be encountered _in vivo_.

The former can be illustrated by considering the equilibrium between the iron(III) chelates of DFB, a hexadentate ligand (Figure 2), and its bidentate analog, acetohydroxamic acid:

$$DFB + Fe^{3+}\left[\begin{matrix} O^{-} - NH \\ | \\ O = CMe \end{matrix}\right]_3 \underset{}{\overset{K}{\rightleftharpoons}} Fe(DFB) + 3MeCONHOH \qquad (12)$$

$$\frac{[Fe(DFB)]}{[Fe(MeCONHO)_3]} = \frac{K[DFB]}{[MeCONHOH]^3} \qquad (13)$$

From stability constant measurements, log K = 2.3 $mol^2 \cdot \ell^{-2}$ (32,33). The equilibrium therefore favors the hexadentate chelate for all but extreme ligand concentrations.

The second component of the chelate effect, the concentration factor, can be illustrated by considering the ability of a drug (D) to sequester transferrin-bound iron:

$$TF\text{-}Fe + nD \rightleftharpoons TF + D_n\text{-}Fe \qquad (14)$$

$$\text{i.e.,} \quad \frac{[TF\text{-}Fe]}{[D_n\text{-}Fe]} = \frac{K^{TF}_{eff}[TF]}{K^{D}_{eff}[D]^n} \qquad (15)$$

For a bidentate drug n = 3, while for a hexadentate drug n = 1. Given a plasma concentration of 4×10^{-5} M for the transferrin-iron complex, and assuming that the same plasma concentration of the drug can be attained, one may calculate the stability constant of the drug (K^{D}_{eff}) which is required if 50% of the transferrin-bound iron is to be sequestered by the drug. If the drug is hexadentate, this is achieved when $K^{D}_{eff} = K^{TF}_{eff}$. However, if the drugs are tridentate or bidentate, their stability constants must be 10^4 and 10^8 times that of K^{TF}_{eff}, respectively.

The same property is illustrated by the data in Table II which compares the degrees of dissociation of tetracoordinate complexes with 0, 2, and 3 chelate rings. A chelate effect of 10^2 per chelate ring is assumed as the basis for the arbitrary stability constants listed. The superior properties of the metal chelate in dilute solution are dramatically illustrated by a simple calculation of the degrees of dissociaton in 1.0 molar and

TABLE II
Degrees of Dissociation of Complexes and Chelates in Dilute Solution

			1.0 M Complexes		$1.0x10^{-3}$ M Complexes	
No. of chelate rings	Formation Constants	Value	Free [M]	% Disso-ciation	Free [M]	% Disso-ciation
0	$\frac{[MA_4]}{[M][A]^4}$	10^{18}	1×10^{-5}	1×10^{-3}	1×10^{-4}	10
2	$\frac{[MB_2]}{[M][B]^2}$	10^{22}	3×10^{-8}	3×10^{-8}	5×10^{-9}	5×10^{-4}
3	$\frac{[ML]}{[M][L]}$	10^{24}	1×10^{-12}	10^{-10}	3×10^{-14}	3×10^{-9}

1.0×10^{-3} molar solutions.

It has been pointed out by Adamson (34) and others (35,36) that the entropy-related chelate effect, as manifested in the stability constants, disappears when unit mole fraction replaces unit molality as the standard state of solutes in aqueous systems. On this basis the stability constants assumed for the model compounds in Table II (20) would have to be equivalent in magnitude regardless of the number of chelate rings formed. On the other hand the relative degrees of dissociation of the model compounds in Table II remain an experimental fact, with the larger concentration unit giving smaller numerical concentrations for the solutions illustrated, thus compensating for the disappearance of the chelate effect in the numerical values of the stability constants.

Some other factors which must be taken into account in designing ligands with maximum stability and selectivity are summarized in Table III (20) and a specific example of the adverse effect of increasing ring size on ΔH and ΔS is shown in Table IV (37). Mutual coulombic repulsions between donor groups in the metal chelate are important, and the extent to which these repulsions are partially overcome in the free chelating ligand relative to analogous unidentate ligands is a manifestation of the enthalpy-based chelate effect. This property, which greatly increases stability constants, is developed to an even higher degree in macrocyclic and cryptate ligands that hold the donor groups at geometric positions relatively close to the positions that they would assume in the chelate. Thus stability and specificity would be increased in all types of multidentate ligands by synthesizing structures in which the freedom of the donor groups to move away from each other is decreased as much as

TABLE III
Factors Influencing Solution Stabilities of Complexes

Enthalpy Effects	Entropy Effects
Variation of bond strength with electronegativities of metal ions and ligand donor atoms	Number of chelate rings
Ligand field effects	Size of the chelate ring
Enthalpy effects related to the conformation of the uncoordinated ligand	Arrangement of chelate rings
Steric and electrostatic repulsions between ligand donor groups in the complex	Changes of solvation on complex formation
Other coulombic forces involved in chelate ring formation	Entropy variations in uncoordinated ligands
	Effects resulting from differences in configurational entropies of the free ligand and the ligand in complex compounds

TABLE IV
Variation of Thermodynamic Constants as a Function of Chelate Ring Size

$$Ca^{2+}(aq) + (^{-}OOCCH_2)_2N\text{-}(CH_2)_n\text{-}N(CH_2COO^{-})_2 \rightleftharpoons [Ca\ chelate]^{2-}$$

	K	$-\Delta H°$(kcal/mole)	$\Delta S°$(cal/degree mole)
n = 2	10.7	6.55	26.6
3	7.28	1.74	27.4
4	5.66	0.9	29.7
5	5.2	-	-
8	4.6	-	-

possible. One of the obvious ways to achieve this objective is to synthesize organic ligands having rigid molecular frameworks, as may be achieved by use of unsaturated linkages and aromatic rings in the bridging groups of the ligands. Increase in rigidity of the ligand would also minimize or remove completely the unfavorable entropy effects related to the decrease of vibrational and rotational freedom of ligand atoms that generally occurs in metal ion coordination.

It is obvious that many of the factors listed in Table III (e.g., number of rings, chelate ring shape) are not sensitive to properties of the metal ion and therefore will not provide differences in metal-ligand interaction. Selectivity in metal complex formation, as measured by differences in stability constants, requires the use of factors that are very sensitive to the nature of the metal ion. The most effective ways to achieve selectivity in chelate formation with multidentate ligands is to change the nature of the donor group in such a way as to change the degree of covalency of the metal ligand bonds formed. Thus the matching of the a and b character of the metal ion and ligand donor atoms would take advantage of differences in electronegativity and polarizability of the metal ions under consideration. Those factors that are sensitive to only the size of the metal ion (e.g., size of the chelate rings) would probably be less effective in achieving selectivity for open-structured ligands, but may be much more effective for macrocyclic anc cryptate ligands. For metal ions differing in ionic charge and/or coordination number there is no difficulty in achieving high degrees of selectivity, even with relatively simple chelating ligands.

Geometric Constraints of Chelate Design

There are often geometric restrictions on the construction of compact, multidentate chelating agents which prevent the most efficient utilization of donor groups or atoms. Several authors (38,39) have discussed the geometric constraints and pointed out the unique ability of the nitrogen atom (and the other Group V elements) to act as both a donor atom and as a linear or bifurcate link between other coordinating groups, so permitting the construction of compact chelate structures. EDTA and its analogs are classic examples of bifurcate chelators. Unfortunately, oxygen does not perform this dual function and DFB is an example of an oxygen based chelator where creation of the optimal octahedral geometry of a hexadentate complex can only be achieved at the expense of introducting two essentially superfluous, eleven-carbon chains (Figure 2). Therefore, if one is to utilize the strong affinity of oxygen ligands for iron(III), it is necessary either to incorporate nitrogen into the chelate system or to mimic DFB by using large and superfluous chain links. Three interesting examples of synthetic hexadentate chelators based on catechol and designed to mimic the natural microbial agent enterobactin are shown in Figure 5 (9,40,41,42,43).

Bioavailability and Biostability

The efficacy of an iron chelating drug which is administered orally will be dependent on the extent to which it is absorbed in the gastrointestinal tract. In man the gastrointestinal pH varies from *ca*. 2 in the stomach, to *ca*. 6.5 in the small intestine and 8 in the colon, while the pH of blood plasma is 7.4. Passage through the gastrointestinal epithelium is, for most drugs, a passive diffusion process which is more facile for the more lipo-

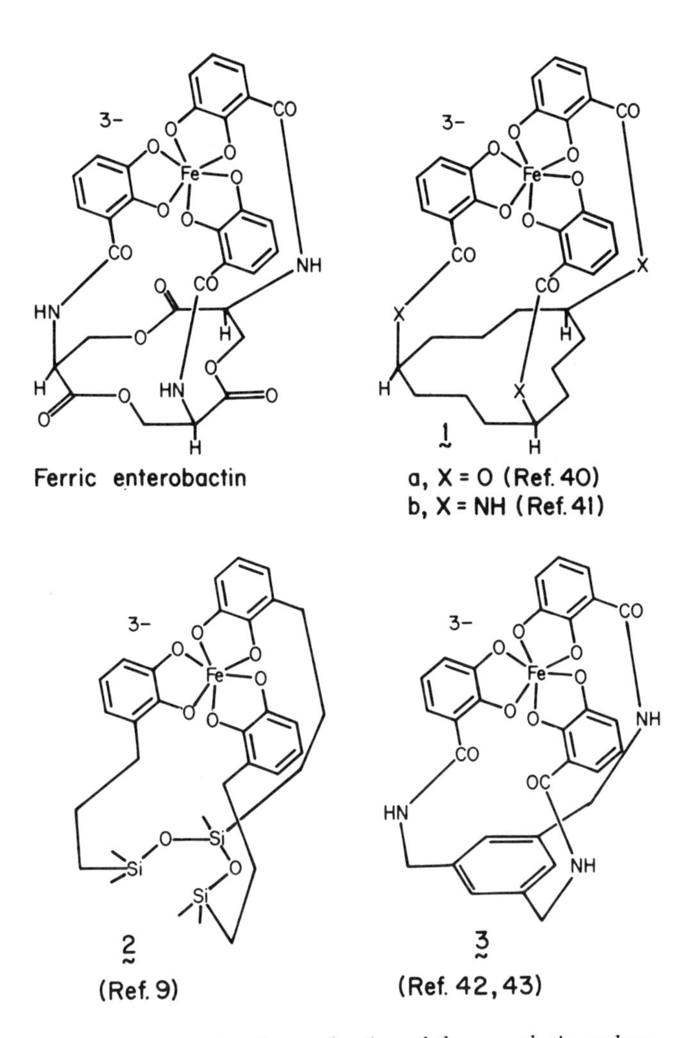

Figure 5. Ferric salts of enterobactin and three synthetic analogs

philic drug. Thus, a drug which is fully ionized in the pH range of the gastrointestinal tract will not pass through the epithelial membrane unless some specific carrier is available. A drug which is partially ionized will pass through this lipo-protein membrane in its unionized form and will be partitioned between the plasma and the gastrointestinal tract to an extent which is dependent on its pK_a and the gastrointestinal pH (44,45). In the case of weak acids

$$\frac{[\mathrm{Drug}]_{\mathrm{plasma}}}{[\mathrm{Drug}]_{\mathrm{GI}}} = (1 + 10^{7.4-pK_a})(1 + 10^{pH-pK_a}) \qquad (16)$$

For these reasons, absorption of an iron chelating drug will be favored if (a) the drug is lipophilic, and (b) the pK_a of any acidic group(s) in the molecule is >3 and the pK_a of any basic group(s) in the molecule if <8, or (c) a specific carrier mechanism is available.

The lipophilicity of a drug is relatively easily manipulated synthetically, without interfering with the iron binding capacity, e.g., by introduction of long chain alkyl substituents. The pK_a is less easily manipulated because it is generally the acidic or basic group which is involved in iron binding; modification of the pK_a may then diminish iron binding capacity. For this reason it is probably desirable to protect strongly acidic and basic groups by temporary conversion to derivatives which regenerate the active functional group after the absorption process. For example, the use of an alkyl ester of a strong acid will eliminate ionization and increase lipophilicity, and subsequent action of plasma esterases will serve to regenerate the free acid. Various examples of this technique, know as "drug latentiation", have been documented by Harper (46) and more recently by Sinkula and Yalkowsky (47). Interestingly, Catsch (48) has suggested that the improved efficacy of certain EDTA-like chelating agents in promoting heavy metal excretion may be related to their ability to form internal esters with enhanced bioavailability; e.g.,

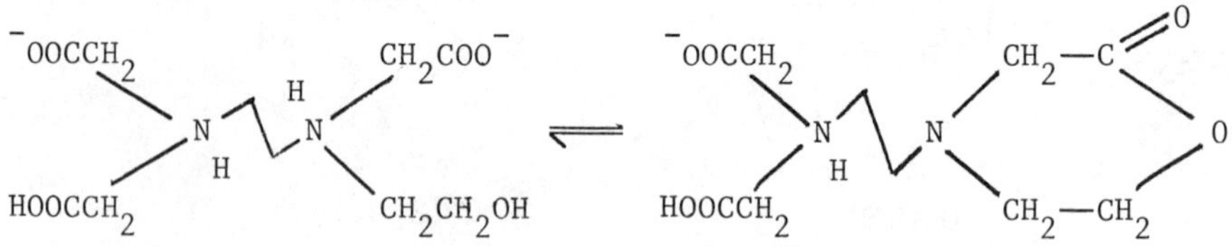

$^{-}OOCCH_2$, $HOOCCH_2$ — N(H) — CH_2CH_2 — N(H) — CH_2COO^{-}, CH_2CH_2OH $\rightleftharpoons$ $^{-}OOCCH_2$, $HOOCCH_2$ — N(H) — CH_2CH_2 — N — (CH_2—C(=O)—O—CH_2—CH_2, ring)

The oral efficacy of at least one iron chelating drug, EHPG (see Figure 1), is markedly enhanced by temporarily blocking the carboxylic acid groups as the alkyl ester.

Following its absorption from the gastrointestinal tract, an iron chelating drug will be transported to the liver *via* the portal vein. Here the drug must survive "detoxification" by microsomal oxidation, conjugation (in the biochemical sense), and excretion, prior to entering the peripheral circulatory system. For example, a number of potentially useful iron-chelating drugs are derivatives of phenols, carboxylic acids, and amines. Carboxylic acids, being a metabolic end product, are stable to microsomal oxidation but are subject to conjugation. Phenols are similarly subject to conjugation and also to microsomal hydroxylation. Amines are subject to conjugation (if primary or secondary), and to oxidation and N-dealkylation.

Conjugation may be blocked to some degree by the same procedure employed to enhance absorption, i.e., administration of the iron chelating drug in the form of a derivative which has transitory *in vivo* stability and reverts to the free drug in the circulatory system (44).

The loss of drug due to microsomal metabolism may also be reduced by molecular modification of the drug. However, it is more difficult to accomplish because of the difficulty in predicting *a priori* what the major metabolic pathway of a specific drug will be. If this can be determined experimentally, substituents can often be introduced (or removed) to block an undersirable metabolic process. The blockage of microsomal hydroyxlation of steroids at C-6 by introduction of a substituent at C-6, or at C-16 by introduction of a substituent at C-17, are cases in point (49). Studies of the metabolism of three iron chelating drugs, EHPG, HBED and N-methyl-N-(2-hydroxybenzyl)glycine, are in progress and will hopefully provide some information on what factors determine their relative efficacy *in vivo*, and how their efficacy may be improved (50).

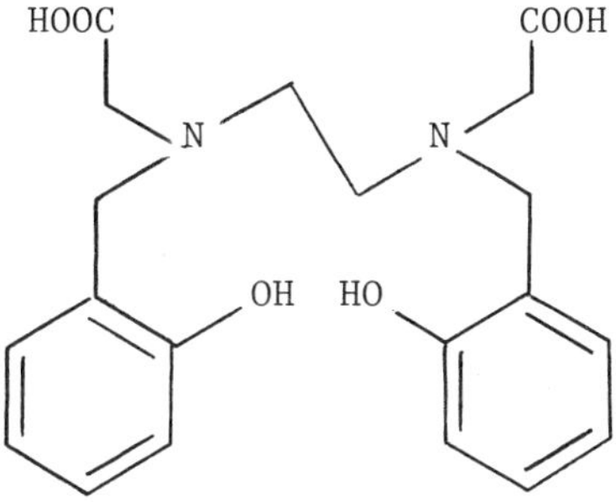

HBED

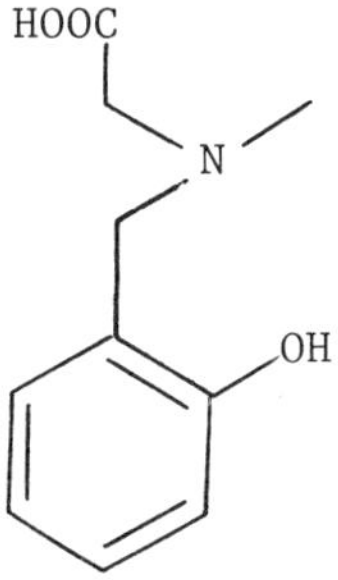

N-methyl-N-(2-hydroxybenzyl)-glycine

The requirement that the chelating agent and its iron(III) chelate not interfere with cellular biochemistry must be considered. Drugs which can pass through the gastrointestinal epithelium are also likely to penetrate other cell membranes. A drug which is suitable for oral administration may therefore have undesirabel side effects unless its lipophilic character is rapidly reduced in vivo. This may be possible if its lipophilicity is associated with an ester function which is rapidly cleaved by plasma esterases, and if the drug acquires ionic charge on binding iron.

In some cases it is possible to control the tissue distribution of the chelating agent DFB has been derivatized as the long chain fatty acyl amide at the terminal amino group and this modification has been shown to enhance biliary excretion of the drug (51). The principal of enterohepatic chelating agents has been proposed (52). Here the basic features which must be exploited in the design of such agents are those which (1) make the chelating agent large and sufficiently non-polar to prevent renal filtration and/or (2) provide it with a structure which permits it to participate in the enterohepatic circulation of the bile acids and their derivatives. The circulation ceases when the chelator acquires charge and greater hydrophilicity on binding Fe(III), and reabsorption in the intestine is no longer favorable. Advantages of this approach are the use of chelating agents more likely to be orally active, avoidance of nephrotoxicity, and more effective localization in the liver, an organ where iron overload can be acute. Cholylhydroxamic acid (53) may be an example of an iron chelating agent which functions by this mechanism.

Some long chain alkyl derivatives of chelating agents have been prepared with the idea of promoting localization in the myocardium (54). This has obvious relevence to iron chelation therapy because of the fatal susceptibility of the heart to iron overload.

Liposomal encapsulation of the chelating agent prior to injection is an alternative means of localizing it in the liver and spleen, an approach which is already being utilized successfully for the administration of drugs to treat leichmaniasis (55). DTPA has been encapsulated with liposomes and in this form is reported to be more effective than the non-encapsulated chelator in removing plutonium in mice (56).

The attachment of iron chelating ligands to polymers is an alternative means of modifying bioavailability. DFB has been covalently bonded to poly(acrolein) and other synthetic polymers and shown to have some potential for use in extracorporeal detoxification of acute iron overloaded plasma (57). Poly(N-methacryloyl-β-alanine hydroxamic acid), a polydentate polymer obtained by derivatization of poly(acrylic acid) with pendant hydroxamic acid groups, has shown significant iron chelation activity in vivo (58), a result which is possibly related to the longer retention of polymeric species in the circulatory system.

Finally, it should be mentioned that continuous infusion of DFB has proven far more effective than single injections for the same dose (59,60). This may be because DFB is only able to remove iron from certain body pools, which are replenished relatively slowly by iron from other body pools. Continuous infusion serves to maintain an effective plasma level of DFB, which would otherwise be reduced by the rapid metabolism of the drug.

Evaluation of Iron Chelating Agents Using Animal Models of Iron Overload

A significant amount of information on the relative efficencies of iron chelating drugs *in vivo* has been collected recently using two animal models to simulate the condition of iron overload (7,61,62,63). Both animal models utilize ip injections of heat damaged red cells to achieve overload. In one screen rat is the test animal (7,61), drug administration and transfusions are concurrent, and efficacy is measured by the iron excreted in the urine and feces. The second screen uses the mouse (62,63), drug administration is initiated two days after transfusions are complete, and efficacy is based on percent iron depleted from the liver and spleen, plus urinary iron excretion. Both screens can be extended to include evaluation of other tissues.

Published results (53,61,63,64) obtained using these animal screens are shown in Tables V-VII, and DFB serves as a useful standard in both screens. It is apparent that there are some differences in the results of the two screens; in particular, 2,3-dihydroxybenzoic acid and rhodotorulic acid are active in the rat screen but not in the mouse screen. This discrepancy possibly arises because of the difference in the transfusion-drug administration sequence in the two screens, for it is known that iron becomes progressively less accessible to chelation as the time between transfusion and drug administration increases.

Most of the compounds listed in Tables V-VII are known or can be expected to have a high and selective affinity for iron(III), with β and K_{eff} values of at least 10^{28} and 10^{20}, respectively. Despite this, many of the compounds tested are only weakly active while others show relative activities not predicted by their stability constants. Of the compounds shown in Figure 1, 3-isopropyltropolone, 1,8-dihydroxynaphthalene and acetohydroxamic acid are essentially inactive, while 2,3-dihydroxynaphthalene-6-sulfonic acid and 8-hydroxyquinoline are active in both screens. EHPG ($\beta = 10^{33.9}$, $K_{eff} = 10^{16.3}$) is more effective than HBED ($\beta = 10^{39.7}$, $K_{eff} = 10^{20.8}$) which is the reverse order of their iron(III) stability constants. These irregularities must be attributed to the importance of factors already referred to, such as bioavailability, biostability, and the kinetics of iron chelation. The increase in the oral efficacy of EHPG on conversion to the dimethyl ester (63) has already been noted. The loss of activity on sulfonation of 8-hydroxyquinoline and 1,8-dihydroxynaphthalene is another illustration of the importance of the ionic form of the chelator.

Hydroxamic acids representative of all of the major structural families of microbial iron chelates (65) have been evaluated. With the possible exception of rhodotorulic acid (*vide supra*) and triacetylfusarinine C, none show activity approaching that of DFB. Several bidentate hydroxamic acids increase iron levels in the

TABLE V
Changes in Hepatic, Splenic and Urinary Iron in Transfused Mice Treated i.p. with Potential Iron Chelators

Compound and Chemical Family	Dose mg/kg	Surviving Animals	% Iron Change vs. Control[a]		
			Spleen	Liver	Urine
Microbial hydroxamic acids					
1. Desferrioxamine B (DFB)	250	10[b]	0	-32[b]	+400
2. Rhodotorulic acid (deferri)	300	10	+19	0	+229
3. Schizokinen (deferri)	300	10	0	0	+42
4. Fusarinine (deferri)	300	10	0	0	+73
5. Mycobactin P (deferri)	400	10	0	0	0
6. Sodium thioformin	300	5[c]	+27	+27	+77
7. Triacetylfusarinine C (deferri)	300	10	0	-18	+247
Synthetic Hydroxamic acids					
8. N-Phenylbenzohydroxamic acid	600	8[c]	+434[d]	+41	0
9. Benzohydroxamic acid	600	8[c]	+194	+17	-39
10. Micotinohydroxamic acid	500	10	+87	0	+18
11 DL-Norleucine hydroxamic acid	500	10	+133	+18	-21
12. L-Histidine hydroxamic acid	300	10	+37	0	0
13. L.Lysine hydroxamic acid hydrochloride	600	10	+41	+18	0
14. N-Hydroxysuccinimide	200	10	-27	0	+84
15. N-Hydroxysuccinamic acid	200	8[c]	0	-30	-64
16. N-Methylacetohydroxamic acid	100[e]	10	+21	-13	-50
17. 2-Hydroxyphenylacetohydroxamic acid	600	10	+9	-8	+7
18. 4-Hydroxyphenylacetohydroxamic acid	600	10	+23	+28	-45
19. Propionohydroxamic acid	300	10	+42	0	0
20. 3,3',3"-Nitrilotris(propionohydroxamic acid) hydrochloride	500	8[c]	+33	+27	0
21. 3,3',3"-Nitrilotris(N-methylpropionohydroxamic acid)	400	7[c]	-7	+1	-2
22. N,N'-Dimethyladipodihydroxamic acid	290	10	0	-10	+33

TABLE V (Continued)

Compound and Chemical Family	Dose	Surviving Animals	% Iron Change vs. Control[a] Spleen	Liver	Urine
23. DL-Phenylalanine hydroxamic acid	265	10	-32	-10	-59
24. Poly(N-methacryloyl-β-alanine hydroxamic acid)[f]	300	10	-20	-52	+380
Phenols					
25. Ethylenediamine-N,N'-bis(2-hydroxyphenylacetic acid) (EDHPA)	25[e]	10	+14	-33	+285
26. EDHPA, dihydrochloride	50[e]	10	0	-54	+1439
27. EDHPA, disodium salt	50[e]	10	0	-26	+191
28. EDHPA, dimethyl ester, dihydrochloride	100[e]	10	-12	-31	+370
29. Ethylenediamine-N,N'-bis(2-hydroxy-5-methylphenylacetic acid)	100	9	0	-60	+698
30. Ethylenediamine-N,N'-bis(5-chloro-2-hydroxyphenylacetic acid)	25[e]	9	+179	-50	+70
31. N,N'-Bis(2-hydroxybenzyl)ethylenediamine, dihydrochloride	150	9	0	0	0
32. N,N'-Bis(2-hydroxybenzyl)ethylenediamine-N,N'-diacetic acid	200	10	0	-22	+119
33. N,N-Bis(2-hydroxybenzyl)propylene-1,3-diamine-N,N-diacetic acid	100	9	0	-35	+88
34. N,N-Bis(2-hydroxybenzyl-glycine methyl ester, hydrochloride	250	9[g]	+11	0	-58
35. N,N-Bis(2-hydroxy-5-sulfobenzyl)glycine trisodium salt	200	10	0	0	+142
36. 8-Hydroxyquinoline	100	7[d]	-50	-20	-60
37. 8-Hydroxyquinoling-5-sulfonic acid	500	5[d]	+25	-16	+308
38. 1,8-Dihydroxynaphthalene	200	8[d]	0	-18	-53
Catechols					
39. 2,3-dihydroxybenzoic acid	100[e]	8[c]	0	0	0
40. N-(2,3-dihydroxybenzoyl)glycine	100	10	0	0	+254
41. 2,3-Dihydroxynaphthalene-6-sulfonic acid, sodium salt	400	10	-39	-26	0

TABLE V (Continued)

Compound and Chemical Family	Dose	Surviving Animals	% Iron Change vs. Control[a]		
			Spleen	Liver	Urine
42. Tetramethyl-1,2-bis(3'-[2',3'-diacetoxyphenyl]-propyl)disiloxane	150	9	0	0	-67
Tropolones					
43. 3-Isopropenyltropolone	150	9[g]	0	+22	0
44. 3-(1'-Methylprop-1'-enyl)tropolone	150[e]	10[c,g]	0	+19	0
45. Tropolone-5-sulfonic acid, ammonium salt	150	10[d]	0	29	+49
46. 4-Isopropyltropolone	100[e]	10	-20	0	0
Miscellaneous Compounds					
47. Di(pyridine-2-carboxaldehydo)azine	50	10	0	0	+222
48. 1-Hydroxyethane-1,1-diphosphonic acid, disodium salt	100	7[c]	+55	+55	+190
49. Diethylenetriaminepentaacetic acid (DTPA)	200	9	0	0	+131

[a] Difference significant at P <.05-.001. [b] Mean of 14 individual bioassays. [c] Predominant toxic sign was central nervous system inhibition. [d] Splenomegaly. [e] Toxic effects at this or higher dose. [f] Submitted for test by Dr. Anthony Winston (Winston and McLaughlin, 1976). [g] Predominant toxic sign was central nervous system stimulation.

TABLE VI
Urinary and Fecal Iron Excretion Promoted by Chelating Agents Administered by Iron Overloaded Rats

Compound	LD 50[a]	Route of[b] Adminis-tration	No. of Animals	Relative Iron Excretion Urine	Feces
Primary aliphatic hydroxamic acids					
Hexanohydroxamic acid	>400 mg/kg	i.p.	6	1	0
Octanohydroxamic acid	>800 mg/kg	i.p.	6	1	1
Decanohydroxamic acid	>800 mg/kg	i.p.	7	1	0
Glycylhydroxamic acid	800 mg/kg	p.o.	9	0	1
Glycylhydroxamic acid	800 mg/kg	i.p.	4	1	1
Histidylhydroxamic acid	3.7 g/kg	i.p.	6	1	1
Acetohydroxamic acid	>800 mg/kg	i.p.	6	0	0
Cholylhydroxamic acid	>1.6 g/kg	p.o.	26	0	3
Primary aromatic hydroxamic acids					
Salicylohydroxamic acid	>400 mg/kg	i.p.	12	1	0
o-Aminobenzhydroxamic acic	>800 mg/kg	i.p.	10	0	1
Nicotinohydroxamic acid	2.3 g/kg	i.p.	6	1	1
2,3-Dihydroxybenzhydroxamic acid	>800 mg/kg	i.p.	5	0	1
2,5-Dihydroxygenzhydroxamic acid	>800 mg/kg	i.p.	6	0	1
Secondary Aliphatic hydroxamic acids					
Desferrioxamine	>800 mg/kg	i.p.	6	1	2
δ-N-acetyl-δ-N-hydroxyornithine	>100 mg/kg[c]	p.o.	12	0	0
Rhodotorulic acid	>800 mg/kg	p.o.	12	1	0
Rhodotorulic acid	>800 mg/kg	i.p.	12	2	3
Derivatives of salicylic acid					
Salicylic acid	1.3 g/kg[d]	p.o.	14	0	0
Acetylsalicylic acid	1.75 g/kg[d]	p.o.	8	0	0
3-Methylsalicylic acid	>300 mg/kg	p.o.	12	0	0
5-Methylsalicylic acid	>300 mg/kg	p.o.	6	0	0

TABLE VI (continued)

Compound	LD 50[a]	Route of[b] Adminis-tration	No. of Animals	Relative Iron Excretion	
				Urine	Feces
Derivatives of 2,3-dihydroxybenzoic acid					
2,3-Dihydroxybenzoic acid	>3 g/kg[d]	p.o.	35	1	0
2,3-Diacetoxyvenzoic acid	>800 mg/kg	p.o.	9	1	0
2,3-Dihydroxybenzoic acid, methyl ester	≃600 mg/kg	i.p.	10	0	2
2,3-Dihydroxybenzoylglycine	>800 mg/kg	p.o.	9	1	2
2,3-Dihydroxybenzoylglycine, ethyl ester	>800 mg/kg	p.o.	11	1	0
2,3-Dihydroxybenzoylglycine, ethyl ester	>800 mg/kg	i.p.	6	0	2
2-Hydroxy-3-methoxybenzoic acid	2.1 g/kg[e]	p.o.	9	1	1
2,3-Dimethoxybenzoic acid	>800 mg/kg	p.o.	12	1	0
Other dihydroxybenzoic acids and derivatives					
2,5-Dihydroxybenzoic acid	>800 mg/kg	p.o.	6	0	0
2,4-Dihydroxybenzoic acid	>800 mg/kg	p.o.	9	1	0
2-Hydroxy-4-acetoxybenzoic acid	>800 mg/kg	p.o.	10	1	0
2,6-Dihydroxygenzoic acid	>600 mg/kg	p.o.	6	1	0
3,4-Dihydroxybenzoic acid	>800 mg/kg	p.o.	11	1	0
3,4-Dihydroxybenzamide	>800 mg/kg	p.o.	8	0	0
3,4-Diacetoxybenzoic acid	>800 mg/kg	p.o.	22	1	0
3-Methoxy-4-hydroxybenzoic acid	>600 mg/kg	p.o.	12	1	0
3-Hydroxy-4-methoxybenzoic acid	3 g/kg[e]	p.o.	1]	1	1
3-Methoxy-4-acetoxybenzoic acid	>800 mg/kg	p.o.	12	1	0
3,4-Dimethoxybenzoic acid	>800 mg/kg	p.o.	12	1	0
3,4-Methylenedioxybenzoic acid	>800 mg/kg	p.o.	9	0	0
3,4-Dietmhoxybenzoic acid, ethyl ester	>600 mg/kg	p.o.	11	1	1
3-Methoxy-4-hydroxybenzoic acid, ethyl ester	5 k/kg	p.o.	12	1	1
3,4-Dimethoxybenzamide	>800 mg/kg	p.o.	6	0	0
3,5-Dihydroxybenzoic acid	>800 mg/kg	p.o.	6	1	0
Derivatives of trihydroxybenzoic acids					
2,3,4-Trihydroxybenzoic acid	>800 mg/kg	p.o.	6	0	0
2,4,6-Trihydroxybenzoic acid	>800 mg/kg	p.o.	6	1	0

TABLE VI (continued)

Compound	LD 50[a]	Route of[b] Adminis-tration	No. of Animals	Relative Iron Excretion	
				Urine	Feces
3,4,5-Trihydroxybenzoic acid	>800 mg/kg	p.o.	20	0	1
Derivatives of phenylacetic acid					
o-Hydroxyphenylacetic acid	>800 mg/kg	p.o.	6	0	1
3,4-Dihydroxyphenylacetic acid	>800 mg/kg	p.o.	12	1	0
Derivatives of β-phenylpropionic acid					
3-(3,4-Dihydroxyphenyl)-propionic acid	>800 mg/kg	p.o.	15	0	1
Derivatives of cinnamic acid					
3,4-Dihydroxycinnamic acid	>800 mg/kg	p.o.	16	0	2
Miscellaneous iron chelators					
Quercetin	>800 mg/kg	p.o.	6	0	0
Quercetin	>800 mg/kg	i.p.	6	0	0
Morin	>800 mg/kg	p.o.	6	0	0
Morin	>800 mg/kg	i.p.	6	0	0
Rutin[d]	950 mg/kg	p.o.	10	1	0
Rutin	950 mg/kg	i.p.	9	1	1
Tropolone	≈150 mg/kg	p.o.	6	0	1
Tropolone	≈150 mg/kg	i.p.	12	0	2
Purpurogallin	≈400 mg/kg	i.p.	6	0	0
D-(-)-Penicillamine	>800 mg/kg	i.p.	6	0	1
N-acetyl-DL-penicillamine	>800 mg/kg	i.p.	6	0	1
2-(o-Hydroxyphenyl)-benzoxazole	>800 mg/kg	i.p.	6	0	0
2-(o-Hydroxyphenyl)-benzothiazole	>800 mg/kg	i.p.	6	0	0
8-Hydroxyquinoline-5-sulfonic acid[d]	>800 mg/kg	p.o.	10	1	0
8-Hydroxyquinoline-6-sulfonic acid	>800 mg/kg	i.p.	6	1	1
2,3-Dihydroxynaphthalene-6-sulfonic acid	>800 mg/kg	p.o.	6	0	0
L-Histidine	>800 mg/kg	p.o.	11	1	2
Cysteamine-HCl	≈500 mg/kg	i.p.	6	0	0
2,3-Dimercaptosuccinic acid	>800 mg/kg	p.o.	6	0	0

TABLE VI (continued)

Compound	LD 50[a]	Route of[b] Adminis-tration	No. of Animals	Relative Iron Excretion	
				Urine	Feces
2,3-Dimercaptosuccinic acid	>800 mg/kg	i.p.	6	0	1
Ellagic acid	>800 mg/kg	p.o.	9	0	2
Kojic acid	1.2 g/kg	p.o.	12	0	0
Maltol	>800 mg/kg	p.o.	9	0	0
Hesperidin	>600 mg/kg	p.o.	12	1	0
α,α-Dipyridyl	200 mg/kg	i.p.	6	1	1
1,10-Phenanthroline	75 mg/kg	i.p.	6	1	0

[a] Intraperitoneally in mice except as noted. [b] All drugs were administered at a dose of 100 mg/kg/day for 5 days except α,α-dipyridyl (40 mg/kg/day); 1,10-phenanthroline (15 mg/kg/day); and tropolone (20 mg/kg/day). [c] Orally in mice. [d] Orally in rabbit. [e] Intraperitoneally in rat.

TABLE VII
Compounds Which Failed to Show Significant Activity in Transfused Mouse Screen

Hydroxamic Acids

Ferrichrome A (deferri), potassium salt
Hadacidin
Acetohydroxamic acid
DL-Tyrosine hydroxamic acid
Isobutyrohydroxamic acid
Glutamohydroxamic Acid
Aerobactin
Salicylohydroxamic acid
Glycine acid
DL-Serine hydroxamic acid
DL-Trypophan hydroxamic acid
Phenylacetohydroxamic acid
DL-Threonine hydroxamic acid

Phenols

N,N-Bis(2-hydroxybenzyl)propylamine
N-(2-hydroxybenzyl)glycine)
1,8-Dihydroxynaphthalene-3,7-disulfonic acid, sodium salt
N,N,N',N'-Tetrakis(2-hydroxybenzyl)ethylenediamine
N,N,N'-Tris(2-hydroxybenzyl)ethylenediamine-N'-acetic acid, dihydrochloride

Miscellaneous

Tropolone
1'-Oxo-2'-formylethylbenzene-1'-thiosemicarbazone-2'-oxime
2'-Oxopropanal-2'-thiosemicarbazone-1'-oxime
4-Chloro-1'-oxo-2'-formylethylbenzene-1'-thiosemicarbazone-2'-oxime
Di(thiophene-2-carboxyldehydo)azine
4-Chloro-1'-oxo-2'-formylethylbenzene-2'-thiosemicarbazone-2'-oxime
1,1,3,5,5-Pentamethyl-1,3,5-tris[2-(2,3-diacetoxyphenyl)propyl]-trisiloxane

liver and spleen of the mouse, suggesting an ability to mobilize iron in the body but subsequent metabolic degradation of the iron chelate occurs prior to excretion. Cholyhydroxamic acid is unique amongst the bidentate class in promoting significant iron excretion after oral administration. The activity of the polymeric hydroxamic acid (#24, Table V) has already been noted.

The tropolones and catechols which have been evaluated show only marginal activity, including the hexadentate ligand corresponding to structure 2 in Figure 5. The lack of activity of the latter may be a result of the susceptability of the catechol group to oxidation, whereas the 1,2-dihydroxybenzoyl group of the microbial agent enterobactin is stabilized by the presence of the α-carbonyl group. Despite the affinity of the phenolic group for iron(III), its presence does not assure activity and the two hexadentate variants of HBED in Figure 6 are ineffective. More extensive tests with EHPG and its derivatives are in progress to optimize its oral activity and this agent appears to offer a potential improvement over DFB.

The best possibility of identifying other effective chelators would appear to depend on devising new hexadentate structures, particularly cryptate systems. An improvement in the understanding of the metabolism of these chelating agents and their iron complexes will also help optimization of the structure.

Abbreviations

DFB	desferrioxamine B
DHG	N,N-bis(2-hydroxyethyl)glycine
DTPA	diethylenetriaminepentaacetic acid
EDTA	ethylenediaminetetraacetic acid
EHPG	ethylene-N,N'-bis-2-(o-hydroxyphenyl)glycine
HBEA	N,N,N'-tris(2-hydroxybenzyl)ethylenediamine-N'-acetic acid
HBED	N,N'-bis(o-hydroxybenzyl)ethylenediamine-N,N'-diacetic acid
HEDTA	N-(2-hydroxyethyl)ethylenediamine-N,N',N'-triacetic acid
HIMDA	N-(2-hydroxyethyl)iminodiacetic acid
NTA	nitrilotriacetic acid
TEA	triethanolamine
TF	transferrin
THBE	N,N,N',N'-tetrakis(2-hydroxygenzyl)ethylenediamine

Abstract

The starting point for the development of iron chelating drugs is the identification of ligands which have a high and preferential affinity for iron(III). Affinity *in vivo* may be defined in terms of an effective stability constant, K_{eff}, which can be estimated from the stability constant (β) of the iron(III) complex measured under optimum *in vitro* conditions by correcting for the physiological pH and competing endogenous metals and ligands.

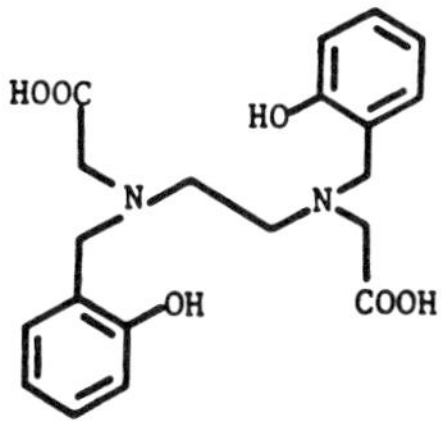

HBED

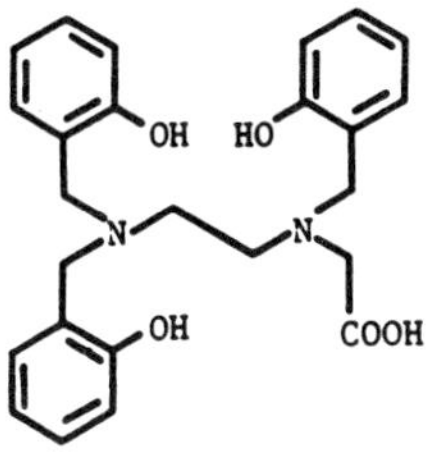

HBEA

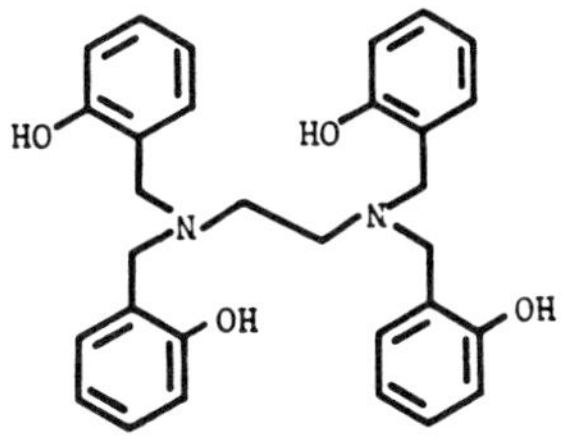

THBE

Figure 6. Two analogs of HBED that failed to show significant activity in transfused mouse screen

Other factors which influence in vivo efficacy are the chelate effect, iron solubilizing ability, kinetic and steric effects, bioavailability and stability, and the chemical state and location of the iron which is to be removed. Using these criteria, the chelating agents with the highest affinity and specificity for iron(III) in vivo should be polydentate and relatively acidic oxy-donors, e.g., derivatives of phenol, catechol, 8-hydroxyquinoline, hydroxamic acid, tropolone, and carboxylic acids such as salicylic acid and glycine. The results of screening such compounds using animal models of iron overload are discussed with the above principles in mind.

Acknowledgement

The authors express their thanks and appreciation to Dr. R. J. Motekaitis of Texas A&M University for carrying out the equilibrium calculations.

Literature Cited

1. Muller-Eberhard, U; Miescher, P.A.; Jaffe, E.R. "Iron Excess. Aberrations of Iron and Porphyrin Metabolism", Grune and Stratton, New York, 1977.
2. Walker, R.J.; Williams, R. "Haemochromatosis and Iron Overload" in "Iron in Biochemistry and Medicine", Jacobs, A; Worwood, M. Eds., Academic Press, New York, 1974.
3. Zaino, E.C. Ed. "Third Conference on Cooley's Anemia", New York Acad. Sci., 1974, p.232.
4. Anderson, W.F.; Hiller, M.C. Ed. "Development of Iron Chelators for Clinical Use", DHEW Publication No. (NIH) 77-994, Bethesda, Md., 1975.
5. Waxman, H.S.; Brown, E.B. Prog. Hematology, 1969, 6, 338.
6. Byers, B.R.; Arceneaux, J.E.L.; Gaines, C.G.; Sciortino, C.V. in "Symposium on the Development of Iron Chelators for Clinical Use", Anderson, W.F.; Hiller, M.C. Eds., DHEW Publication No. (NIH) 77-994, Bethesda, Md., 1975, pp.213-228.
7. Cerami, A.; Graziano, J.H.; Grady, R.W.; Peterson, C.M. in "Symposium on the Development of Iron Chelators for Clinical Use", Anderson, W.F.; Hiller, M.C. Eds., DHEW Publication No. (NIH) 77-994, Bethesda, Md., 1975, pp.261-268.
8. Crumbliss, A.L.; Palmer, R. A.; Sprinkle, K.A.; Whitcomb, D.R. in "Symposium on the Development of Iron Chelators for Clinical Use", Anderson, W.F.; Hiller, M.C. Eds., DHEW Publication No. (NIH) 77-994, Bethesda, Md., 1975, pp.175-205.
9. Pitt, C.G.; Gupta, C. in "Symposium on the Development of Iron Chelators for Clinical Use", Anderson, W.F.; Hiller, M.C. Eds., DHEW Publication No. (NIH) 77-994, Bethesda, Md., 1975, pp.137-174.
10. Martell, A.E.; Smith, R. M. "Critical Stability Constants" Vols.1-4, Plenum, New York, 1977.

11. Lippard, S.J.; Schugar, H.; Walling, C. Inorg. Chem., 1967, 6, 1825.
12. Assa, R.; Malstrom, B.G.; Saltman, P.; Vanngard, T. Biochim. Biophys. Acta, 1963, 75, 203.
13. Schubert, J. in "Iron Metabolism" Gross, F. Ed., Springer-verlag, Berlin, 1964, p.466.
14. May, P.M.; Linder, P.W.; Williams, D.R. J. Chem. Soc. Dalton, 1977, 588.
15. Ringbom, A. J. Chem. Ed., 1958, 35, 282.
16. Schwarzenbach, G. "Die Komplexometrische Titration" F. Einke, Stuttgard, 1956, p.17.
17. Luk, C.K. Biochemistry, 1971, 10, 2838.
18. Peters, G.; Keberle, H.; Schmid, K.; Brunner, H. Biochem. Pharm., 1966, 15, 93.
19. Reilly, C.H.; Schmidt, R.W.; Sadek, F.S. J. Chem. Ed., 1959, 36, 555.
20. Martell, A.E. Pure Appl. Chem., 1978, 50, 813.
21. Doran, M.; Martell, A.E. unpublished studies.
22. Bogucki, R.F.; Martell, A.E. J. Am. Chem. Soc., 1968, 90, 6002.
23. Schugar, H.; Walling, C.; Jones, R.B.; Gray, H.B. J. Am. Chem. Soc., 1967, 89, 3712.
24. Bates, G.W.; Billups, C.; Saltman, P. J. Biol. Chem., 1967, 242, 2816.
25. Pollack, S.; Vanderhoff, G.; Lasky, F. Br. J. Hematology, 1976, 34, 231.
26. Pollack. S.; Vanderhoff, G.; Lasky, F. Biochim. Biophys. Acta, 1977, 497, 481.
27. Morgan, E.H. Biochim. Biophys. Acta, 1977, 499, 169.
28. Margerum, D.W. in "Coordination Chemistry" Vol.2, ACS Monograph 174, Martell, A.E. Ed., American Chemical Society, Washington, D.C., 1978, pp.174-194.
29. Pape, L.; Multani, J.S.; Stitt, S.; Saltman, P. Biochem., 1968, 7, 613.
30. Harrison, P.M.; Hoy, T.G.; Macara, I.G.; Hoare, R.J. Biochem. J., 1974, 143, 445.
31. Crichton, R.R.; Roman, F. J. Mol. Catalysis, 1978, 4, 75.
32. Anderegg, G.; L'Eplattenier, F.; Schwarzenbach, G. Helv. Chim. Acta, 1963, 46, 1400.
33. Schwarzenbach, G.; Schwarzenbach, K. Helv. Chim. Acta, 1963, 46, 1390.
34. Adamson, A.W. J. Am. Chem. Soc., 1954, 76, 1578.
35 Jameson, R.F. in "An Introduction to Bio-Inorganic Chemistry" Williams, D.R. Ed., C.C. Thomas, Springfield, 1976.
36. Munro, D. Chem. Britain, 1977, 13, 100.
37. Anderegg, G. Helv. Chim. Acta, 1964, 47, 1801.
38. Black. D.St.; Hartshorn, A.J. Coord. Chem. Rev., 1972, 9, 219.
39. Lyons, F. Rev. Chem. Progr., 1961, 22, 69.

40. Collins, D.J.; Lewis, G.; Swan, J.M. Aust. J. Chem., 1974, 27, 2593.
41. Corey, E.J.; Hurt, S.D. Tetrahedron Lett., 1977, 3923.
42. Venuti, M.C.; Rastesster, W.H.; Neilands, J.B. J. Med. Chem. 1979, 22, 123.
43. Weitl, F.L.; Raymond, K.N. J. Am. Chem. Soc., 1979, 101, 2728.
44. Schanker, L.S.; Tocco, D.J.; Brodie, B.B.; Hogben, C.A. J. Pharmacol. Exp. Ther., 1958, 123, 81.
45. Goldstein, A.; Aronow, L.; Kalman, S.K. "Principles of Drug Action", Wiley, New York, 1974.
46. Harper, N.J. Progr. Drug. Res., 1962, 4, 221.
47. Sinkula, A.A.; Yalkowsky, S.F. J. Pharm. Sci., 1975, 64, 181.
48. Catsch, A. Fed. Proc., 1961, 20, Suppl.10, pt. II, 206.
49. Deghenghi, R.; Munson, A.J. in "Medicinal Chemistry" Burger, A. Ed., Wiley-Interscience, New York, 1970, p.900.
50. Cook C.E. Research Triangle Institute, unpublished studies.
51. Meyer-Brunot, H.G.; Keberle, H. Amer. J. Physiol, 1968, 214, 1193.
52. Jones, M.M.; Pratt, T.H.; Mitchell, W.G.; Harbison, R.D.; McDonald, J.S. J. Inorg. Nucl. Chem., 1976, 38, 613.
53. Grady, R.W.; Graziano, J.H.; White, G.P.; Jacobs, A.; Cerami, A. J. Pharmacol. Exp. Ther., 1978, 205, 757.
54. Eckelman, W.C.; Karesh, S.M.; Reba, R.C. J. Pharm. Sci., 1975, 64, 704.
55. Alving, C.R.; Steck, E.A.; Hanson, W.L.; Loizeaux, P.S.; Chapman, Jr., W.L.; Waits, V.B. Life Sciences, 1978, 22, 1021.
56. Rahman, Y.E.; Rosenthal, M.W.; Cerny, A.E. Science, 1973, 180, 300.
57. Ramirez, R.S.; Andrade, J.D. J. Macromol. Sci. Chem., 1976, A10, 309.
58. Winston, A.; McLaughlin, G.R. J. Poly. Sci., 1976, 14, 2155.
59. Copper, B.; Bunn, H.F.; Propper, R.D.; Nathan, D.G.; Rosenthal, D.S.; Maloney, W.C. Amer. J. Med., 1977, 63, 958.
60. Propper, R.D.; Cooper, B.; Rudo, R.R. N. Engl. J. Med., 1977, 297, 418.
61. Graziano, J.H.; Grady, R.W.; Cerami, A. J. Pharmacol. Exp. Ther., 1974, 190, 570.
62. Gralla, E.J. in "Symposium on the Development of Iron Chelators for Clinical Use", Anderson, W.F.; Hiller, M.C. Eds., DHEW Publication No. (NIH) 77-994, Bethesda, Md., 1975, pp.229-254.
63. Pitt, C. G.; Gupta, G.; Estes, W.E. et al. J. Pharmacol. Exp. Ther., 1979, 208, 12.
64. Grady, R. W.; Graziano, J.H.; Akers, H.A.; Cerami, J. Pharmacol. Exp. Ther., 1976, 196, 478.
65. Neilands, J.B. 178th National Meeting, American Chemical Society, Washington, D.C., Sept.9-14, 1979, Paper No. Inorg. 36.

RECEIVED May 22, 1980.

18

The Synthesis, Thermodynamic Behavior, and Biological Properties of Metal-Ion-Specific Sequestering Agents for Iron and the Actinides

KENNETH N. RAYMOND[1], WESLEY R. HARRIS, CARL J. CARRANO, and FREDERICK L. WEITL

Department of Chemistry and Materials and Molecular Research Division, Lawrence Berkeley Laboratory, University of California, Berkeley, CA 94720

Certain types of anemia require regular transfusions of whole blood, since the victims of these diseases cannot correctly manufacture their own. One such disorder, β-thalassemia, is described in more detail elsewhere in this publication (1). Because the body lacks any mechanism for excreting significant amounts of serum iron, the iron contained in transfused blood (∿ 250 mg per pint) can accumulate to lethal levels. At this time the only treatment for Cooley's anemia is continual transfusion and so some way must be found to efficiently remove this excess iron. Thus there is an obvious need for a drug which can selectively bind iron *in vivo* and facilitate its excretion.

Anderson has described the current use of desferrioxamine B for clinical iron removal in the treatment of β-thalassemia (1). Desferrioxamine B belongs to the class of compounds called siderophores (2), which were discussed by Neilands (3). Siderophores are produced by microorganisms for the purpose of binding exogenous ferric ion and facilitating its transport across the cell membrane. These compounds generally utilize either hydroxamic acid or catechol groups to bind ferric ion and form very stable, high spin, octahedral complexes. Desferrioxamine B is one of the hydroxamate type siderophores (Figure 1). It is a linear molecule consisting of alternating units of succinic acid and 1,5-diaminopentane, which combine to give three hydroxamic acid groups. Although the effectiveness of desferrioxamine therapy has been improved, there appear to be fundamental limitations to its potential for iron removal which are probably inherent to hydroxamates in general.

This led to our interest in the second major type of siderophores, the catechols. The best known member of this class of compounds is enterobactin, a cyclic triester of 2,3-dihydroxybenzoylserine, shown in Figure 2. Although there were indications that enterobactin formed very stable ferric complexes (4, 5), the formation constant of ferric enterobactin had never been

[1]Author to whom correspondence should be addressed.

0-8412-0588-4/80/47-140-313$05.00/0

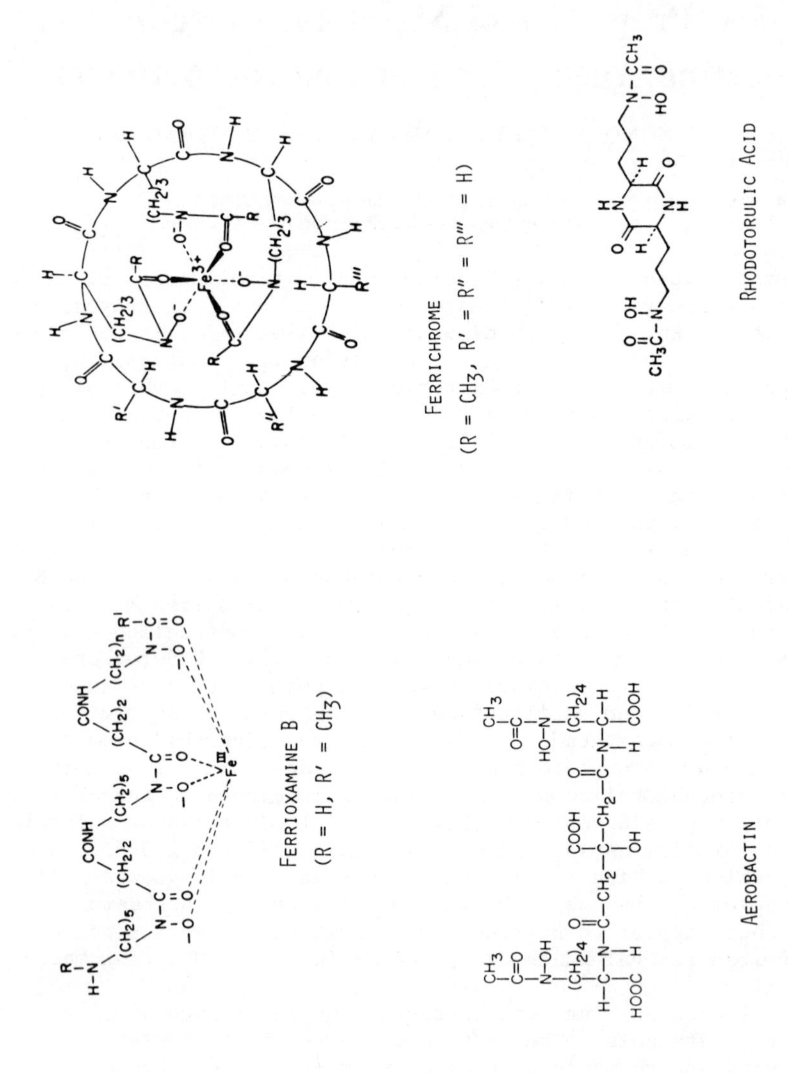

Figure 1. Structural formulas of representative types of hydroxamate siderophores

determined. Therefore, our first priority was to determine the iron affinity of enterobactin.

Solution Equilibria of Ferric Enterobactin

The potentiometric titration curve of ferric enterobactin, shown in Figure 3, has a sharp inflection after the addition of six equivalents of base. Such a break indicates that the six phenolic oxygens from the three dihydroxybenzoyl groups are displaced by ferric ion in the ferric enterobactin complex. This interpretation is further supported by the absorbance maximum at 490 nm (ε 5600), which is very similar to simple tris(catecholato)iron(III) complexes (6, 7). The very low pH at which complexation of enterobactin occurs, with virtually complete complex formation by pH 6, is a strong indication of a very stable complex. However, the titration is prematurely terminated at pH 3.8 by the precipitation of a purple neutral iron complex (whose composition and structure will be discussed later) which makes it impossible to determine the stability constant of ferric enterobactin from potentiometric data alone.

Instead, the stability constant of enterobactin has been determined spectrophotometrically by competition with EDTA, as described by the equation:

$$Fe(ent)^{3-} + EDTA^{4-} + 6H^{+} \overset{K_x}{\rightleftharpoons} Fe(EDTA)^{-} + H_6ent \qquad (1)$$

It is necessary to take advantage of the strong pH dependence of Eq. 1. At neutral or basic pH, this equilibrium lies completely on the side of ferric enterobactin. At pH 5, however, a measurable distribution of ferric ion is obtained with less than a tenfold excess of EDTA. The intense charge transfer band of ferric enterobactin provides a convenient way of determining the concentration of $Fe(ent)^{3-}$, and the remaining concentrations are obtained from mass balance considerations. Using the literature value for the formation constant of ferric EDTA (8), one can calculate a value of the proton-dependent equilibrium constant

$$K_6^* = \frac{[Fe(ent)^{3-}][H^+]^6}{[Fe^{3+}][H_6ent]} = 10^{-9.7(2)} \qquad (2)$$

In order to convert K_6^* into the conventional, i.e. proton independent, formation constant, it is necessary to know the six ligand protonation constants of enterobactin. Unfortunately, the ester groups of enterobactin are extremely susceptible to base-catalyzed hydrolysis, which precludes independent measurement of ligand pKa's. However, we have measured the protonation constants of the bidentate ligand 2,3-dihydroxy-N,N-dimethylbenzamide as log K_1^H = 12.1 and log K_2^H + 8.4. By using these values as estimates for the enterobactin protonation constants, we have estimated the

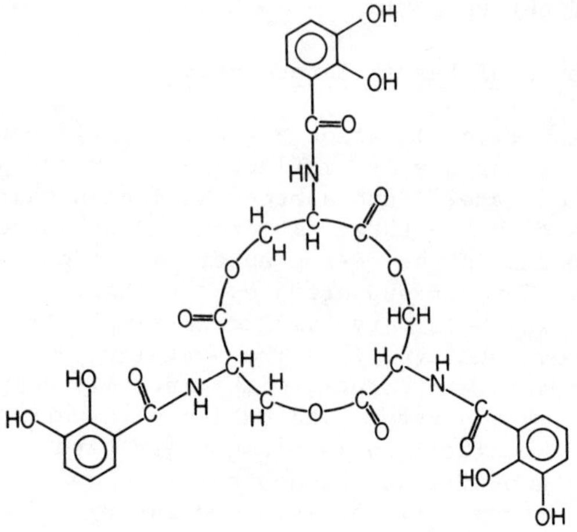

Figure 2. Structural formula of the tricatecholate siderophore enterobactin

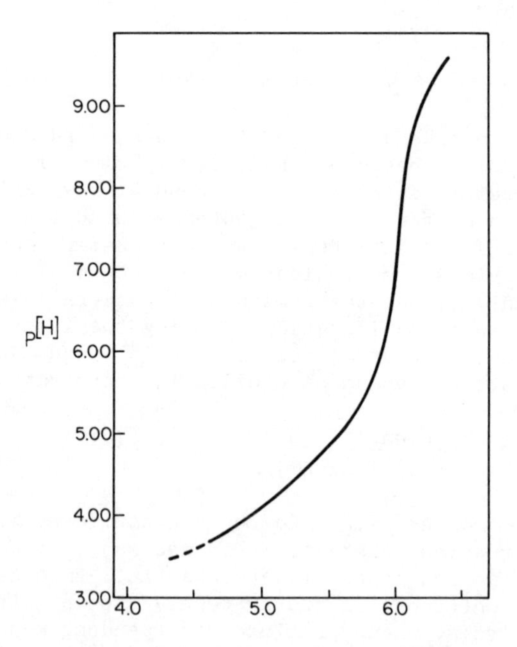

Figure 3. Potentiometric equilibrium curve of a 1:1 solution of ferric enterobactin: T = *25°C;* μ = *0.10*M *(KNO_3);* a = *moles of base per mole of iron*

overall formation constant of ferric enterobactin to be

$$K_{ML} = \frac{[Fe(ent)^{3-}]}{[Fe^{3+}][ent^{6-}]} \cong 10^{52} \quad (3)$$

This is the largest formation constant ever reported for a ferric complex and indicates the exceptional stability of ferric enterobactin.

Because of the very weak acidity of the phenolic oxygens of enterobactin, the full impact of log K_{ML} would be realized only above pH 12. What we are really interested in is the ligand's ability to sequester ferric ion at physiological pH. We also need some method of comparing the relative effectiveness of various ligands which can take into account changes in ligand protonation constants, the hydrolysis of some ferric complexes, and the formation of polynuclear species in some systems. Thus we have chosen to compare ligands by calculating the equilibrium concentration of free hexaaquoiron(III) in a pH 7.4 solution which is 1 μM in iron and 10 μM in ligand. The results are expressed as pM values (pM = - log $[Fe(H_2O)_6{}^{3+}]$), with a larger pM indicative of a more stable complex under the prescribed conditions. This gives a direct measurement of the relative iron binding affinity, since in a competition between two ligands under the conditions stated, the ligand with larger pM will dominate. Table I lists pM values of all the siderophores for which formation constants are known. [Note that the pM value for enterobactin is determined directly from the observed equilibrium constant $K_6{}^*$ (Eq. 2) and not from the estimated K_{ML} (Eq. 3).] Although the trihydroxamic acid siderophores such as ferrioxamine B and ferrichrome form very stable complexes, it is clear that enterobactin not only has a much larger formation constant, but is also many orders of magnitude more effective at sequestering ferric ion at physiological pH. Although enterobactin itself is not suitable for chelation therapy (due in part to the facile hydrolysis of its ester linkages at physiological pH) it does represent a uniquely promising model on which to base the design of new synthetic ferric ion sequestering agents.

Synthetic Analogues of Enterobactin

With these results in mind, we have prepared and evaluated a number of catecholate type ligands. The structural formulas for these ligands are shown in Figure 4. Like enterobactin, all these ligands can bind ferric ion via six phenolic oxygens from three catechol groups. Unlike enterobactin, they are also hydrolytically stable over normal pH ranges.

The sulfonation of these types of ligands is designed to serve a number of purposes. It stabilizes the catechol groups against oxidation to the corresponding quinone, and also increases the otherwise very low water solubility of these ligands. In

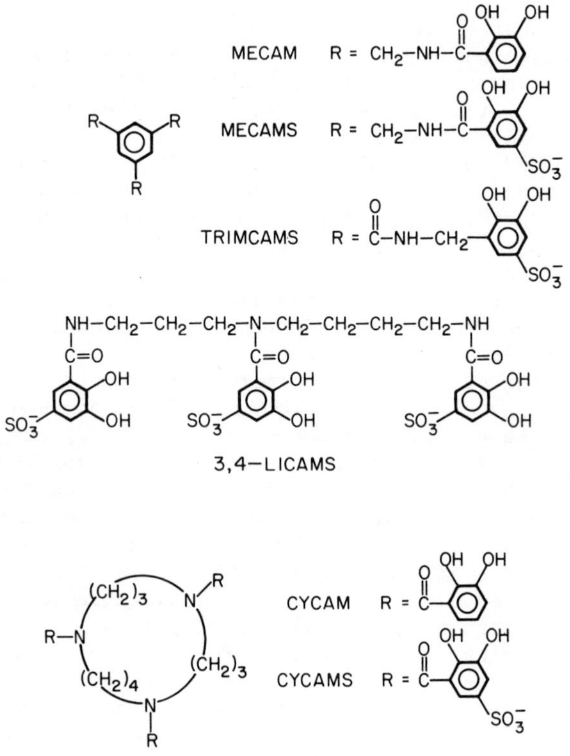

Figure 4. Structural formulas and acronyms of synthetic tricatecholate ligands

addition, sulfonation substantially lowers the ligand protonation constants. Because of the very weak acidity of the catechol ligating groups, competition from hydrogen ion is a significant interference to metal complexation, even at physiological pH. By decreasing the affinity of the ligating groups for hydrogen ion, it is possible to enhance metal complexation at neutral pH.

The formation constants of the ferric complexes of these synthetic catecholate ligands have been determined spectrophotometrically by competition with EDTA, as described above for enterobactin. The first three (most acidic) ligand protonation constants have been determined by potentiometric titration of the free ligand. The second, more basic, set of protonation constants are too large to be determined readily potentiometrically. Thus the proton-dependent stability constant is expressed as

$$K_3^* = \frac{[FeL][H]^3}{[Fe][H_3L]} \tag{4}$$

Such constants are valid over the pH range in which the final three phenolic protons are essentially undissociated, i.e. up to $\sim$ 10.5. As with enterobactin, one can estimate the higher ligand protonation constants based on values reported for simple catechols and then convert K_3^* into K_{ML} (K_{ML} = [FeL]/([Fe][L])), which is written in terms of the fully deprotonated form of the ligand. Values of log K_3^* and log K_{ML} are listed in Table II.

It is much easier to make direct comparisons between these catecholate compounds and other classes of ligands such as the hydroxamates in terms of pM values rather than log K_{ML}. Therefore, the K_3^*'s have been used to calculate pM values under the same conditions prescribed above for the siderophores: pH 7.4, 1 μM total Fe^{3+}, and 10 μM total ligand. These pM values are also listed in Table II.

The pM values of 3,4-LICAMS, 3,4,3-LICAMS, MECAM, and MECAMS are all exceptionally high, ranging from 28.5 to 31.0. Although not as powerful as enterobactin, these ligands are clearly superior to the usual amino acid type ligands such as diethylenetriaminepentaacetic acid and to the hydroxamate based siderophores. In particular, desferrioxamine B has a pM value of 26.6, so that the ligands described above are up to 10,000 times more effective at sequestering Fe(III) <u>at pH 7.4</u> than is desferrioxamine B.

The CYCAM type ligands (Figure 4) are significantly less effective than their linear analogues, such as 3,4-LICAMS, or the other platform type ligand, MECAMS. It appears that the combination of having the two ends of an aliphatic amine linked to form a cyclic group <u>and</u> having the amide nitrogens actually contained within this ten-membered ring leads to rather severe strain when the molecule is configured to fully encapsulate a ferric ion. In MECAM and enterobactin, the amide nitrogens are appended to, rather than contained within, the central rings. This additional flexibility appears to be necessary for an effective ligand. In

Table I. Relative ferric ion complexing abilities at neutral pH for human transferrin and several microbial iron sequestering agents.

Ligand[a]	pM[b]	log K_{ML}	Ref.
Enterobactin	35.5	52	6
Desferrioxamine E	27.7	32.5	9
Desferrioxamine B	26.6	30.6	9
N-acetyl-desferrioxamine B	26.5	30.8	9
Desferrichrysin	25.8	30.0	9
Desferrichrome	25.2	29.1	9
Aerobactin	23.3	23.1	10
Rhodotorulic acid	21.9	–[c]	11
Human transferrin	23.6	–	6

[a]The structural formulas of several of the siderophores are shown in Figures 1 and 2, for the other siderophore structures see Ref. 2. The "des" prefix explicitly denotes the iron free siderophore ligand.

[b]Calculated for pH 7.4, 10 μM ligand, and 1 μM Fe^{3+}.

[c]Exists in solution solely as an Fe_2L_3 dimer under these conditions.

Table II. Relative ferric ion complexing abilities at neutral pH for synthetic tricatechol sequestering agents.

Ligand[a]	pM[b]	log K_3*	log K_{ML} (est.)
3,4,3-LICAMS	31.1	8.51(3)	43
MECAMS	29.4	6.6(1)	41
MECAM	29.1	9.5(3)	46
3,4-LICAMS	28.5	6.40(9)	41
TRIMCAMS	25.1	4.4(1)	41
3,4,3-CYCAMS	24.9	3.4(1)	38
3,4,3-CYCAM	23.0	1.5(1)	38

[a]For the structural formulas of these compounds see Figure 4.

[b]As defined in Table I.

addition, the triester ring of enterobactin is flexible, as opposed to the rigidly planar benzene ring of MECAMS. This extra conformational freedom is probably a contributing factor to the enhanced stability of enterobactin over MECAM.

The Mode of Coordination of Enterobactin and Analogous Tricatechols

At high pH, the tricatecholate ligands bind iron via the six phenolic oxygens, just as observed for the tris complexes of simple catechols such as 2,3-dihydroxybenzamide, tiron, or catechol itself. In the MECAM system, this red $[Fe(MECAM)]^{3-}$ complex has an absorbance maximum at 492 nm with $\varepsilon = 4700\ M^{-1}\ cm^{-1}$. As the pH is lowered, the λ_{max} shifts to longer wavelengths and an isosbestic point is formed at 542 nm, as shown in Figure 5. Such data may be analyzed using the equation

$$\varepsilon_{obsd} = \frac{1}{K_{MH_nL}} \frac{(\varepsilon_{ML} - \varepsilon_{obsd})}{[H]^n} + \varepsilon_{MH_nL} \tag{5}$$

in which ε_{obsd} is the absorbance observed at a given pH divided by the analytical iron concentration, and ε_{ML} and ε_{MH_nL} are the molar extinction coefficients of the $[Fe(MECAM)]^{3-}$ and $[Fe(H_nMECAM)]^{(3-n)-}$ species. The exponent $\underline{n}$ is the stoichiometric coefficient of hydrogen ion in the reaction:

$$Fe(MECAM)^{3-} + nH^+ \xrightleftharpoons{K_{MH_nL}} Fe(H_nMECAM)^{(3-n)^-} \tag{6}$$

$$K_{MH_nL} = \frac{[Fe(H_nMECAM)]}{[Fe(MECAM)][H]^n} \tag{7}$$

Linear plots of ε_{obsd} vs $(\varepsilon_{ML} - \varepsilon_{obsd})/[H]^n$ are obtained only for $\underline{n} = 1$, which establishes the equilibrium of Eq. 6 is a one-proton reaction. The slope of this plot gives log K_{MHL} + 7.08(5). As the pH is lowered further, a second one-proton reaction occurs (with an isosbestic point at 588 nm) for which log $K_{MH_2L} = 5.6(1)$. Below pH 4.8, a dark purple complex precipitates, which is the neutral $Fe(H_3MECAMS)$ complex. Thus ferric MECAM reacts in a series of 3 one-proton steps.

The solid-state IR spectra of free H_6MECAM, $K_3Fe(MECAM)$, and $Fe(H_3MECAM)$ are shown in Figure 6. There is an amide carbonyl band for the free H_6MECAM ligand at 1635 cm^{-1} which shifts to 1620 cm^{-1} upon formation of the red $K_3Fe(MECAM)$ complex. In the spectrum of the neutral, triply-protonated complex, $Fe(H_3MECAM)$, the carbonyl band is absent, and has apparently moved underneath the phenyl ring modes at 1580 and 1540 cm^{-1}. Such a shift is

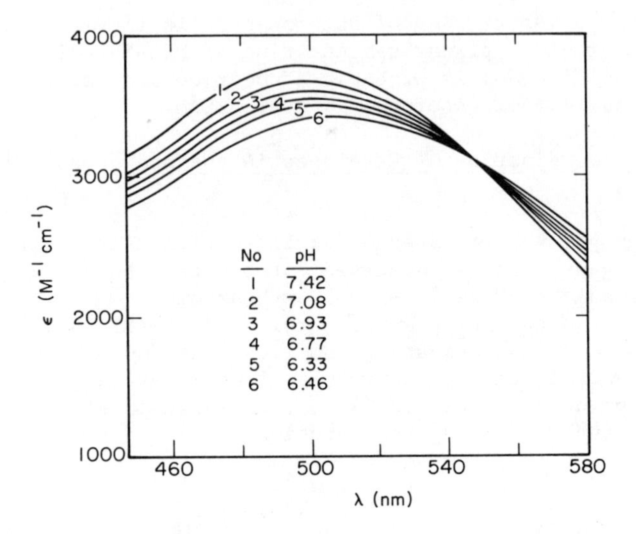

Figure 5. Visible spectra of ferric MECAM, as a function of pH, from pH 6.5 to 7.5: [Fe(MECAM)$^{3-}$] = 2 × 10^{-4}; μ = 1.0 (KNO_3); T = *25°C*

Figure 6. IR spectra of KBr pellets of (– · – · – ·) MECAM, (– – –) [Fe-(MECAM)$^{3-}$], and (——) Fe(H_3MECAM)

characteristic of a metal-bound carbonyl, and indicates a shift in the mode of bonding from a "catecholate-type" in which coordination is by the two phenolic oxygens to a "salicylate type" in which coordination is by one phenol and the carbonyl oxygen, as shown below.

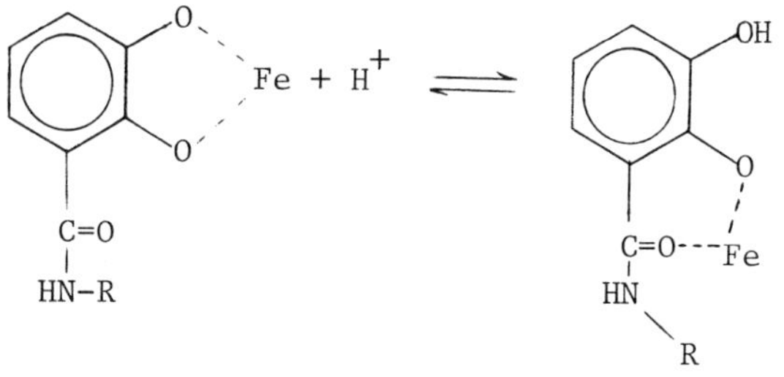

$$\text{(catecholate: ring–O,O–Fe; C=O, HN–R)} + H^+ \rightleftharpoons \text{(salicylate: ring–OH, O--Fe--O=C; HN–R)}$$

Spectrophotometric data indicate the presence of analogous, sequential one-proton reactions for the ferric complexes of enterobactin and all the synthetic tricatecholate ligands <u>except</u> TRIMCAMS (Figure 4), in which the amide carbonyl groups have been relocated *α* to the central benzene ring, and are not a substituent of the catechol rings as in MECAMS. Thus the carbonyl is no longer available to form a six-membered chelate ring with the ortho phenolic oxygen. The protonation reaction of TRIMCAMS thus provides a powerful test of the catecholate-salicylate mode of bonding equilibrium that we propose.

The visible spectra of ferric TRIMCAMS as a function of pH are shown in Figure 7. There is a single, sharp isosbestic point at 540 nm which remains throughout the addition of <u>two</u> equivalents of hydrogen ion to $[Fe(TRIMCAMS)]^{6-}$. A plot of Eq. 6 is linear for *n* = 2, with

$$K^{2}_{MH_2L} = \frac{[Fe(H_2TRIMCAMS)^{4-}]}{[Fe(TRIMCAMS)^{6-}][H^+]^2} = 10^{13.7} \qquad (8)$$

These results thus support the model proposed above in which one-proton reactions are due to a shift from a catecholate to a salicylate mode of bonding.

Removal of Iron from Human Transferrin

Most normal body iron is contained either in hemoglobin or in the storage proteins ferritin and hemosiderin. Certainly for the heme proteins it is unlikely that chelating agents will be able to remove significant amounts of iron — and the iron in the storage proteins is also quite inaccessible. However, the high-spin Fe(III) in the iron transport protein transferrin is

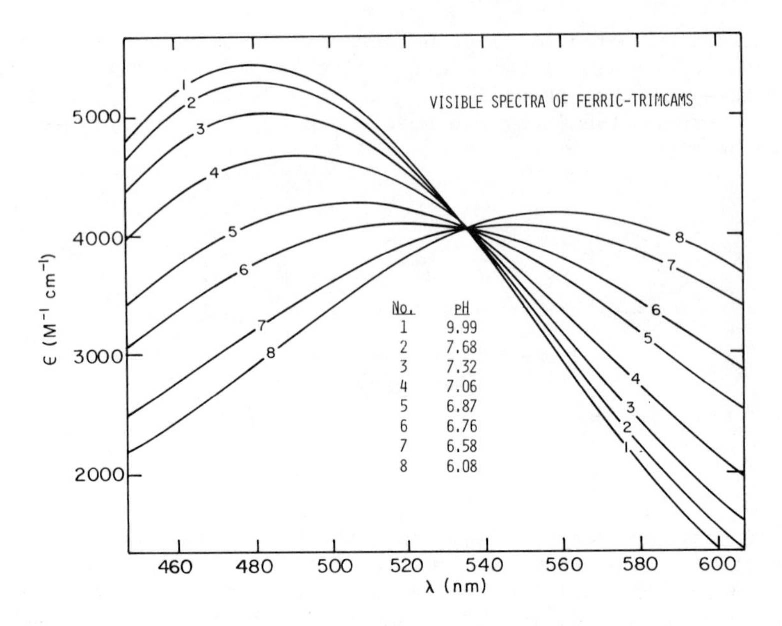

Figure 7. Visible spectra of ferric TRIMCAMS as a function of pH from pH 6 to 10

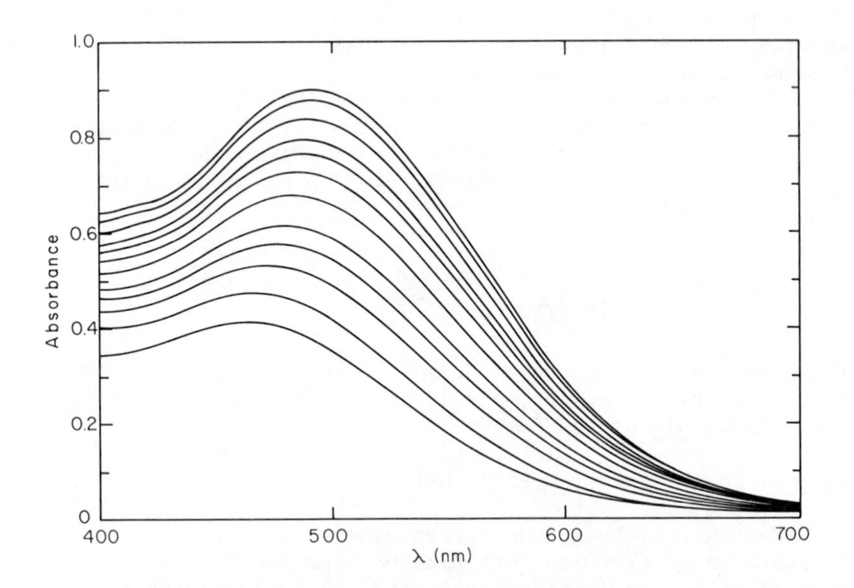

Figure 8. Spectral changes accompanying iron removal from transferrin. The bottom curve represents the unreacted transferrin and the top curve the final product, ferric 3,4-LICAMS.

relatively labile, and apotransferrin is able to obtain iron from ferritin. Thus if we could remove and excrete transferrin iron, we may then allow the apotransferrin to mobilize stored iron. The key factor is obviously the ability to remove ferric ion from transferrin. This is even more of a kinetic problem than a thermodynamic one. Thus, while the hydroxamates such as desferrioxamine B are thermodynamically capable of removing iron from transferrin, kinetically they are able to do so at a useful rate only in the presence of other ligands.

The pM values of most of the catecholate ligands are well above that of transferrin, indicating that iron removal is thermodynamically favored. The question remains, however, as to the rate of this exchange reaction. Therefore we have investigated the kinetics of iron removal from transferrin by these types of catecholate ligands. The addition of 3,4-LICAMS to diferric transferrin results in the series of spectra shown in Figure 8. The absorbance maximum shifts smoothly from the 470 nm λ_{max} of diferric transferrin (ε = 2500/Fe) to the 495 nm λ_{max} of ferric 3,4-LICAMS (ε = 5500). With excess 3,4-LICAMS, plots of $\ell n[(A - A_\infty)/(A_0 - A_\infty)]$ vs time are linear over three half-lives (Figure 9).

Previous results have indicated that iron removal from transferrin might involve the formation of a ternary iron-transferrin-ligand intermediate, which dissociates into FeL and apotransferrin (12, 13). Such a scheme is outlined in Eq. 9.

$$\text{FeTr} + \text{L} \underset{k_{-1}}{\overset{k_1}{\rightleftharpoons}} \text{FeTrL} \xrightarrow{k_2} \text{FeL} + \text{Tr} \tag{9}$$

Such a mechanism predicts a hyperbolic relationship between k_{obsd} and the concentration of the competing ligand, and Figure 10 shows that such a relationship does pertain. One can express k_{obsd} as

$$k_{obsd} = \frac{k_2[\text{L}]K_{eq}}{2.3 + 2.3\,K_{eq}[\text{L}]} \tag{10}$$

where $K_{eq} = k_1/k_{-1}$. The data for 3,4-LICAMS were refined by a nonlinear least-squares fit of k_{obs} vs [L] to give values k_2 = 0.066(4) min^{-1} and K_{eq} = 4.1(6) x 10^2 $\ell\ mol^{-1}$. Similar results are obtained with MECAM and enterobactin as competing ligands, although the low solubility of these ligands limits the ratio of ligand:transferrin. Table III lists the percentage of iron removed in 30 min at a specified ligand concentration. While desferrioxamine B can remove less than 5% of the iron at a 100:1 excess of desferrioxamine, 3,4-LICAMS removes 50% of transferrin iron at only a 40:1 ratio. Thus the catecholate type ligands represent an effective combination of both a high affinity for ferric ion coupled with the ability to remove iron from transferrin at a reasonable rate.

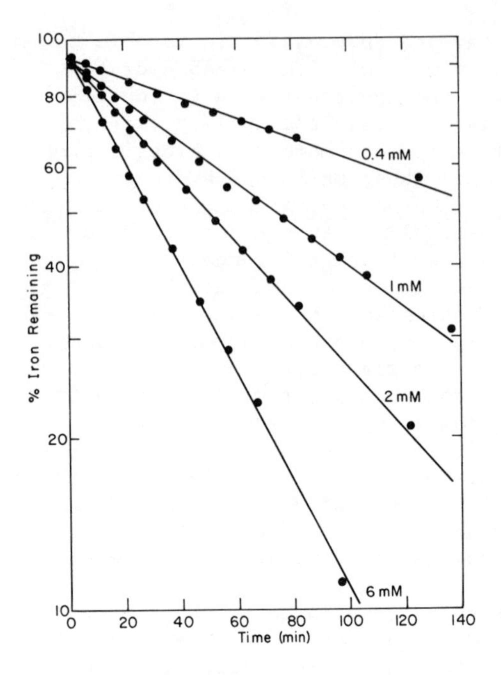

Figure 9. Iron removal from ~ 0.2mM diferric transferrin by various concentrations of 3,4-LICAMS

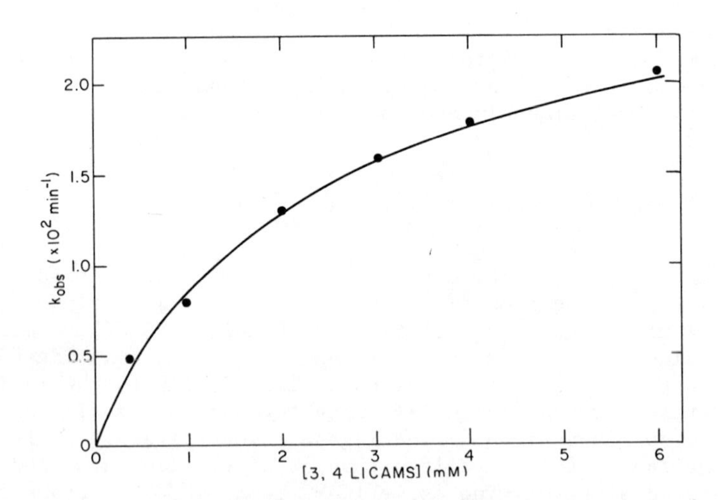

Figure 10. Plot of the observed rate constant for iron removal from transferrin (~ 0.4mM) vs. the concentration of 3,4-LICAMS. The points represent the experimental data, the line is calculated from the derived rate constants.

Table III. Relative kinetics of iron removal from human transferrin by several iron sequestering agents.

Ligand	[L]/[Tr][a]	% Fe Removed[b]
3,4-LICAMS	40	50
3,4,3-LICAMS	100	60
Desferrioxamine B	100	5
EDTA	2500	37
MECAM	1	13
Enterobactin	1	6
3,4-LICAMS	1	6
EDTA	1	0

[a] Ratio of ligand to transferrin concentration; [L] = ligand, [Tr] = diferric transferrin.

[b] After 30 minutes.

Specific Sequestering Agents for Actinide(IV) Ions

In addition to the problems posed by acute and chronic iron-overload poisoning, we are also interested in the problems posed by transuranium actinide contamination. There are a number of similarities in the coordination chemistry of Fe(III) and Pu(IV). Indeed, the great biological hazard of plutonium is because, once in the body, it is associated with the iron binding proteins transferrin and ferritin and is deposited essentially irreversibly in iron storage sites. Using enterobactin as a model, we can predict that catechol-based ligands should form very stable plutonium complexes — especially when incorporated in large multidentate ligands.

In order to satisfy the preferred higher coordination number of the actinides, we have synthesized several potentially octadentate, tetracatecholate compounds with the structures shown in Figure 11. Of these, we have studied the solution chemistry of 3,4,3-LICAMS with Th(IV) as a model for Pu(IV). The potentiometric equilibrium curve of a 1:1 ratio of TH(IV):3,4,3-LICAMS is shown in Figure 12. Complexation is complete by ∿ pH 8, which indicates fairly strong complexation. However, the sharp inflection at 7.75 equivalents of base per Th(IV) shows that the thorium species formed under these conditions is polynuclear, with at least three and possibly more thorium ions per molecule. Polynuclear complexes of thorium are quite common with ligands having fewer than eight donor groups (14, 15, 16, 17). In particular, Th(IV) forms a ligand-bridged dimeric complex with tiron (15) (1,2-dihydroxy-3,5-disulfobenzene), even in the presence of a large excess of ligand.

The coordination of all four catechol groups of 3,4,3-LICAMS to Th(IV), displacing the eight catechol protons, would result in a break in the titration curve at 8 equivalents of base, compared to the observed break at 7.75 equivalents. The titration of fewer than the expected number of protons, plus the polymeric nature of the Th(IV):3,4,3-LICAMS complex both suggest that all four dihydroxybenzoyl groups are not coordinating, thus leaving coordination sites available for hydroxyl bridging groups.

Unfortunately, the complexity of the Th(IV):3,4,3-LICAMS system precludes the calculation of formation constants. However, the relative sequestering abilities of 3,4,3,-LICAMS and DTPA have been investigated by direct competition between these two ligands using difference ultraviolet spectroscopy to measure the fraction of thorium bound to 3,4,3-LICAMS. At pH 6.5, the Th(IV) is bound predominantly to DTPA. The fraction of Th(IV) bound to 3,4,3-LICAMS then increases rapidly with increasing pH, leveling off around pH 7.5 with ∿ 80% of the thorium associated with 3,4,3-LICAMS. Thus at physiological pH, the 3,4,3-LICAMS complex is slightly favored thermodynamically over Th(IV)-DTPA.

Several linear and cyclic tetracatecholates have been evaluated *in vivo* for their ability to enhance the excretion of Pu(IV) from mice (18). Between 20-30 μmole/Kg of the ligand was administered 1 hr after the injection of plutonium citrate. The mice were sacrificed after 24 hr and the distribution of Pu(IV) in the tissues and excreta were measured. The percentages of plutonium excreted are listed in Table IV. About 60-65% of the Pu(IV) is excreted with 3,4,3- and 4,4,4-LICAMS, which is roughly comparable to the effectiveness of DTPA at this dose level. The length of the middle bridging alkane group in the LICAMS series appears to be critical, since the presence of a propyl group at this position reduces the amount of Pu(IV) excreted to ∿ 40%. The cyclic catecholates are less effective than the linear compounds. The unsulfonated ligand 3,3,3,3-CYCAM binds plutonium and restricts its deposition in the liver and skeleton, which is where 80% of injected plutonium is deposited in the control. However, the 3,3,3,3-CYCAM complex apparently dissociates in the kidneys, so the Pu(IV) is simply concentrated in this organ.

Although 3,4,3-LICAMS and DTPA removed roughly equal amounts of plutonium under the conditions described above, 3,4,3-LICAMS offers several advantages over DTPA. The catecholate ligands do not form strong complexes with most divalent metal ions and thus the depletion of essential metals such as Zn, Mn, Cu and Co, (which is a severe problem in DTPA therapy) is greatly reduced. In addition, dose-response studies show that the 3,4,3-LICAMS dose can be reduced by a factor of 100 (to 0.3 μmoles/Kg) and still effect ∿ 50% excretion of Pu(IV). In contrast, DTPA therapy at this level is virtually ineffective at this dose level (19).

A

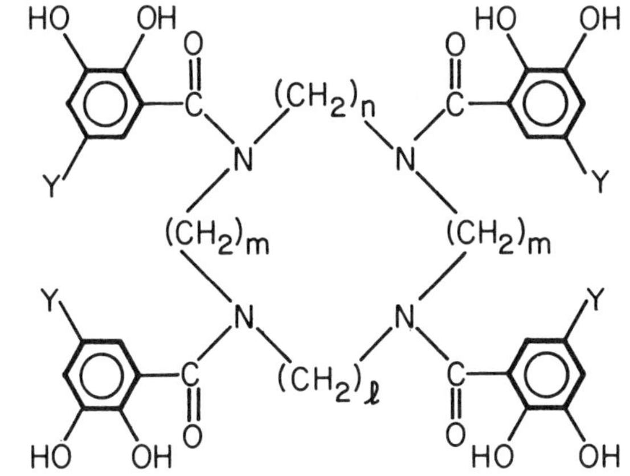

$n = m = \ell = 3$; $Y = H$ 3,3,3,3-CYCAM

$n = m = \ell = 3$; $Y = SO_3Na$ 3,3,3,3-CYCAMS

B

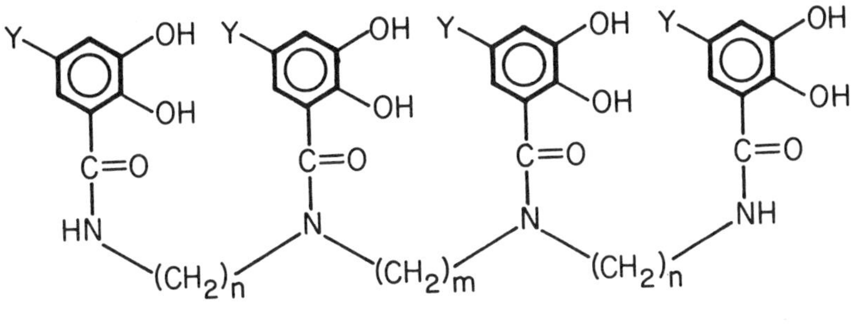

$n = m = 4$; $Y = H$ 4,4,4-CYCAM

$n = m = 4$; $Y = SO_3Na$ 4,4,4-CYCAMS

$n = 3$; $m = 4$; $Y = H$ 3,4,3-CYCAM

$n = 3$; $m = 4$; $Y = SO_3NA$ 3,4,3-CYCAMS

Figure 11. Structural formulas and acronyms for synthetic tetracatecholate ligands

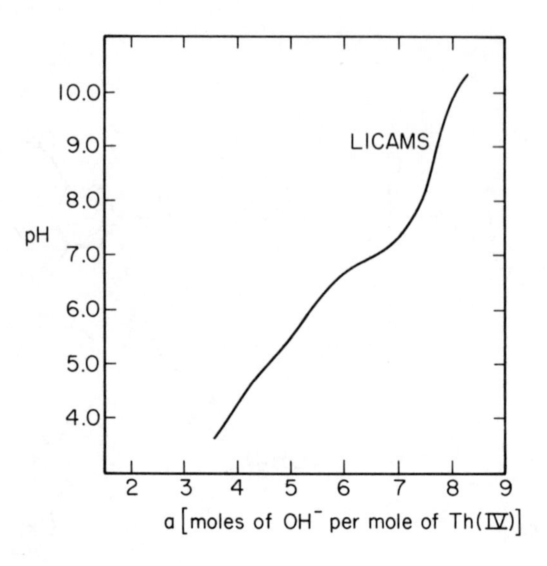

Figure 12. Potentiometric equilibrium curve of a 1:1 ratio of Th(IV) 3,4,3-LICAMS: [*Th*] = [*LICAMS*] = 2×10^{-3}M, T = 25°C, μ = 0.10 (KNO_3)

Table IV. Relative efficacy of Pu removal from mice by several tetracatechol ligands.

Ligand[a]	% Pu Excreted (24 hrs)
3,4,3-LICAMS (i.v.)	64.6
4,4,4-LICAMS (i.v.)	59.3
3,3,3-LICAMS (i.v.)	44.1
4,3,4-LICAMS (i.v.)	43.4
2,3,2-LICAMS (i.p.)	37.0
2,3,3,3-CYCAMS (i.v.)	38.6
3,3,3,3-CYCAMS (i.v.)	35.3
3,3,3,3-CYCAM (i.m.)[b]	11.1
$CaNa_3DTPA$ (i.p.)	63.2
Desferrioxamine B (i.p.)	47.8
Control	6.1

[a] See Figure 11 for structural diagrams.

[b] Note that this is the unsulfonated derivative.

Finally, 3,4,3-LICAMS is substantially more effective than DTPA for removing Pu which has been deposited in the skeleton. The properties and chemistry of these compounds are described in greater detail in a monograph resulting from another symposium of the 1979 Washington ACS meeting (20).

Summary

The results described above show that these new catecholate compounds do indeed form exceptionally stable complexes with very hard Lewis acids such as Fe^{3+}, Th^{4+} and Pu^{4+}. The sulfonation of these ligands increases their water solubility, stabilizes the catechol groups against air oxidation, and enhances their ability to sequester metal ions at neutral and slightly acidic pH. The iron complexes are among the most stable ever characterized, and these types of ligands are thermodynamically able to compete with transferrin for serum iron. In contrast to the hydroxamates, our kinetic studies have shown that the catechol-based ligands are in fact capable of removing substantial amounts of iron from transferrin within a few minutes.

With Th(IV), the tetracatecholates form complex polynuclear species. Competition studies *in vitro* indicate these complexes are slightly more stable than the monomeric Th:DTPA complex at pH 7.4. *In vivo* studies on plutonium decorporation from mice show that 3,4,3-LICAMS is as effective as DTPA at fairly high doses. However, dose-response curves show that the 3,4,3-LICAMS retains 80% of its effectiveness at very low concentrations, whereas the DTPA activity drops to essentially zero under the same conditions.

Acknowledgment

We are pleased to acknowledge the continuing collaboration of Dr. P. W. Durbin (Donner Laboratory, Lawrence Berkeley Laboratory) and her co-worker S. Sarah Jones. We thank the National Institutes of Health for supporting the biological iron chelation chemistry reported here. The actinide sequestering project is supported by the Division of Nuclear Sciences, Office of Basic Energy Sciences, U.S. Department of Energy under Contract No. W-7405-Eng-48.

Literature Cited

1. Anderson, W. F. "Iron Chelation in the Treatment of Cooley's Anemia"; this publication.
2. Neilands, J. B., Ed. "Microbial Iron Metabolism"; Academic Press: New York, 1974.
3. Neilands, J. B. "High Affinity Iron Transport in Microorganisms"; this publication.

4. O'Brien, I. G.; Gibson, F. Biochim. Biophys. Acta 1970, 215, 393-402.
5. Cooper, S. R.; McArdle, J. V.; Raymond, K. N. Proc. Natl. Acad. Sci. U.S.A. 1978, 75, 3551-54.
6. Harris, W. R.; Carrano, C. J.; Cooper, S. R.; Sofen, S. R.; Avdeef, A. E.; McArdle, J. V.; Raymond, K. N. J. Am. Chem. Soc. 1979, 101, 6097-6104.
7. Avdeef, A. E.; Sofen, S. R.; Bregante, T. L.; Raymond, K. N. J. Am. Chem. Soc. 1978, 100, 5362-70.
8. Martell, A. E.; Smith, R. M. "Critical Stability Constants"; Plenum Press: New York, 1977; Vol. I.
9. (a) Schwarzenbach, G.; Schwarzenbach, K. Helv. Chim. Acta 1963, 46, 1390-1400.
(b) Anderegg, G.; L'Eplattenier, F.; Schwarzenbach, G. Helv. Chim. Acta 1963, 46, 1409.
10. Harris, W. R.; Carrano, C. J.; Raymond, K. N. J. Am. Chem. Soc. 1979, 101, 2722-27.
11. Carrano, C. J.; Cooper, S. R.; Raymond, K. N. J. Am. Chem. Soc. 1979, 101, 599-604.
12. Bates, G. W.; Billups, C.; Saltman, P. J. Biol. Chem. 1967, 242, 2810-15.
13. Bates, G. W.; Billups, C.; Saltman, P. J. Biol. Chem. 1967, 242, 2816-21.
14. Courtney, R. C.; Gustafson, R. L.; Chabarek, S.; Martell, A. E. J. Am. Chem. Soc. 1957, 80, 2121-28.
15. Murakami, Y.; Martell, A. E. J. Am. Chem. Soc. 1960, 82, 5605-07.
16. Boguki, R. F.; Martell, A. E. J. Am. Chem. Soc. 1958, 80, 4170-74.
17. Boguki, R. F.; Martell, A. E. J. Am. Chem. Soc. 1968, 90, 6022-27.
18. Durbin, P. W.; Jones, E. S.; Raymond, K. N.; Weitl, F. L. Radiation Research 1979, in press.
19. Durbin, P. W.; Jones, E. S.; Raymond, K. N.; Weitl, F. L. Unpublished results.
20. Raymond, K. N.; Smith, W. L.; Weitl, F. L.; Durbin, P. W.; Jones, E. S.; Abu-Dari, K.; Sofen, S. R.; Cooper, S. R. "Actinide Chemistry"; ACS Symposium Series, Edelstein, N., Ed.; American Chemical Society: Washington, D.C., 1979.

RECEIVED April 7, 1980.

CHELATION THERAPY FOR REMOVAL OF TOXIC OVERLOADS OF OTHER METAL IONS

19

Chelate Therapy for Type b Metal Ion Poisoning

MARK M. JONES and MARK A. BASINGER

Department of Chemistry and Center in Environmental Toxicology,
Vanderbilt University, Nashville, TN 37235

The fact that type b/soft metal ions (for our purposes those that form stronger coordinate bonds to S than to O) have their own distinctive set of coordination preferences has clear consequences for both the distribution of these metals in the mammalian body and the most effective reagents for use in facilitating their removal. Their toxicity is presumably due to the combined action of their coordination preferences and their stereochemistry in the determination of those biological binding sites which are preferentially attacked and subsequently inactivated. For purposes of the present discussion we will use a combination of the classification schemes of Ahrland, Chatt and Davies (1) and Pearson (2). Of the elements which fall in this broad category, many are so uncommonly encountered clinically that human cases are extraordinarily rare. The elements for which some information is available on body distribution and decorporation will be the ones on which our attention will be centered. These include Hg(II), Au(II), As(III), Sb(III), Cu(I), and Pt(II).

It is necessary to note that there are differences in the organ distribution of these elements and that the processes involved in their metabolism are generally complex and imperfectly understood. However, type b metals do, by and large, show a tendency to accumulate in the kidneys and liver; renal failure (nephrotic syndrome) seems to be a common result of their action. When the "natural" half-life of these elements in the mammalian body is examined, it is generally found that the rate processes correspond to the loss of the element from two or more compartments (3,4) in which the stability of the metal to biological molecule bonds vary greatly. As a result, it is somewhat misleading to quote a single half-life as characteristic of the loss of the metal from the whole animal. Furthermore, the relative amount of a metal which goes into these different compartments can vary with both the rate of loading and the time interval between the metal loading and the analysis.

This brings into somewhat clearer focus the problem of metal-binding sites in whole animals. The sites with the shortest

0-8412-0588-4/80/47-140-335$05.00/0

half-lives seem to be those in which the hard/soft properties of the metal and donor species are rather poorly matched. Thus sites in which Hg(II) is bonded to O would be expected to release the Hg(II) more readily than those in which it is bonded to S. As time goes by, that part of the metal which remains in the body can bind to sites with which it forms successively more stable complexes. As this occurs, the removal of the metal from the body may become increasingly more difficult. This particular problem is of great importance in dealing with chronic metal poisoning, especially among industrial workers exposed to lead and cadmium (4). Very often these cases become apparent only after extended exposure, during which time a large part of the total body burden has been transferred to sites from which its removal is extremely difficult or impossible. In lead poisoning the accumulation is in the bones; in cadmium poisoning the accumulation is in the kidneys and perhaps the testes.

In many types of metal poisoning, there appears to be an approximately maximum allowable time interval between intoxication and the initiation of chelate therapy. When treatment is begun after this interval, the extent and irreversibility of the damage is such that the death of the organism can be assumed to follow no matter how much antidote is administered; for less than toxic doses an analogous phenomenon is the occurrence of permanent damage.

The chemical basis of this time period may not be uniform for all of the type b metals. For mercury(II) it appears to be related to the occurrence of a minimum amount of what is literally the destruction of tissue and proteins. In the case of cadmium it appears to have a different basis at least in part. Subsequent to the injection of an otherwise lethal dose of a cadmium compound there is a relatively short period when it can be complexed and removed and the animal saved. As the interval between the cadmium injection and that of the antidote (e.g. CaEDTA or CaDTPA) increases, the cadmium becomes more and more difficult to mobilize and a point is soon reached where the antidote is without effect (5). In the same fashion, aged cadmium deposits are apparently resistant to mobilization by EDTA (6).

General Procedures

With this as a background, we may now examine the procedures available for the removal of such metals. The normal route consists in the reaction of the deposited metal with a chelating agent with which it forms a water soluble complex. This metal complex is then excreted through the kidneys into the urine. Because the kidneys are themselves a delicate and complex organ this process is attended by a certain inherent danger. It is in the matching up of these metal ions with the most appropriate chelating agents that coordination chemists can make a significant contribution. Let us first look at some of the requirements which these chelating agents must meet. Desiderata for ideal therapeutic chelating agents include the following:

1. They must possess an ip LD50 ≥ 400 mg/kg.
2. They must form very stable complexes with the metal ion whose removal is desired.
3. Any selectivity which they exhibit in forming complexes is usually advantageous if other adverse properties are not simultaneously enhanced.
4. The complexes formed should be water soluble.
5. The passage of the complexes through the kidneys should be accompanied by a minimum of damage.
6. It is probably helpful if the chelating agents undergo no significant metabolism.

It is necessary to note that these complexes may be mixed complexes involving the chelating agent the metal and some biological molecule and the mixed complex is then the metal containing species which passes through the kidneys (7).

The toxicity of chelating agents is an area in which the general features are becoming known in great detail as more data is collected. To begin, the daily administration of appreciable amounts of almost any good chelating agent is ultimately toxic or lethal because of the enhanced urinary excretion of essential trace elements which it provokes. Some chelating agents, such as the aminopolycarboxylates, which undergo little if any metabolic changes in the human body, appear to act as toxic materials almost exclusively via their ability to upset metal ion equilibria of various sorts in the human or animal body. When administered as calcium or zinc complexes, they have a very modest toxicity indeed, much reduced from that of the parent compound. For chelating agents which can be metabolized, the toxicity arises from the depletion of essential trace metals plus the metabolic activity of the chelating agent. Sulfhydryl groups can interact with biological redox systems and derange them. For such compounds a high lipid solubility is often associated with a high toxicity. This high lipid solubility also allows the metal complexes which are formed to intrude into areas normally protected by lipid barriers, such as the brain (8). It is for this reason that the "best" of such chelating agents for our purposes are those which possess a very high water solubility (and a correspondingly low lipid solubility) for both the parent compound and the metal complexes which are formed in the detoxification process. The structures of some of the therapeutic chelating agents which are used for the type b/soft ions are shown in Table I. These have one or more sulfur donor groups plus other groups to enhance the water solubility of both the compound and its metal complexes. This enhancement is to be understood in the relative sense that the water solubility is greater than in the corresponding derivatives without such groups. The absolute solubility of the metal complexes formed may still be quite small. The toxicities of these, of some other therapeutically useful chelating agents and some chelating agents which are not so used are listed in Table II. A chelating agent will not always use all of its donor groups when interacting with a metal!

Table II
LD50 Values of Therapeutic and Non-therapeutic Chelating Agents

Chelating Agent	LD50 mg/kg	Conditions	Lit.
$Na_2CaEDTA$	3800	rat, i.p.	a
Na_4EDTA	380	mouse, i.p.	a
D-penicillamine	334	mice, i.p.	b
N-Acetyl-D,L-penicillamine	1000	rats, i.p.	b
Ethylenediamine	427	mouse, s.c.	a
2,3-dimercaptopropanesulfonate	2000	mouse, s.c.	c
Triethylenetetramine	4340	rat, oral	a
BAL	105	rat, i.v.	a
Salicyclic Acid	891	rat, oral	a
Acetylacetone	1000	rat, oral	a
Deferrioxamine	329	rat, i.v.	d
2,3-dimercaptosuccinate sodium	5000	mouse, oral	e
Dimethylglyoxime	250	rat, oral	a

a. "Toxic Substances 74" U.S. Dept. HEW, Rockville, Md.
b. Aposhian, H. V.; Aposhian, M. M. Pharmacol. Exptl. Ther., 1959, 126, 131.
c. Klimova, L. K. Farmakol. i Toksilsol., 1958, 21, 53.
d. Goldenthal, E. I. Tox. Appld. Pharmac., 1971, 18, 185.
e. Stohler, H. R.; Frey, J. R. Ann. Trop. Med. Parasitol., 1964, 58(4), 431.

One of the strange problems of metal ion toxicity is the lack of a direct relationship between the stability constant of a metal-chelate complex and the ability of that chelating agent to offset the acute toxicity of that metal ion. There is a good theoretical model of the relationship between metal chelate stability constants and the decorporation process (9,10,11) for groups of closely related chelating agents. Part of this problem lies in the fact that the stability constants for many metal-chelate combinations of interest simply have not been determined with the required degree of accuracy. Nevertheless, it is generally accepted that if a chelating agent is to be a successful antidote for a given toxic metal it must form complexes with that metal which are more stable than the complexes which that metal forms with the biological binding sites. Given this then, how does one guarantee that the chelating agent can gain access to the bound metal perhaps within the cell? This is a difficult question to answer but it does often seem that the cell walls are in fact far more permeable to water soluble chelating agents than has previously been supposed. The fact is that these compounds can remove metals which are bound to structures *inside* the cells.

There are really two easily envisaged mechanisms by which this might occur. In the first, (Figure 1) the chelating agent passes through the membrane, in one manner or another, comes in direct contact with the metal ion, pulls it off the biological binding site, transports it back through the cellular membrane to the intracellular fluid and then carries it through the kidneys and into the urine. In the second mechanism (Figure 2) the chelating agent need not actually penetrate the cell wall if a system involving appropriate carrier molecules is present. Cellular membranes are now recognized as being more permeable to certain types of small molecules and the first mechanism is no longer considered so improbable.

In either mechanism the more effective matching up of the donor/acceptor properties can be seen to lead to an increased efficacy.

Detoxification and Decorporation of Specific Metal Ions

The metal ions considered here are those which bond preferentially to S donors, especially sulfhydryl groups. The bonds to O or N can generally be broken by an appropriate sulfhydryl group, though the peculiar coordination chemistry in these situations does not always lead to either chelate rings containing two sulfur donors or a complete replacement of O or N donors by S (12,13). These ions are not always derived from type b elements in the usual classification scheme. Thus As is put in class a on the basis of the behavior of As(V). But the reducing powers of cellular media may (43) transform As(V) to As(III), whose coordination behavior shows a marked preference for S donors over those of O or N. The same is true of Sb. For copper an analogous process can be used to transform the borderline species Cu(II) to one in which preferences for bonding to S are more marked, i.e. Cu(I) in order to facilitate its excretion.

Mercury. Mercury(II) and the various organic derivatives of mercury do not have the same pattern of either body distribution or response to chelating agents (14). The half-life of mercury (II) in the human body is relatively short (3). The critical period in mercury poisoning is apparently the period immediately following the introduction of the compound into the body. During this period the extent of damage increases with time and the portion of this which represents permanent, apparently irrepairable damage also increases with time. After this has reached a certain point there is a limit to the extent of the healing that will occur. In humans this has the consequence that after certain stages of poisoning the accelerated removal of the mercury from the human body has relatively little effect on the clinical state of the patient (15).

Sulfhydryl bearing compounds of various sorts are by far the best agents yet discovered for accelerating the removal of

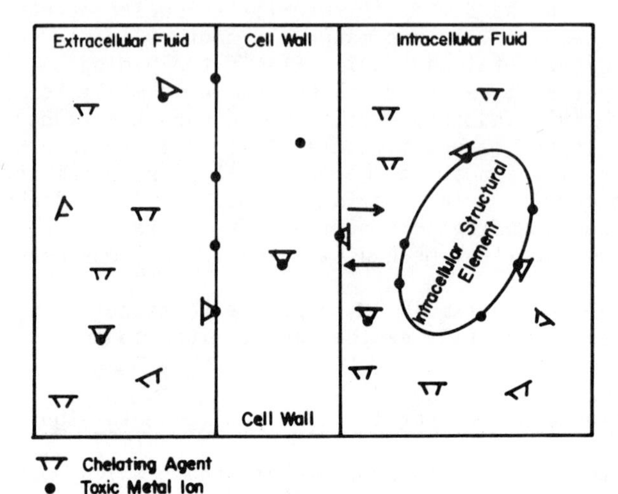

Figure 1. Mechanism A: overall process by which a chelating agent can remove a toxic metal ion from a cell when the cell wall is permeable to the chelating agent

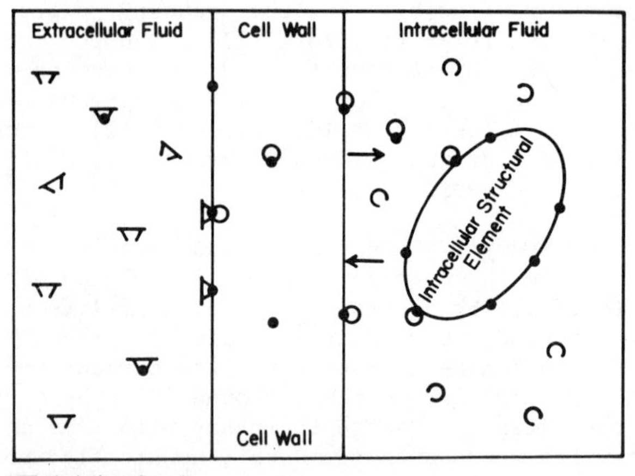

Figure 2. Mechanism B: overall process by which a chelating agent can remove a toxic metal ion from a cell when the cell wall is impermeable to the chelating agent

mercury(II) and organomercury compounds from the mammalian body. The results of a comparative study of the efficacy of some of these compounds in offsetting the lethality due to ip $HgCl_2$ is shown in Table III (16). Of the compounds listed in this table,

Table III

Survival Rates for Various Chelating Agents
All animals given $HgCl_2$ i.p. at a level of 10 mg/kg

Mole Ratio Chelator/$HgCl_2$	Chelator[a]	Survival Ratio
30:1	DMSA	14/20
30:1	DMPS	16/20
30:1	NAPA	15/20
30:1	DPA	17/20
30:1	BAL	1/20
10:1	DMSA	19/50
10:1	DMPS	30/50
10:1	NAPA	30/50
10:1	DPA	26/50
10:1	BAL	19/50

a. DMSA = 2,3-dimercaptosuccinic acid
DMPS = sodium 2,3-dimercaptopropanesulfonate
NAPA = N-Acetyl-D,L-penicillamine
DPA = D-penicillamine
BAL = 2,3-dimercaptopropanol-1

(The survival rate of untreated animals is less than 5%.) BAL may still be the most widely used in human cases of mercury poisoning, but the others would probably be found to be more effective under most conditions, and to produce fewer side effects. The compounds which are listed in Table III other than BAL, all appear to posses a rather low lipid solubility and are rapidly cleared from the human body through the kidenys.

In an attempt to extend the range of donor species useable for mercury poisoning we have carried out an evaluation of the use of the phosphine $(C_6H_5)_2PC_6H_4SO_3Na$ (17). The phosphine was, in fact, able to offset some of the pathological changes characteristic of mercuric chloride poisoning. Its own inherent toxicity was too great, however, and the combined effect of phosphine plus $HgCl_2$ proved more lethal than either alone, Table IV. The possibility of using a water soluble phosphine for mercury poisoning is a remote one in the absence of compounds of this sort possessing a low inherent toxicity.

Analogous data collected on some water soluble dithiocarbamates which have more than one functional group (to assure water solubility) revealed that these are both materials of modest inherent toxicity and potential use in heavy metal poisoning (18). These dithiocarbamates do not possess exceptional chemical

Table IV

Survival Ratios for Mice Given Diphenylphosphinebenzenesulfonate As An Antidote for $HgCl_2$, i.p.

Compounds Administered	Amounts Administered	Survivors Total
$HgCl_2$	8 mg $HgCl_2$/kg	3/10
DPPBS	0.2 mmole DPPBS/kg	8/10
DPPBS	0.3 mmole DPPBS/kg	7/10
$HgCl_2$ + DPPBS	8 mg $HgCl_2$/kg + 0.2 mmole DPPBS/kg	1/15
$HgCl_2$ + DPPBS	8 mg $HgCl_2$/kg + 0.3 mmole DPPBS/kg	0/10

stability, though they are easier to prepare than most of the typical chelating agents which contain sulfur donors. They are not as effective in offsetting lethality in acute mercuric chloride poisoning as the sulfhydryl compounds listed in Table III. The most common of the dithiocarbamates available as a pure laboratory chemical is diethyl dithiocarbamate, a compound which gives lipid soluble metal complexes. This has been tried in a few types of metal poisoning without notable success (19). This compound seems inherently less promising for such antidotal action than analogous compounds which we have prepared which bear additional polar groups.

In spite of the fact that effective chelating agents are available for the removal of mercury from the mammalian body, the chemistry of these interactions, the structures of the complexes formed and many aspects of their mechanism of action are still unknown quantities. One of the key properties of a donor-metal interaction in judging the suitability of the donor as a therapeutic agent for the metal is the stability constant of the complex formed. We have recently completed a study which tried to introduce some system into the information on these parameters for mercury complexes of interest in therapy. Using a three electrode system we determined estimates of the stability constants for many of these (Table V) (20). A striking feature of these are their

Table V

Stability Constants of Mercury(II) Complexes With Sulfhydryl Containing Donors

Donor Molecule	log K_1	log K_2
D-penicillamine	38.3	6.1
N-Acetyl-D,L-penicillamine	35.4	6.2
2,3-Dimercaptopropanesulfonate	42.2	10.9
2,3-Dimercaptopropanol-1	44.8	7.11
2,3-Dimercaptosuccinic Acid	39.4	6.5
Mercaptoacetic Acid	36.5	5.9
Mercaptoethanesulfonate	36	6
N-Acetylcysteine	38.1	7.5

large magnitudes. While it is reasonably certain that these complexes are all more stable than the corresponding Hg-EDTA complex, it is possible that their stability constants are overestimated here because of some systematic error. The results are in agreement with some earlier studies on this same problem (21,22) involving mercury(II) complexes with cysteine and HS^-.

Arsenic. Arsenic poisoning is generally regarded as involving, in a key step, the reaction of arsenious acid or a derivative with sulfhydryl groups in critical enzymes (23), e.g. pyruvate oxidase. Because arsenic(III) favors coördination to sulfur over that to nitrogen or oxygen, the situation here has some similarities to that found with mercury. An important difference here however is the general acceptance of the hypothesis that chelation of the arsenic(III) by two sulfhydryl sulfurs is involved in the action of effective chelating agents. The most effective of the chelating agents are those which contain vicinal sulfhydryl groups: BAL, sodium 2,3-dimercaptopropanesulfonate, and 2,3-dimercaptosuccinic acid. These are capable of competing with the enzymes for the arsenic. They transform it into a water soluble form and facilitate its excretion through the kidneys. Because BAL is so effective in arsenic poisoning and because this kind of poisoning is not as common as formerly, relatively little effort has been expended in the search for new arsenic antidotes over the past twenty years. On the basis of animal studies it appears that any of the chelating agents which contain vicinal sulfhydryl groups and are effective with other heavy metals are effective with arsenic(III).

Antimony. Antimony(III) is the active constituent of a large number of drugs designed to kill human and animal parasites (24, 25). As a consequence of the significant variation in human response to these drugs, most antimony poisoning appears to be iatrogenic in origin. Because the coordination preferences of trivalent antimony and arsenic are so similar, the same group of vicinal dimercaptides that are used with arsenic poisoning also furnish the most effective therapeutic agents for antimony poisoning, i.e. 2,3-dimercaptosuccinic acid (26), 2,3-dimercaptopropanesulfonate (27), and 2,3-dimercaptopropanol-1 (28). The principal therapeutic use of antimony complexes lies in the treatment of diseases typical of tropical zones. As a result iatrogenic antimony poisoning is rather uncommon in the United States or Europe.

Gold(I). Gold(I) compounds are widely used in the treatment of arthritis (29). These compounds are generally ones in which the gold is coördinated to a sulfur atom (e.g. as in $Na_2[Au(S_2O_3)_2]$ or in complexes with organic mercaptides). Gold(I) coordination preferences run as: O < N < S, with S donors apparently able to replace O or N with ease. There are significant individual variations in the response of patients to gold therapy inasmuch as the immune system is involved. The occurrence of side effects involving the kidneys are not unknown and it is occasionally necessary to enhance the excretion of the gold which has been administered. Clinically, all of the compounds listed as used in mercury poisoning have also found use in the treatment of gold poisoning in arthritic patients. Little information seems to be available on the comparative efficacy of the various compounds which have been used for this purpose. D-penicillamine and N-Acetyl-D,L-penicillamine appear to be preferred in the United States and Western Europe because of their low toxicity and the ease with which they can be obtained and given; both of these compounds can be administered orally. The incidence of side reactions in gold therapy is fairly high (∿20%) and a small percentage of these is very serious. Like other class b or soft metals, gold can accumulate in and cause damage to the kidneys.

Copper. Copper(II) is one of the borderline species whose behavior falls between that of type a and type b species. It is easily reduced to copper(I) in the presence of sulfhydryl groups and this reaction is a part of the overall process when copper excretion is enhanced by sulfhydryl containing chelating agents (30). Its excretion can also be accelerated by species which are not capable of reducing copper(II) to copper(I), such as triethylenetetramine (31). As a result of the fundamental long-term studies of Walshe (32) on the treatment of hereditary copper-accumulating disorders, very effective treatments are now available for the treatment of chronic copper poisoning. The most effective of these use D-penicillamine to enhance urinary excretion of copper. This involves an initial reduction to copper(I) followed by the urinary excretion of the copper(I) complex. A complication may arise in individuals who are allergic to penicillin - these also have an allergic response to D-penicillamine. It is for these individuals that Walshe developed the use of triethylenetetramine as an alternative compound capable of enhancing the urinary excretion of copper.

Platinum(II). The recent upsurge in the use of platinum metal complexes in cancer chemotherapy, has led to an increased appreciation of their toxicity. The most widely used of these materials, cis[$Pt(NH_3)_2Cl_2$] is a material which is quite toxic, but also comparatively insoluble. As a consequence it is administered as a dilute solution, often via iv drip. Since platinum(II) is in many ways a typical class b ion, it can generally be mobilized by the same group of chelating agents which have been found to be effective for mercury(II) or gold(I). The nephrotoxicity which is a common feature of the use of this compound (32,33,34) can often be reduced by appropriate hydration of the patients (35). Reports of the use of chelating agents to offset the toxic effects of cis[$Pt(NH_3)_2Cl_2$] are not common but D-penicillamine has been reported to reduce the nephrotoxicity of this compound (36). Unfortunately this also limits the action of the cis complex against the cancer cells. There is no reason to believe that the other complexing agents shown in Table I would not be just about as effective as D-penicillamine in limiting the nephrotoxicity of the cis$[Pt(NH_3)_2Cl_2]$.

"Odd" Oxidation States. The toxicities of the type b elements are quite markedly dependent upon their oxidation states and mode of combination. These factors affect both the mode of distribution of these elements within the mammalian body and the manner in which they interact chemically with enzymes and other critical biomolecules. In this general area we find problems posed by organomercurials, organoarsenicals and compounds such as arsine and stibine. The toxicology of these compounds tend to exhibit some striking differences when compared with the compounds of these elements usually encountered in aqueous solution. To begin with, they often possess an elevated solubility in lipids and a correspondingly increased ability to pass through cell walls and to pass over the so-called "blood brain barrier". As a consequence some of these substances are extremely toxic and are also rather difficult to remove from the human body using water soluble chelating agents.

A very clear account of these problems may be found in the report of Petrunkin and his co-workers describing their search for an effective antidote for *arsine* poisoning (37). They found that typical water soluble chelating agents of the sort which are very effective in the treatment of poisoning by arsenite, were quite ineffective in the treatment of arsine poisoning. They ascribed this to the poor lipid solubility of these compounds which prevented them from gaining access to the critical sites at which arsine acted. They solved this problem by the design and synthesis of appropriate lipid soluble compounds which contained vicinal mercapto groups. Of these, the compound which they found most effective in animal studies on arsine poisoning was

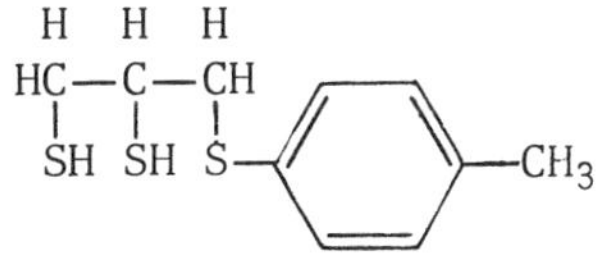

```
 H   H   H
HC — C — CH
 |   |   |
SH  SH   S — C6H4 — CH3
```

Mixed Ligand Chelate Therapy. It is well established that metal chelate complexes consisting of two or more chelating agents possess an additional stability over those containing just a single molecule of one of these chelating agents. This has led to the prediction that the appropriate combination of chelating agents should be more effective than a single one in removing certain metals from the mammalian body. The experimental information on the behavior of mixed chelate systems in acute cadmium poisoning is contradictory (38,39), but the effect may be present in certain studies (40) involving other elements.

One of the difficulties attending such studies is the problem of the *a priori* selection of the appropriate combination of chelating agents. The question has been discussed briefly in the literature but there seems to be no effective experimental testing of the suggestions.

In any discussion by a chemist of the therapeutic usage of chelating agents it is possible to loose sight of two important points. The first is that it is possible for a compound to serve as an antidote for say mercury poisoning even though it does not have a particularly notable ability to enhance the *excretion* of mercury (41). The second is that after a sufficiently long time interval between the introduction of the metal, the rapid enhancement of metal excretion from the human body does not necessarily lead to any significant clinical improvement (15).

Summary

The most striking feature of a survey of this sort is to see that the optimum chelate therapy for the metals in class b is invariably based upon the use of one of a small group of mercaptide bearing chelating agents (shown in Table I). Furthermore, on the basis of the work done in this area so far, it seems reasonable to predict that these same compounds would be effective in the treatment of types of class b toxicity which have not yet been studied experimentally. Thus toxicity due to compounds of Pd(II), Ag(I), Rh(III), and Ir(III) should be offset by the timely administration of these compounds. At the present time there is no evidence to suggest that a single one of these compounds will be found to be the optimum agent for all of these ions or even that the sequence of effectiveness will be the same when these compounds are tested against the different metal ions in this class. There is however little reason to doubt that the compounds shown in Table I will be at least partly effective in offsetting the toxicity of *any* of these ions. These, incidentally, are among those ions classified as nitrogen/sulfur seekers by Nieboer and Richardson (42) on the basis of X-ray structural information.

Acknowledgement

This work was carried out under the auspices of the Center of Environmental Toxicology, Department of Biochemistry, Vanderbilt University School of Medicine, Nashville, Tennessee, 37235 and supported by Grant 2R01ES01018-4 of the National Institutes of Environmental Health Sciences. I also wish to thank Neil H. Weinstein of the University of Florida for his stimulating and critical comments.

Literature Cited

1. Ahrland, S.; Chatt, J.; Davies, N. R. Quart. Rev., 1958, 12, 265.
2. Pearson, R. G., Ed. "Hard and Soft Acids and Bases"; Dowden, Hutchison & Ross:Stroudsburh, PA, 1973.
3. Friberg, L.; Vostal, J., "Mercury in the Environment", CRC Press:Cleveland, OH, 1972; pp. 70-88.
4. Tsuchiya, K., "Cadmium Studies in Japan: A Review", Kodansha Ltd.:Tokyo, 1978; pp. 76-79.
5. Voight, G. E.; Skold, G. Z. Ges. Exper. Med., 1963, 136, 326.
6. Friberg, L. A.M.A. Arch. Ind. Health, 1956, 13, 18.
7. Walshe, J. M. Clin. Sci., 1964, 26, 461.
8. Berlin, M.; Lewander, T. Acta Pharmacol., 1964, 22, 1.
9. Schubert, J. Atompraxis, 1958, 4, 393 and the references cited therein.
10. Catsch, A., "Dekorpierung radioaktiver und stabiler Metallionen", Karl Thiemig:Munich, 1968; pp. 7-14.
11. Heller, H.-J.; Catsch, A. Strahlentherapie, 1959, 109, 464.
12. Canty, A. J. in "Organometals and Organometalloids", Brinckman, F. E.; Bellama, J. M., Eds., ACS Symposium Series 82, American Chem. Soc.:Washington, D. C., 1978; pp. 327-338.
13. Carty, A. J. in "Organometals and Organometalloids", Brinckman, F. E.; Bellama, J. M., Eds., ACS Symposium Series 82, American Chem. Soc.:Washington, D. C., 1978; pp. 339-358.
14. Swensson, A.; Ulfvarson, J. Int. Arch. fur Gewerbepath. u. Gewerbehygiene, 1967, 24, 12.
15. Bakir, F.; Al-Khalidi, A.; Clarkson, T. W.; Greenwood, R. Bull. W.H.O., 1976, 53, 87, Suppl. "Conf. on Intoxication Due to Alkylmercury Treated Seed".
16. Jones, M. M.; Basinger, M. A.; Weaver, A. D.; Davis, C. M.; Vaughn, W. K. Res. Comm. Chem. Path. and Pharm. In press.
17. Mitchell. W. M.; Holy, N. L.; Jones, M. M.; Basinger, M. A.; Vaughn, W. K. Tox. Appl. Pharm. In press.
18. Jones, M. M.; Burka, L. T.; Hunter, M. E.; Basinger, M. A.; Campo, G.; Weaver, A. D. J. Inorg. Nucl. Chem., 1979, in press.
19. Sunderman, F. W. Jr.; White, J. C.; Sunderman, F. W.; Lucyszym, G. W. Amer. J. Med., 1963, 34, 875.
20. Casas, J. S.; Jones, M. M. J. Inorg. Nucl. Chem., 1979, in press.
21. Stricks, W. A.; Kolthoff, I. M. J. Am. Chem. Soc., 1953, 75, 5673.
22. Schwarzenback, G.; Widmer, M. Helv. Chim. Acta, 1963, 46, 2613.
23. Peters, R. A. "Biochemical Lesions and Lethal Synthesis", The Macmillan Co.:New York, 1963; pp. 40-58.
24. Katz, M. Advances in Pharmacol. Chemother., 1977, 14, 1.
25. Jordan, P.; Randall, K. Trans. Roy. Soc. Trop. Med. Hyg., 1962, 56, 136.

26. Chi, C.-T. Farmakol. i Toksikol., 1959, 22(1), 94. Chem. Abstr., 1959, 53, 20578.
27. Stevenson, D. S.; Suarez, R. M. Jr.; Marchard, E. J. Puerto Rico J. Pub. Health Trop. Med., 1948, 23, 533.
28. Walz, D. T.; Dimartino, M. J.; Suttin, B. M. in "Antiinflammatory Agents", Scherrer, R.; Whitehouse, M. M., Eds., Academic Press:New York, 1974; Vol. 1, pp. 209-239.
29. McCall, J. T.; Goldstein, N. P.; Randall, R. V.; Gross, J. B. Amer. J. Med., 1967, 254, 35/13.
30. Walshe, J. Quart. J. Med. N. S., 1973, XLII, 441.
31. Walshe, J. Brain, 1967, 90, 149.
32. Madias, N. E.; Harrington, J. T. Amer. J. Med., 1978, 65, 307.
33. Dentino, M.; Luft, F.; Yun, M. N.; Williams, S. D.; Einhorn, L. H. Cancer, 1978, 41, 1271.
34. Gozalez-Vitale, J. C.; Hayes, D. M.; Cvitkovic, E.; Sternberg, S. S. Cancer Treat. Rep., 1978, 62, 693.
35. Stark, J. J.; Howell, S. B. Clin. Pharmacol. Therap., 1978, 23, 461.
36. Osieka, R.; Bruntsch, U.; Gallmeier, W. M.; Seeber, S.; Schmidt, C. G. Deutsch. Med. Wochschr., 1976, 101, 192.
37. Mizyukova, I. G.; Petrunkin, V. E.; Lysenko, N. M. Farmakol. Tosikol (Moscow), 1971, 34(1), 70. Chem. Abstr., 1971, 74, 97223.
38. Schubert, J; Derr, S. K. Nature, 1978, 275, 311.
39. Jones, M. M.; Basinger, M. A. Res. Commun. Chem. Path. Pharmacol., 1979, 24, 525.
40. Volf, V.; Seidel, A.; Takada, L. Health Physics, 1977, 32, 155.
41. Gabhard, B. Arch. Toxicol., 1976, 35, 15.
42. Neiboer, E.; Richardson, D. H. S. Environmental Pollution (Series B), 1979. In press.
43. Ginsberg, J. M. Am. J. Physiol., 1965, 208, 832.

RECEIVED April 7, 1980.

20

Gold-Binding to Metallothioneins and Possible Biomedical Implications

C. FRANK SHAW III

Department of Chemistry and The Laboratory for Molecular Biomedical Research, The University of Wisconsin–Milwaukee, Milwaukee, WI 53201

Our group has undertaken an exploration of the biological and inorganic chemistry which control the metabolism of gold(I) in vivo and which must, therefore, form the basis for the mechanism of action of gold in treating rheumatoid arthritis (1,2,3,4). This article will proceed from a brief review of chrysotherapy and gold biochemistry to the interaction of gold with metallothionein and its role in the gold biochemistry.

Although the data and conclusions presented are drawn primarily from our own research, D. A. Gerber at SUNY Downstate Medical Center, B. M. Sutton and colleagues at Smith, Kline and French Corporation, Philadelphia, C. J. Danpure, P. J. Sadler, D. H. Brown and their respective coworkers, in England, Munthe and Jellum in Norway, and Mogilnicka and Piotrowski in Poland are making important contributions to understanding the biochemistry of gold and should be acknowledged.

Background

A. Chrysotherapy. The term chrysotherapy refers to the treatment of rheumatoid arthritis, RA (5,6) and recently pemiphigus (7) with gold complexes. Chrysotherapy began with a 1929 report by the Frenchman Forestier (8) and is thus a long established mode of treatment. Although its popularity waned after the introduction of cortisones, the problems associated with cortisones and the realization that gold can arrest the course of RA have led to renewed interest in chrysotherapy. In fact, Kaye and Pemberton in reviewing various treatments for RA state that the results of chrysotherapy "can be spectacular" (9). Commonly used drugs in the United States are gold sodium thiomalate (Myochrysin), I, and gold thioglucose (Solganal) II. Recently Sutton and co-workers (10,11) at Smith, Kline and French Corporation have developed a new drug, Auranofin, S-2,3,4,6-tetraacetyl-1-β-D-thioglucose(triethylphosphine)gold(I), III. This drug which can be administered orally circumvents some of the problems associated with gold therapy and is currently undergoing clinical

0-8412-0588-4/80/47-140-349$06.00/0

testing. The older compounds Na_2AuTM and AuTG must be administered in weekly to monthly intramuscular injections.

I Na_2AuTM II AuTG III $Et_3PAupATG$

At present, the sight and mode of action of chrysotherapy are uncertain, but a number of viable hypotheses, including the possibility of interactions with the immune system are under active consideration (5,6).

B. Gold Metabolism. The metabolism of gold follows the general sequence shown below during uptake, distribution and excretion:

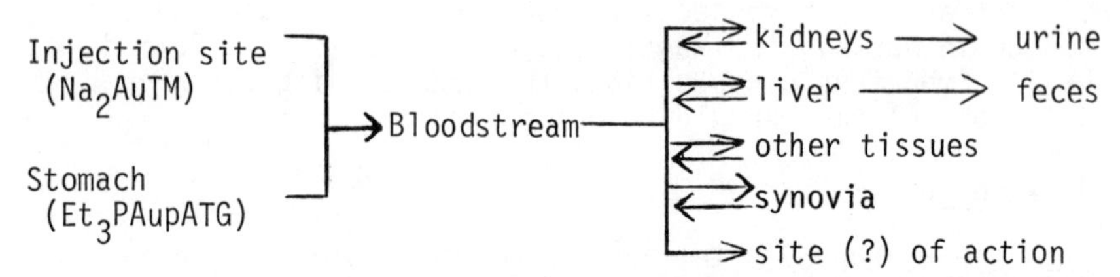

Gold(I) is adsorbed into the bloodstream from which it is transferred in substantial quantities to the kidney and liver, which are major sites of deposition as well as the routes of excretion, and in smaller amounts it is deposited in virtually all tissues of the body. At present, the site and mode of action of chrysotherapy are uncertain although gold is known to enter the synovial fluid of the joints (1,5). The deposition of gold is long-term, although the equilibration of recently injected $[^{195}Au]$-Na_2AuTM with accumulated gold takes place (12).

C. Gold Biochemistry. The biological chemistry has been reviewed recently (1,13) and will be briefly summarized here. An early experiment which provides an important clue to the biochemical fate of gold(I) was D. A. Gerber's demonstration that Na_2AuTM bonds to the single free sulfhydryl residue, cys-34, of bovine serum albumin (14). In the binary reaction, the thiomalate carrier ligand is retained:

$$\text{Protein-SH} + \text{AuTM}^{-2} \longrightarrow \text{Protein-S-Au-TM}^{-3} + \text{H}^{+}$$

However, recent experiments by Jellum and Munthe suggest that the reactions of Na_2AuTM *in vivo* and in blood serum *in vitro* lead to the displacement of thiomalate carrier ligand from gold (15,16). The exchange of thiol ligands at gold(I) occurs very rapidly as demonstrated by averaging of free and bound lignad signals in proton and carbon-13 nmr experiments (18,19). The dissociation of the thiomalate *in vivo* may result from displacement of the carrier ligand from the initially formed protein-gold-thiomalate complex by an endogenous thiol such as glutathione or cysteine (1):

$$\text{Protein-S-Au-TM}^{-3} + \text{RSH} \longrightarrow \text{Protein-S-Au-SR}^{-} + \text{HTM}^{-2}$$

In fact, gold(I) reactions in the cell cytoplasm are dominated by the gold sulfhydryl interactions. It is expected from the HSAB principle that gold(I) will have a strong affinity for sulfur ligands, but it is unique in comparison to other soft acids such as Pt(II), Hg(II) and Cd(II) in that it has an unusually low affinity for nitrogen bases (19).

Figure 1 compares the gold and thiol concentrations of rat renal coritcal cytosol from gold injected rats after fractionation by gel-exclusion chromatography. Four gold peaks are observed which correspond to the thiol maxima (2). These substances , labelled peaks I through IV may be identified as follows (2,3): I corresponds to the excluded fraction and includes proteins of molecular weight $\geq$110,000 Daltons; II is a protein of moderate size which has not been identified; III, a protein-bound species of approximately 10,000 Daltons, is now clearly established as gold bound to a metallothionein-like protein from the work described in this paper; IV is a mixture of low molecular weight complexes, predominantly glutathione and cysteine bound gold. Similar patterns are observed after administration of Et_3PAuCl orally, and $NaAuCl_4$ and AuTG intraperitoneally (Figure 2) suggesting that the pattern is independent of the carrier ligand used to administer the gold (2).

The correspondence of the gold of the cytosol fractions with the thiol groups suggests that the gold may distribute itself in a thermodynamic equilibrium among protein and non-protein thiol groups. Confirmation of this hypothesis is provided by Figures 3 and 4 which demonstrate the results of fractionating cytosol from control rats after incubating with Na_2AuTM and the effect of adding a thiol reagent such as cysteine or glutathione to the cytosol of a gold treated rat. The fact that *in vitro* addition of gold produces the same pattern (peaks I through IV) as *in vivo* administration is strong evidence for a thermodynamically--rather than biologically--controlled distribution. The ability of

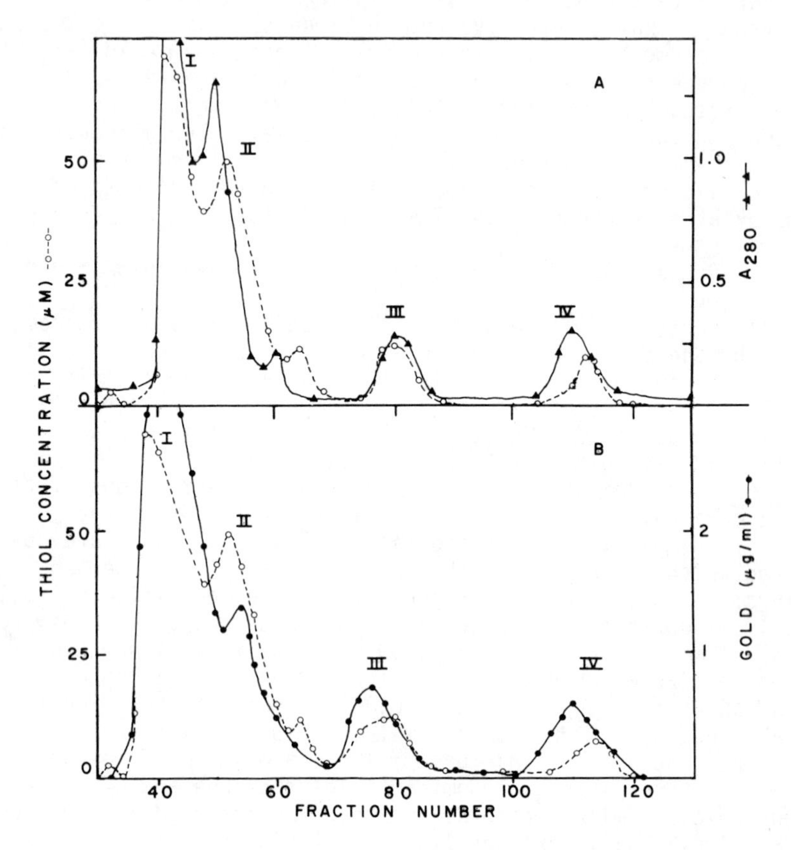

Bioinorganic Chemistry

Figure 1. Relationship of rat renal cytosolic thiol groups to gold distribution: a, [SH] and protein (A_{280}) for a control animal; b, [Au] and [SH] for an Na_2AuTM injected rat (2). Preparation of cytosol, fractionation on Sephadex G-100, and collection were performed under nitrogen to prevent air oxidation of the thiol groups. Conditions: Tris–HCl buffer 40mM pH 8.6; 2.5 × 60 cm column; flow rate 20 mL fractions.

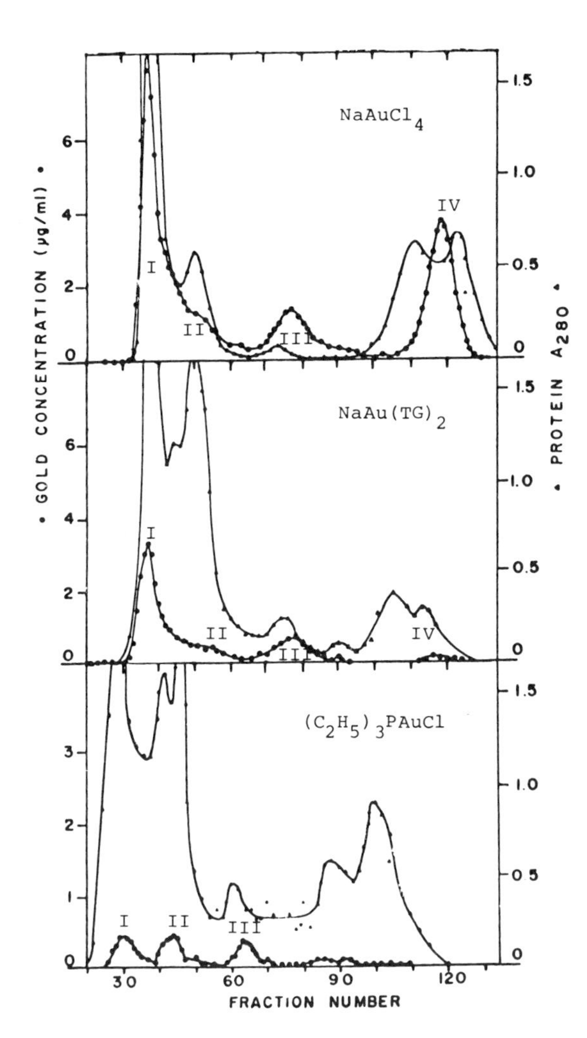

Bioinorganic Chemistry

Figure 2. Cytolsolic gold distribution following in vivo administration of: a, $NaAuCl_4$; b, $NaAu(TG)_2$; and c, $(C_2H_5)_3PAuCl$ to rats (2). In each case peaks I, II, and III are observed. Peak IV appears only at higher gold loadings (3). Conditions as in Figure 1. $NaAu(TG)_2$ was prepared by the in situ reaction of NaTG and AuTG in aqueous solution.

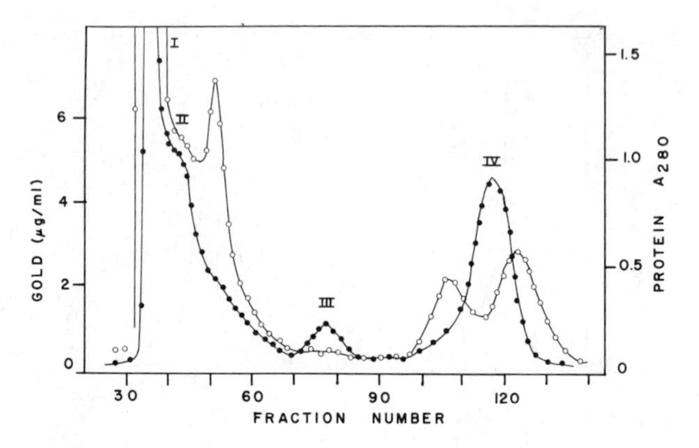

Bioinorganic Chemistry

Figure 3. The gold distribution following in vitro addition of Na_2AuTM (2.6μm) to renal cytosol from a control animal is qualitatively identical to that from in vivo administration, implying that it is a thermodynamically—not biologically—determined distribution (2)

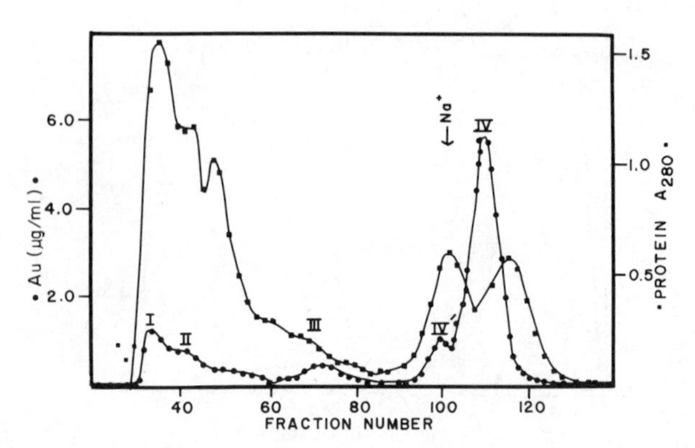

Inorganic Chemistry

Figure 4. Addition of cysteine to renal cytosol from a goldtreated rat at a 100/L thiol/gold ratio shifts the gold from Peaks I, II, and to a lesser extent, III into the low molecular weight fraction, Peak IV, which is $Au(cys)_2^-$ (3). The small peak, IV', is a glutathione complex of gold. Conditions as in Figure 1.

cysteine to compete for and displace protein-bound gold(I) (primarily from peaks I and II into peak IV) further supports this hypothesis.

The data of Figure 3 lead to another important conclusion, that the observed distribution of gold among cytoplasmic protein does not require *de novo* protein synthesis to explain the results, since the same pattern is observed *in vitro* as *in vivo*.

D. Metallothionein. Metallothioneins, MT, are a family of ubiquitous metal binding proteins isolated mainly from the livers and kidneys of mammals, but also found in other tissues and species. It was first reported in 1957 by Vallee and co-workers (20,21,22). It is a very small protein, 6,100 Daltons, (excluding metal ions), which can bind up to seven metal ions, primarily zinc and cadmium but also including trace to significant quantities of copper and traces of other metal ions (20,21). In gel-exclusion chromatography it appears to have a molecular weight of *ca*. 10,000, apparently due to a prolate ellipsoidal shape with a 6/1 axial ratio (20). A truly unique aspect of this protein is the presence of twenty cysteine residues among 61 amino acids (23). The structure of the equine renal protein, which was used in the chemical studies described here, is given in Figure 5. The protein as isolated from horse kidney is metal saturated--that is, there are no free thiol groups; they are all involved in metal mercaptide linkages. The average stoichiometry of the metal-sulfur binding sites must be 3SH/metal, although bridging sulfurs may lead to $[M^{+2}S_4]^-$ binding sites as proposed by Sadler (24). The biosynthesis of MT in response to cadmium exposure in various organisms and the high specificity of cadmium for metallothionein have lead to the hypothesis that MT may have a heavy metal detoxifying function (25,26,27,28). Recent evidence, however, demonstrates that increased hepatic Zn-MT levels are observed in response to nonheavy metal physiological stresses, including bacterial infection, partial hapectomy, CCl_4 exposure and burns (29,30,31). These observations suggest that metallothionein may have a role in zinc metabolism. Indeed, we have observed low levels of a zinc, copper binding protein with metallothionein-like properties in Erlich-Ascites tumor cells in the absence of heavy metal exposure (32,33).

E. Metallothionein-Gold Reactions. A number of laboratories (2,34,35,36,37) have observed and interpreted the occurrence of "10,000 Dalton" gold containing species in the cytosol of renal and hepatic tissue, as summarized in Table I. It is clear that various interpretations of gold-metallothionein binding and the ability of gold to induce MT biosynthesis have been promulgated. Therefore a comprehensive study of gold binding to MT and the related question of zinc, copper and cadmium antagonisms, which may relate to the mechanisms of gold therapy and toxicity, were undertaken in our laboratory (38)

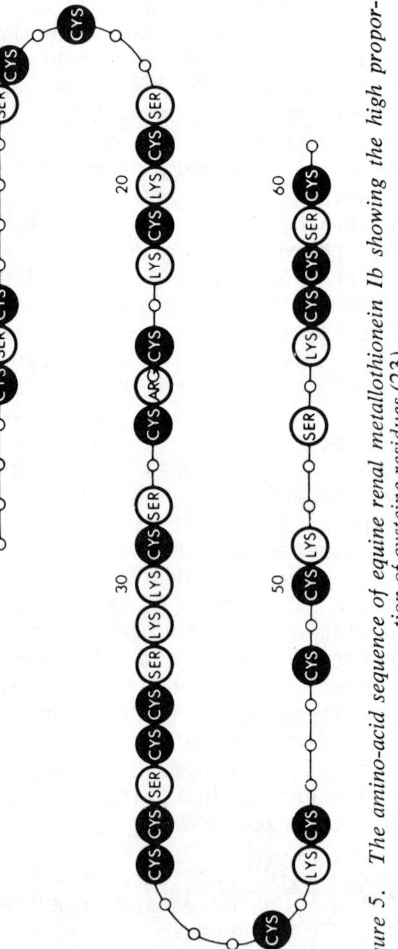

Figure 5. The amino-acid sequence of equine renal metallothionein Ib showing the high proportion of cysteine residues (23)

TABLE I. Previously Reported "10,000 Dalton" Gold-Binding Proteins

Compound	Dose in mg gold	Tissue	% of gold MT-bound	Attributed to MT-like protein	Au stimulated biosynthesis	Ref.
Na_2AuTM	7x16 mg/kg/day	rat kidney	13*	probably not	no	2
NaAuTG	7x16 mg/kg/day	rat kidney	18*	probably not	no	2
Et_3PAuCl	14x10 mg/kg/day	rat kidney	27*	probably not	no	2
$AuCl_4^-$	7x16 mg/kg/day	rat kidney	15*	probably not	no	2
$AuCl_4^-$	7x44 mg/kg/2 days	rat kidney	20%	yes	?	30
$AuCl_4^-$	2x2.5 mg/kg/day	rat liver	†	†	yes	31
Na_2AuTM	40 mg/kg	rat liver	33%*	---	---	28
Na_2AuTM	17x40 mg/kg	mouse liver	---	+	no	29
Na_2AuTM	17x40 mg/kg + $ZnCl_2$	mouse liver	small	yes +	no	29

*Planimeter estimates from published figures [28] or the original graphs [2].

†Increased levels of ZnMT were observed following gold(III) administration.

+No gold was observed in the MT fractions unless it was stimulated by $ZnCl_2$ administration.

and will be described here.

We have employed a variety of systems to study Au(I)-MT interactions including reactions of Na_2AuTM with highly purified equine renal MT (DEAE-Sephadex and Sephadex G-75); *in vivo* administration of $CdCl_2$ and Na_2AuTM to rats; and *in vitro* addition to rat liver cytosol. Because this is an inorganic symposium, the chemical aspects will be emphasized and the *in vivo* results only briefly discussed. [*Added note:* After the original version of this manuscript was submitted, a paper appeared demonstrating that $NaAuCl_4$ induces the biosynthesis of a metallothionein-like protein in rat liver and kidney to which some of the gold in these tissues is bound (43).]

Experimental

More complete details will be published in papers submitted elsewhere (38,39). All metal analyses (Au, Cd, Zn, Cu) were performed on a Perkin-Elmer 360 Atomic absorption spectrophotometer calibrated with serial dilutions of commercial standards. Column chromotography was carried out at 4°C on Glencoe columns packed with Sephadex-G-50 or G-75 using a peristaltic pump to maintain constant flow rates. Details regarding buffers, column size, etc., are given in the figure legends. Fractions were collected automatically and subsequently enalyzed for metal content by AA and absorption at 250 nm (when pure MT was used) or 280 nm (when cytosol was fractionated).

Metallothionein was isolated from horse kidney tissue by adaptation of standard methods and purified by Sephadex G-75 and DEAE-Sephadex chromotography (38). The reactions of purified MT with various amounts of Na_2AuTM were conducted by incubating samples at 4°C in Tris-HCl buffer, pH 7.8 for 2 hours and fractionating the mixture on Sephadex G-50. Fractions were analyzed for free and bound metals. The kinetics of reaction between MT and DTNB was studied by mixing DTNB and MT under N_2 in anaerobic cuvettes containing 10 mm Tris-HCl/0.1M KCl, pH 8.6 at 25°C. The reaction was followed by the appearance of 5-thio-2-nitrobenzoate which absorbs at 412 nm. Pseudo-first order conditions were maintained with excess (10 to 100 fold) DTNB.

Cytosol was prepared from the kidney and liver tissue of rats by homogenizing the tissue in buffered sucrose media containing 2-mercaptoethanol and centrifuging the homogenate at 5,000 *xg* and subsequently at 40,000 *xg* to yield a particulate free supernatant. At this point, a heat treatment step, in which the cytosol was heated in boiling water bath to 60°C, held at 60°C for 60 sec and cooled in an ice bath, was applied, after which the cytosol was recentrifuged at 40,000 *xg* (20 min) to remove denatured proteins. The heat treated cytosol was then fractionated on Sephadex G-75 and the metal content and A_{280} absorbance measured on all fractions. In various experiments, cytosol was isolated from the livers of rats injected intraperitoneally with

Na_2AuTM (7 x 8 mg Au/kg daily) or chronically exposed to cadmium (34.4 μg Cd^{+2} in the drinking water for 134 days). In some cases, (mixed in viro/in vitro experiments) the heat-treated cytosol from Cd-treated rats was incubated for 24 hours with Na_2AuTM and, conversely, heat-treated cytosol from Au-treated rats was incubated with $CdCl_2$. In other experiments (co-administration of $CdCl_2$ and Na_2AuTM), four rats were given $CdCl_2$ (2 mg Cd/Kg) and two days later Na_2AuTM (10 mg Au/kg), then sacrificed 24 or 96 hours later.

Results

A. In vitro Metallothionein/Na_2AuTM Reactions. Highly purified equine renal MT was allowed to react with various amounts of Na_2AuTM and then fractionated on Sephadex G-50 to separate the protein-bound and free metals. Because the protein does not have any UV-visible chromophores independent of the metal mercaptide linkages, the protein was "quantitated" by its cadmium content (measured by atomic absorption spectroscopy) and results are reported for various Au/Cd ratios of the reactants. Aliquots of a reaction mixture having a 10/1 Au/Cd ratio were fractionated after 2, 24 and 48 hours incubation time to insure complete reaction. In fact, the reaction had gone to completion within two hours (Figure 6), and all subsequent reactions were carried out for two hours before fractionation.

TABLE II. Na_2AuTM/Metallothionein Reactions†

Au/Cd ratios	0	1	2	3	5	25
Metallothionein						
Zn	0.66	0.26	0.38	-	-	-
Cd	1.00	1.00	1.00	0.86	0.46	0.17
Au	-	0.84	1.89	2.08	3.10	6.88
Low molecular weight species						
Zn	-	0.30	1.36	1.11	0.48	0.62
Cd	-	-	-	0.14	0.53	0.83
Au	-	-	-	0.50	1.43	19.34

†Mole ratios of metals in metallothionein and low molecular weight fractions relative to total Cd = 1.0.

When various Au/Cd ratios were employed, the results indicated a dramatic difference between the reactivities of protein-bound zinc and cadmium. Chromatograms for 2/1 and 5/1 ratios are presented in Figure 7 and data for these and other stoichiometries are summarized in Table II. Clearly the zinc is preferentially displaced by gold, in a reaction which goes to completion and is limited by the amount of gold present at the concentrations employed. However, when the zinc is completely displaced, an equilibrium displacement of cadmium is observed, and a decreasing proportion of the gold is MT bound. The chemistry can be represented as follows:

$$\text{Au(I)} + \text{Zn,CdMT} \longrightarrow \text{Au,CdMT} + \text{Zn}^{+2}$$

$$\text{Au(I)} + \text{Au,CdMT} \rightleftharpoons \text{AuMT} + \text{Cd}^{+2}$$

Thus with a 2/1 Au/Cd ratio, all the gold is protein-bound and 78% of the zinc, but no cadmium was displaced. At 3/1 and higher ratios, all of the zinc and progressively more cadmium are displaced from the protein. Even with a large excess of Na_2AuTM, complete displacement of cadmium was not observed (Table 2).

The equations just presented ignore the possible role of the thiomalate carrier ligand in gold-MT binding. Three different binding models are possible: (a) bonding of one AuTM moiety to a single SH group, which requires 3 Au bound/M^{+2} displaced; (b) isomorphous displacement of M^{+2} by Au(I) which requires 1 Au bound/M^{+2} displaced; and (c) chelation of gold by two SH groups, which results in 1.5 Au bound/M^{+2} displaced.

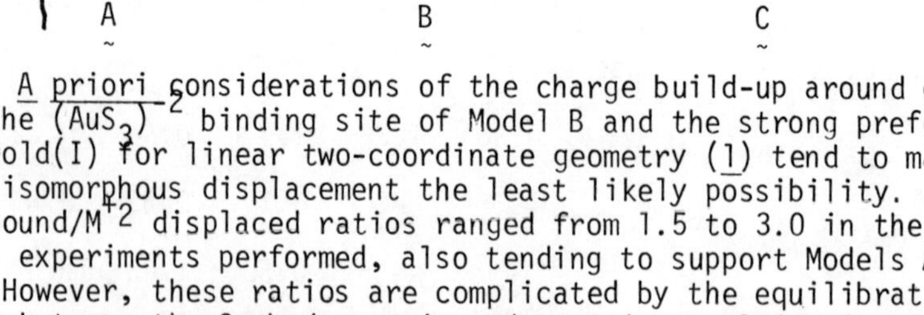

SAuTM
SAuTM
SAuTM
A

S
Au—S
S
B

S — Au — S
C

A priori considerations of the charge build-up around gold in the $(AuS_3)^{-2}$ binding site of Model B and the strong preference of gold(I) for linear two-coordinate geometry (1) tend to make the isomorphous displacement the least likely possibility. The Au bound/M^{+2} displaced ratios ranged from 1.5 to 3.0 in the various experiments performed, also tending to support Models A and C. However, these ratios are complicated by the equilibration of zinc between the Sephadex resin and protein resulting in zinc recoveries between 80 and 160%, and it is not possible to distinguish between Models A and C. Experiments with [^{35}S]-Na_2AuTM are planned to resolve the question.

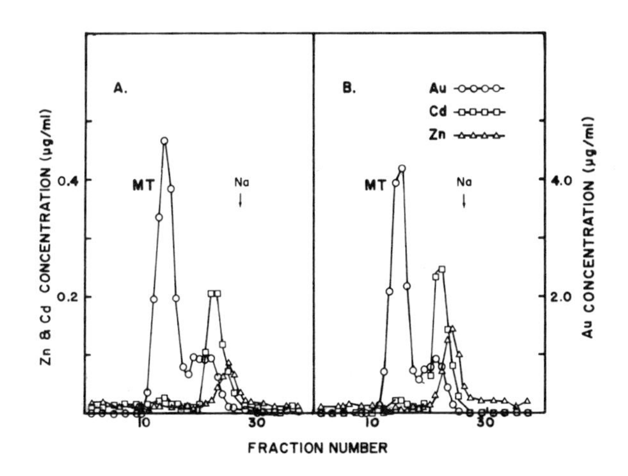

Figure 6. The distribution of metals between MT and the low molecular weight fraction after incubation of Na_2AuTM and equine renal MT at a 10/L Au/Cd ratio for: a, 2 h; and b, 26 h. The reaction is complete within 2 h. Conditions: Sephadex G-50 column (1.5 × 60 cm) eluted with Tris–HCl 40mM, pH 7.8 at 5 mL/h; 3.2 mL fractions.

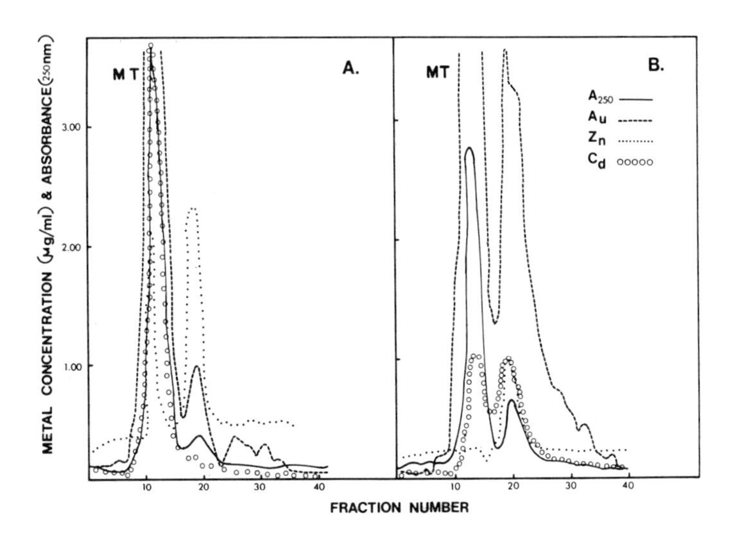

Figure 7. The effect of increasing Au/Cd ratio on metal displacement from MT: a, 2/1 ratio; and b, 5/1 ratio. Conditions as in Figure 6 with 2-h reaction time.

These results unambiguously demonstrate that gold(I) will bind to metallothionein *in vitro* with displacement of zinc and cadmium. However, this fact, even in conjunction with a "10,000" Dalton gold protein species *in vivo* does not prove that the latter peak is MT-bound gold.

B. The DTNB Reaction of Metallothionein. The appearance of reactive (or "free") thiols in the 10,000 Dalton peak of rat renal cytosol, which was previously reported, was apparently inconsistent with the published properties of equine renal metallothionein--namely, that all sulfhydryl groups were present as metal mercaptide linkages. We have subsequently examined the reaction of 5,5'-dithiobis(2-nitrobenzoic) acid (DTNB, a standard agent for measurement of the SH concentrations (40)) with Zn-thionein. A bi-phasic reaction was found in which each step had DTNB dependent and independent terms (Figure 8):

$$k_I = 0.41\ M^{-1}\ sec^{-1}\ [DTNB] + 0.65 \times 10^{-3}\ sec^{-1}$$

$$k_{II} = 0.044\ M^{-1}\ sec^{-1}\ [DTNB] + 0.12 \times 10^{-3}\ sec^{-1}$$

(In these experiments, Zn-thionein was prepared by dialysis against EDTA at low pH, and reconstituted with a single metal ion.) The observed rate constants indicate that both dissociative and associative steps are occurring. In the dissociative step the sufhydryl dissociates from the Zn and reacts with DTNB in a fast step. The mechanism below is proposed to explain the DTNB dependence of the reaction as well as other observations not discussed here (particularly the absence of effects due to EDTA or hydrogen ion on the rates (39)).

COO⁻ ⁻OOC
O_2N–⟨O⟩–S–S–⟨O⟩–NO_2 + Protein—S—Metal ⟶

Metal
Protein--S
S–⟨O⟩–NO_2 + O_2N–⟨O⟩–S ⟶
⁻OOC ⁻OOC

⁻OOC
Protein—S—S–⟨O⟩–NO_2 + Metal

The biphasic nature of the reaction will probably require elucidation of the tertiary structure of MT for a correct explanation.

For our purposes, the significant aspect of this reaction is that it provides an explanation for the reactive SH groups of the 10,000 Dalton fractions of the kidney cytosol: they are the result of metallothionein reacting slowly with the DTNB. The measured rate constants are adequate to explain the 50 μM SH concentration previously observed after fifteen minutes reaction time.

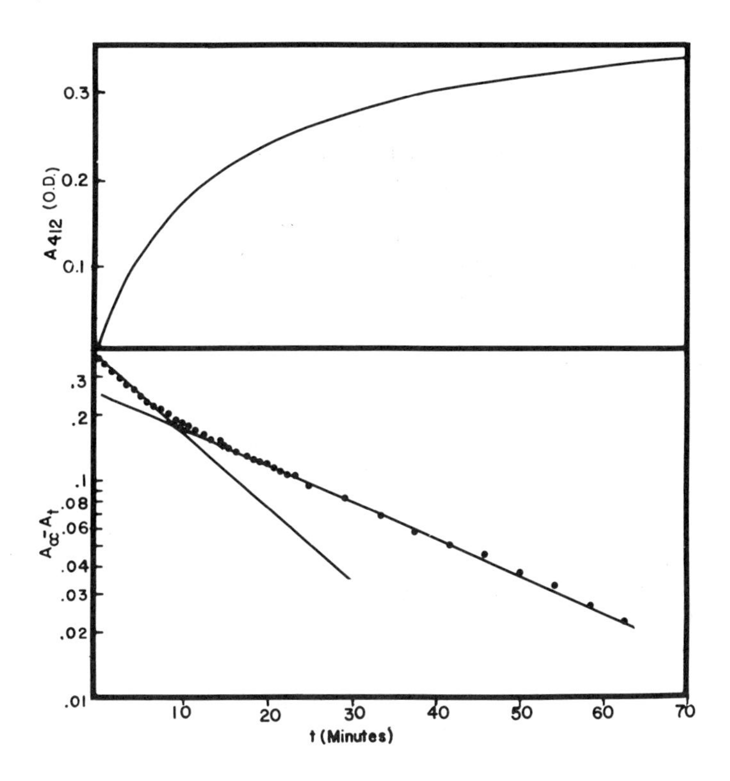

*Figure 8. The reaction of reconstituted equine renal Zn–MT with DTNB (5,5′-dithiobis (2-nitrobenzoic acid)) followed by the absorbance at 412 nm. A two-step reaction in which each has two components is observed. Conditions: 10m*M *Tris–HCl, 0.1*M *KCl, pH 6.8 at 25°C, [DTNB] = 6.7m*M, *$[Zn]_{MT}$ = 17* μM.

C. Co-Administration of $CdCl_2$ and Na_2AuTM *in vivo*. To further explore whether gold and cadmium both bind to metallothionein *in vivo*, $CdCl_2$ and Na_2AuTM were injected into a rat (2 mg Cd/kg and 10 mg Au/kg intraperitoneally 48 hours apart). After sacrificing the animals, the kidneys and livers were isolated and cytosol (100,000 xg supernatant) prepared and fractionated on Sephadex G-75. The cytosol fractions were analyzed for gold, cadmium and zinc by atomic absorption spectroscopy. The chromatograms are presented in Figure 9. As expected, the gold and cadmium peaks in the 10,000 molecular weight region, peak III, coincide thereby providing strong support for the hypothesis that gold binds to a metallothionein-like protein *in vivo*. Peaks I and II are also clearly present, but peak IV, found only with high gold levels (3), does not occur in the liver. This is the first direct demonstration that cadmium and gold co-elute in the "10,000 Dalton" metal binding species.

D. Differences in Au-MT and Cd-MT Specificity. Although the *in vivo* and *in vitro* experiment described above convincingly establish that a metallothionein-like protein can and indeed does bind gold(I) *in vivo*, there is a striking difference in the specificity of metallothionein for gold and cadmium. These differences are well illustrated by a series of experiments using rat liver cytosol. The heat treated cytosol was prepared from liver homogenate and then fractionated after *in vivo* exposure of the rats to Na_2AuTM (7 x 8 mg Au/kg/day intraperitoneal injections) or $CdCl_2$ (34.4 μg Cd/ml in drinking water for 134 days) as shown in Figures 10a and 11a. Table III shows the dramatic changes in the metal/protein ratios and the percentages of metal initially present in going from the homogenate to metallothionein. While almost 70% of the cadmium in the initial homogenates is MT bound, less than 1% of the gold is. Although the doses and methods of administration for Cd and Au used here are not exactly comparable, numerous other experiments under a variety of conditions establish that the results obtained here for cadmium and gold are, in fact, typical of the two elements.

E. Mixed in vivo/in vitro Experiments. The same liver cytosol preparations as above were used to study metal displacements from metallothionein by adding gold to the *in vivo* Cd^{+2} treated cytosol and vice versa. These experiments more nearly reproduce the situation when metallothionein is presented with gold(I) in the cellular milieu than do the experiments with pure equine renal MT. The resulting chromatograms are shown in Figures 10b and 11b, and the metal contents of the metallothionein peaks are given in Table IV.

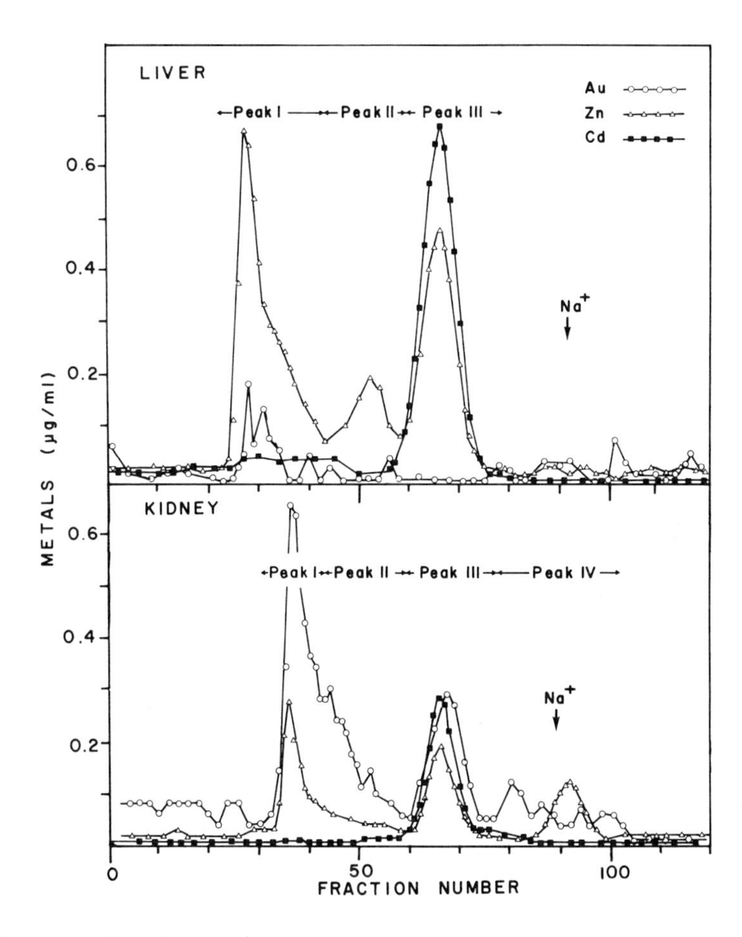

Figure 9. Hepatic and renal cytosol from a rat administered $CdCl_2$ and Na_2AuTM, demonstrating the co-elution of metallothionein bound Cd and Au, Peak III. Conditions: Sephadex G-75 (2.5 × 60 cm) column eluted with 50mM Tris–HCl/0.15M KCl; 3 mL fractions.

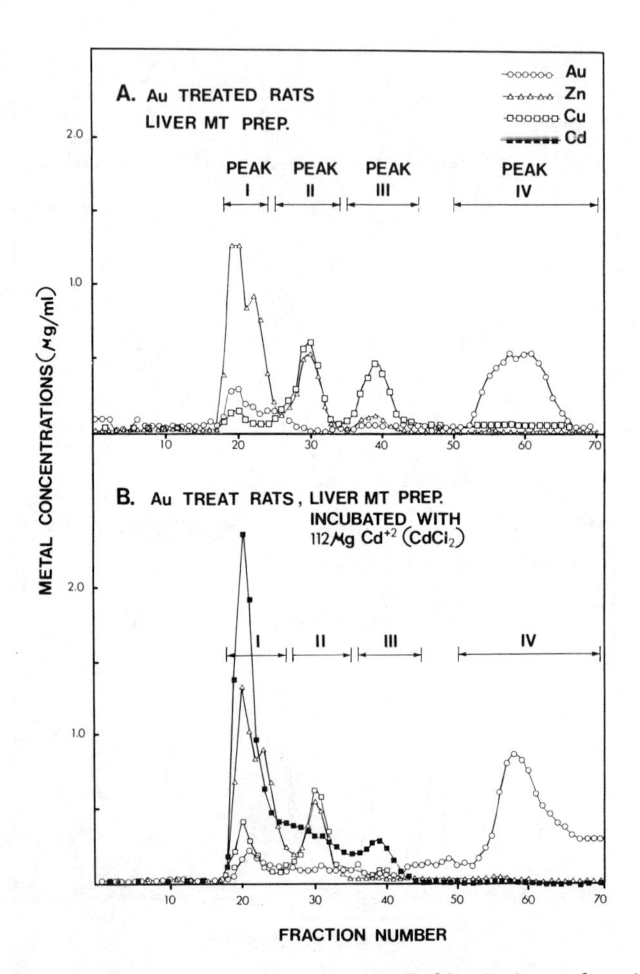

Figure 10. Displacements of metal from metallothionein in rat hepatic cytosol. Fractionation of: a, cytosol from rat administered Na_2AuTM in vivo; b, cytosol from a that was incubated with 112 μg of Cd^{+2} for 24 h. 2-Mercaptoethanol was added to the cytosol to prevent air oxidation. Conditions: Sephadex G-75-120 column (2.5 × 100 cm) eluted with 5mM Tris–HCl, pH 7.8 at 20 mL/hr; 9 mL fractions.

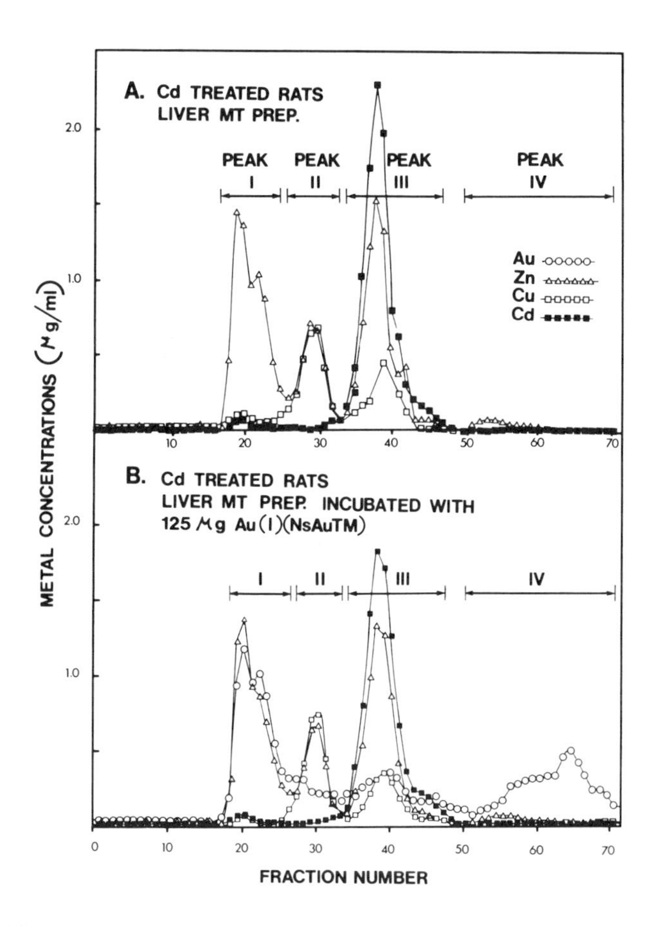

Figure 11. Metal displacement from Cd-induced metallothionein of rat hepatic cytosol. Fractionation of: a, cytosol from Cd exposed rate; and b, cytosol from a to which 125 μg Au(I) was added. Conditions as in Figure 10.

TABLE III

Metal/Protein Ratios of Liver Supernatants

	Cd-treated rats[†]			Au-treated rats*		
	Cd%	Prot.%	μgCd/ mg Prot.	Au%	Prot.%	μgAu/ Prot.
Homogenate	100	100	.0624	100	100	.224
40K Supernatant	75	50	.0939	29	61	.100
Heat-treated Supernatant	72	14	.283	5	14	.078
Metallothionein	68	+	+	>1	+	+

[†]homogenate: 13.4 μg Cd/g wet tissue

*homogenate: 50.4 μg Au/g wet tissue

+the biuret protein assay is not accurate for metallothionein because of the high cysteine content.

TABLE IV

Metal Displacements from Metallothionein in Rat Liver Cytosol by *In Vitro* Addition of Metal

Treatment		nmoles of metal in MT Fraction (%)*			
in vivo	*in vitro*	gold	cadmium	zinc	copper
Cd	Au	118(18)	764(91)	868(40)	292(39)
Cd	-	0(0)	789(95)	926(33)	321(44)
Difference		+118	-25	-58	-29
Au	Cd	33(6)	117(11)	33(3)	55(9)
Au	-	29(5)	0(0)	67(6)	248(40)
Difference		+4	+117	-34	-193

*Percentage of each metal in cytosol which is Mt-bound(%= $\frac{\text{nmol in MT X100}}{\text{nmol in cytosol}}$.

In the liver a substantial amount of copper is metallothionein bound, in contrast to the traces found in kidney tissue. In these experiments a Zn/Cd/Cu ratio of 9/8/3 was observed. When gold is added to the cadmium-treated cytosol, zinc, copper and cadmium are displaced from MT in a 2/1/1 ratio. As with the horse kidney MT, the zinc is preferentially, although not exclusively, displaced by gold. The MT in the gold-treated cytosol has an unusually high copper content, and when cadmium is added it is predominantly the copper and some zinc which are displaced in a 6/1 ratio.

Discussion

The data for the chemical reactions of MT with Na_2AuTM and DTNB allow us to conclude that gold binds to metallothionein *in vitro* and to a metallothionein-like binding protein *in vivo*. The binding of gold(I) to MT can be entirely explained by reaction with pre-existing MT, although the possibility of limited MT biosynthesis can not be eliminated from presently available data. However, it is quite evident that MT plays only a minor and possibly incidental role in the metabolism of gold. In contrast, metallothionein is the principal binding site for cadmium in the cell.

Rajagopalan *et al.* (37) have demonstrated that administration of $AuCl_4^-$, a gold(III) complex, to rats results in increased levels of liver zinc metallothionein (the gold content of the cytosol fractions were not measured). However, gold(III) is much more toxic than gold(I)(1) and the effect may be a non-specific stress induction of MT, similar to that produced by CCl_4, bacterial infections, etc. (29-31). Although the distinction may seem to be semantic, the extent of MT-biosynthesis induced by cadmium versus gold and the consequences for the biochemistry of the metals are quantitatively and qualitatively different.

When Na_2AuTM reacts with purified MT, there is a strong discrimination between zinc and cadmium, the former being much more readily displaced. However when Na_2AuTM is added to the cytosol, from Cd-treated animals, it results in a displacement of zinc, copper and cadmium in a 2/1/1 ratio. When cadmium is added to the cytosol of Au-treated rats in which MT contains zinc, copper and gold only zinc and copper were displaced. From this it can be deduced that the affinity of metallothionein for gold is comparable to that for cadmium and greater than that for zinc:

$$Au \sim Cd > Zn$$

Although the binding of Au(I) to MT *in vivo* may be an incidental aspect of its metabolism, it could have profound consequences affecting the toxicity of gold compounds and their mechanism of action in rheumatoid arthritis. If it is assumed that metallothionein has some role in zinc and possibly copper metabolism--and the stress induction of Zn-MT supports this supposition--then the binding of gold to MT with displacement of zinc and copper may alter zinc and possibly copper metabolism in some inimical fashion, contributing to or complicating other toxic reactions.

A number of treatments which are in general use or clinical testing for rheumatoid arthritis may alter zinc or copper metabolism. These include gold(I) complexes such as Na_2AuTM and $Et_3PAupATG$, penicillamine which is a good metal chelating agent, and salts of copper and zinc themselves. The possibilities of common mechanisms for some of these agents has been discussed by Whitehouse and Walker (41) and Jellum and Munthe (15,16). Evans has previously postulated that zinc/copper antagonisms may occur at metallothionein (42). Our research now expands that to include possible gold antagonisms of copper and zinc occurring at least in part at metallothionein on the molecular level. Continuing research on the biochemical mechanisms and basis of gold metabolism will be necessary to elucidate the role of gold-MT interactions in such processes.

[The paper of Mogilnicka and Piotrowski (Biochem. Pharmacol., 1979, 28, 2625-2531) published after the original submission of this article establishes that some de novo biosynthesis of a metallothionein-like binding protein does occur in response to $AuCl_4^-$ administration to rats. Movement of copper into the kidney was also established. Whether these effects occur to the same extent for gold(I) thiolates remains to be determmined.]

V. Literature Cited

1. Shaw, C. F. III. Inorg. Persp. Biol. Med., 1979, 2, 287-355.
2. Thompson, H. O.; Blaszak, J.; Knudtson, C. J.; Shaw, C. F. III. Bioinorg. Chem., 1978, 9, 375-388.
3. Shaw, C. F. III; Schmitz, G., Thompson, H. O.; Witkiewicz, P. J. Inorg. Biochem., 1979, 10, 317-330.
4. Schaeffer, N. S.; Shaw, C. F.; Thompson, H. O.; Satre, R. W. Arthritis Rheum., 1980, 23, 0000.
5. R. H. Freyberg, revised by Ziff, M.; Baum, J. in J. L. Hollander, Ed., "Arthritis and Allied Conditions," 8th Edition, Lea and Fibiger, Philadelphia, 1972, 455-482.
6. David, P.; Harth, M. J. Rheumatol., 1979, 6, Suppl. 5.
7. Pennys, N. S.; Eaglestein, W. E.; Frost, P. Arch. Dermatol., 1976, 112, 185-187.
8. Forestier, J. Bull. et Mem. Soc. Med. d'Hop. de Paris, 1929, 53, 323.
9. Kaye, R. L.; Pemberton, R. E. Arch. Int. Med., 1976, 136, 1023-1028.
10. Sutton, B. M.; McGusty, E.; Walz, D. T.; DiMartino, M. J. J. Med. Chem., 1972, 15, 1095-1098.
11. Finkelstein, A. E.; Walz, D. T.; Batista, V.; Mizraji, M.; Roisman, F.; Misher, A. Ann. Rheum. Dis., 1976, 35, 251-257.
12. Gerber, R. C.; et al., J. Lab. Clin. Med., 1974, 83, 778-789.
13. Sadler, P. J. Struct. Bonding, 1976, 29, 171-219; Gold Bulletin, 1976, 5, 110-118.
14. Gerber, D. A. J. Pharmacol. Exp. Ther., 1964, 143, 137-40.

15. Jellum, E.; Munthe, E.; Guldal, G.; Aaseth, J. Scand. J. Rheumatology, Suppl. 28, 1979, 28-36.
16. Jellum, E.; Munthe, E. in Willoughby, D. A.; Giround, J. P.; Velo, P., Eds., "Perspectives in Inflammation, Future Trends and Developments," MTP Press Ltd., Lanchester, England, 1976, 578-581.
17. Isab, A. A.; Sadler, P. J. J. Chem. Soc., Chem. Commun., 1976, 1051- 1052.
18. Eldridge, J. E.; Cancro, M. E.; Shaw, C. F. III. Abstract 136, 1979 Great Lakes Regional ACS Meeting, Rockford, Illinois, June 4-6, 1979.
19. Puddephatt, R. J., "The Chemistry of Gold, Elsevier, Amsterdam, 1978, 274 pages.
20. Kägi, J. H.; Himmelhoch, S. R.; Whanger, P. D.; Bethune, J. L.; Vallee, B. L. J. Biol. Chem., 1974, 249, 3537-3542.
21. Kagi, J. H. R.; Vallee, B. L. J. Biol. Chem., 1961, 236, 2435-2442.
22. Margoshes, M.; Vallee, B. L. J. Am. Chem. Soc., 1957, 79, 4813-4814.
23. Kojima, Y.; Berger, C.; Vallee, B. L.; Kägi, J. H. R. Proc. Nat'l. Acad. Sci. USA, 1976, 73, 3413-3417.
24. Sadler, P. J.; Bakka, A.; Benyon, P. J.; FEBS Lett., 1978, 94, 315-318.
25. Kägi, J. H. R.; Vallee, B. L. J. Biol. Chem., 1960, 235, 3460-3465.
26. Webb, M. Biochem. Pharmacol., 1972, 21, 2751-2765.
27. Shaikh, Z. A.; Smith, J. C. Chem. Biol. Interact., 1977, 19, 161-171.
28. Hildago, H. A.; Koppa, V.; Bryan, S. E. Biochem. J., 1978, 170, 219-225.
29. Oh, S. H.; Deagan, J. T.; Whanger, P. D.; Wesig, P. H. Am. J. Physiol., 1978, 234, E282-285.
30. Richards, M. P.; Cousins, R. J. J. Nutr., 1976, 106, 1591-1599.
31. Sobocinski, P. Z.; Canterbury, W. J., Jr.; Mapes, C. A.; Dinterman, R. E. Am. J. Physiol., 1978, 234, E399-406.
32. Koch, J.; Wielgus, S.; Shankara, B.; Shaw, C. F. III; Petering, D. H. Biochem. J., in press.
33. Petering, D. H.; Koch, J.; Wielgus, S.; Shaw, C. F. III. Fed. Proc., 1978, 37, 1352.
34. Lawson, K. J.; Danpure, C. J.; Fyfe, D. A. Biochem. Pharmacol., 1977, 26, 2417-2426.
35. Turkall, R. M.; Bianchine, J. R.; Lerber, A. P. Fed. Proc., 1977, 36, 356.
36. Mogilnicka, E. M.; Piotrowski, J. K. Biochem. Pharmacol., 1977, 26, 1819-1820.
37. Winge, D. R.; Premakumar, R.; Rajagopalan, K. V. Arch. Biochem. Biophys., 1978, 188, 466-475.
38. Schmitz, G.; Minkel, D. T.; Gingrich, D.; Shaw, C. F. III. J. Inorg. Biochem., 1980, 11, 0000.

39. Li, T.-Y.; Minkel, D. T.; Shaw, C. F. III; Petering, D. H. Biochem. J., submitted.
40. Ellman, G. L. Arch. Biochem. Biophys., 1959, 82, 70-77.
41. Whitehouse, M. W.; Walker, W. R. Agents Actions, 1978, 8, 85-90.
42. Evans, G. W.; Majors, P. F.; Cornatzer, W. E. Biochem. Biophys. Res. Commun., 1970, 40, 1142-1148.
43. Mogilnicka, E. M.; Piotrowski, J. K. Biochem. Pharmacol., 1979, 28, 2625-2631.

RECEIVED April 24, 1980.

21

Wilson's Disease and the Physiological Chemistry of Copper

I. HERBERT SCHEINBERG

Albert Einstein College of Medicine, Bronx, NY 10461

Copper, zinc and iron are the only heavy metals known with certainty to be essential to human life and health. As cations all three are similar to other heavy metals, such as mercury and lead, in their capacity to combine with many proteins, precipitating them or impairing their physiological functions. Yet, though present in some tissues in concentrations far in excess of toxic concentration of mercury or lead, copper, iron or zinc toxicity is seen in man only in the presence of a specific genetic defect in the metal's metabolism, as in Wilson's disease; or when the metal enters the body through a parenteral route, as in the chills, fever, and nausea resulting from the inhalation of zinc fumes. Studies of Wilson's disease of the last 30 years suggest that the remarkable innocuousness of essential heavy metals is a consequence of their essentiality.

Wilson's Disease

All newborn infants have an increased concentration of copper in the liver, and a decreased concentration of a specific plasma copper-protein, ceruloplasmin, in comparison to individuals over one year of age (1). The rare infant who has inherited a pair of the specific autosomal recessive gene that causes Wilson's disease exhibits an excess of hepatic copper and a deficiency of ceruloplasmin for life.

Although there are a number of human copper proteins - Table I lists those known to occur in man - ceruloplasmin is the only one known to present an abnormality in Wilson's disease. Ceruloplasmin is a blue, alpha globulin, with a molecular weight of 131,000 daltons, and 6 atoms of prosthetic, tightly bound copper per molecule (2). It is a glycoprotein, with a well-defined sugar moiety, and both cupric and cuprous ions in its structure and exhibits oxidase activity towards a number of polyamines, polyphenols and the ferrous ion. Its phyisological function is unknown. The fact, however, that Wilson's disease, which is chronic copper toxicosis of man, is associated with lifelong deficiency or

0-8412-0588-4/80/47-140-373$05.00/0

Table I
Human Copper Proteins

Protein	Organ or tissue Source	Molecular Weight x 10^{-3}	Copper content (atoms/ molecule)	Color	Biochemical function	Physiological function	Hereditary defect	Comments
Ceruloplasmin (EC 1.12.3.1)	Plasma	132	6	Blue	Polyamine oxidase; ferroxidase	Unknown	Deficient or absent in Wilson's disease, and in Menkes' syndrome	Glycoprotein
Dopamine -mono-oxygenase (EC 1.14.17.1)	Plasma; Cerebrospinal fluid; Pheochromocytoma of adrenal	290	4	Colorless	Dopamine - hydroxylase	Biosynthesis of noradrenaline		Glycoprotein
Cytocupreins (EC 1.15.11) Erythrocuprein Cerebrocuprein Hepatocuprein	 Erythrocytes Brain Liver	31-33	2	Blue-green	Superoxide dismutase	Catalysis of dismutation of O_2^-		Each molecule also contains 2 atoms of zinc
Cytochrome c oxidase EC 1.9.3.1	Liver	200-250	2	Colorless	Cytochrome c: oxygen oxido-reductase	Electron transport	Activity depressed in Menkes' syndrome	Hemoprotein 1 heme/1 copper
Cytosolic copper binding protein; L-6D	Liver	10	1-5	Colorless	Sequestration of cuprous ions	Unknown		
Neonatal hepatic mitochondrocuprein	Liver	6 for a solubilized peptide derivative	2-4	Colorless	Sequestration of cuprous ions	Unknown		Despite its name localized in heavy lysosomes
Tyrosinase EC 1.10.3.1	Skin	(118)	4	Colorless	Hydroxylation of tyrosine to dihydroxyphenylalanine (DOPA)	Melanogenesis	Absent in albinism	Quantitative data on tyrosinase derived from mushroom preparation

absence of ceruloplasmin suggests that the metabolism of ceruloplasmin and the homeostatic mechanism that maintains normal copper balance are in some way connected. No other primary clinical or biochemical abnormality has been reported even in patients with Wilson's disease who appear to have no ceruloplasmin at all.

Experiments with ceruloplasmin whose sugar chains have been freed of their terminal sialic acid residues indicate that desialylated ceruloplasmin is rapidly removed from the circulation (3), transported into hepatocytic lysomes where is is catabolized, and whence at least some of its copper is excreted into the bile (4). The rate of accumulation of the excess bodily deposits of copper in patients with Wilson's disease is compatible with the hypothesis that a block in the excretory pathway of ceruloplasmin-copper constitutes the fundamental metabolic defect in this disorder.

Whether or not a result of abnormal ceruloplasmin metabolism, the patient with this disease steadily accumulates copper in his liver as he ages. The normal concentration of copper in the liver is below 50 micrograms per gram of dry liver, but it is not unusual to find concentrations of 1000 in adolescent patients who may simultaneously be completely asymptomatic and apparently in perfect health (5).

Nevertheless, it is certain from microscopic examinations of liver biopsy samples, and chemical measurements on blood in over a hundred asymptomatic patients with Wilson's disease, that the accumulation of hepatic copper is accompanied by structural and functional damage to the liver. At some time between the middle of the first and fifth decades of life necrosis of the primary functional cells of the liver results from this organ's overload of copper (6).

Death of hepatic cells may be so massive that severe hepatitis may end the patient's life within weeks of the first clinical sympton. Even when the hepatitis is not fatal it is often accompanied by the release into blood of sufficient copper from damaged hepatocytes to cause copper-induced hemolysis of so many erythrocytes that anemia results (7).

Frequently the clinical evidence of copper's toxic effects on the liver are slight, or disregarded. Necrotic hepatocytes may release copper to the blood gradually, resulting in the accumulation of significant amounts of free plasma copper. Whereas in normal individuals virtually all but a few percent of the plasma copper is in ceruloplasmin, untreated patients with Wilson's disease may have more than 50 micograms of copper per 100 ml of plasma - half the normal total plasma copper - which is not bound to ceruloplasmin. Unlike ceruloplasmin-copper, the non-ceruloplasmin, or free copper of plasma is able to diffuse across membranes. One apparently paradoxical result of this is that untreated patients with Wilson's disease generally excrete several hundred micrograms of copper in the urine daily while normal individuals, excreting little more than 10 micrograms daily, do not accumulate toxic amounts of copper as they age. Unfortunately, patients are not kept in zero copper balance by this elevated urinary output, and their free plasma copper diffuses across the blood-brain barrier.

The toxic effects of copper in the brain are disastrous (Fig.1). If the patient with Wilson's disease does not die of the damage copper produces in the liver, he will eventually suffer from severe neurologic or psychiatric disturbances, and will finally die of copper toxicosis of the central nervous system (8).

How can an element that is essential to life be lethal? The answer is not simply that the concentrations of copper are greater in patients with Wilson's disease than in other individuals. In fact, the total concentration of copper in the plasma of patients with Wilson's disease is almost always lower than that of a normal individual, since the elevated level of non-ceruloplasmin-copper in the patient generally does not compensate for his deficient level of ceruloplasmin-copper. Even in the brain of a dead patient, concentrations of copper may be only several-fold higher than normal (9).

Copper is toxic in patients with Wilson's disease because its thermodynamic activity is appreciable. In normal individuals the activity of copper is virtually zero because copper is almost always present only as the prosthetic element of one of another specific protein. A copper ion that is firmly bound to a specific copper-protein is not toxic, as exemplified by ceruloplasmin-copper, which is normally present at a concentration of about 100 micrograms per 100 ml. In the third trimester of pregnancy the plasma concentration of ceruloplasmin-copper normally and innocously rises as high as 300 micrograms per 100 ml. Yet 30 micrograms of non-ceruloplasmin-copper is a characteristic and highly toxic level in an untreated patient with Wilson's disease.

It is paradoxical that in Wilson's disease it seems likely that free copper ions effect their toxicity by combining with sulfhydryl, carboxyl or amino groups of non-copper proteins and impairing their physiological roles.

The validation of much of the foregoing is found in the results of removing copper pharmacologically from patients with Wilson's disease. Although Wilson described the clinical aspects of this disease in 1911 (8), it was not until almost 40 years later that evidence was obtained that implicated copper as the etiologic agent. In 1948 Cumings (10) proposed that the administration of 2,3-dimercaptopropanol might arrest the progression of what was then known to be a uniformly fatal disorder by chelating and excreting copper. The administration of this thiol - known as B.A.L. - to severely ill patients with Wilson's idsease was undertaken by Denny-Brown who showed that copper removal was accompanied by improvement in the clinical condition of the patients (11). B.A.L., however, had to be administered intramuscularly and was painful and disagreeable. In 1956 Walshe showed that administering β,β-dimethylcysteine (D-penicillamine) orally to patients with

Wilson's disease significantly increased the urinary excretion of copper(12).

Although the rarity of Wilson's disease made it commercially unprofitable, it proved possible to induce a pharmaceutical company to manufacture D-penicillamine. In 1963 the United States Food and Drug Administration approved penicillamine as safe and effective in the treatment of Wilson's disease.

The clinical effects of the continual administration of penicillamine in Wilson's disease could hardly have been more astonishingly gratifying. Patients who appeared to be days away from death because of liver or brain disturbances were restored to normal health. And patients who had been asymptomatic, but in whom the unequivocal diagnosis of Wilson's disease was made (usually by the demonstration of both deficiency of plasma ceruloplasmin, less than 20 milligrams per 100 ml of plasma; and an excess of hepatic copper, greater than 250 micorgrams per gram of dry liver) were kept free for decades of any sign or sympton of the illness by the continual administration of penicillamine (13).

Objective evidence has been abundant that these clinical effects have been accompanied by, and have depended upon, the removal of much of the excess copper that had accumulated before therapy was instituted. With penicillamine, uninary copper excretion increased to amounts that assured the physician that a negative copper balance had been achieved, as shown in Fig. 2, where almost 8 milligrams of copper was excreted in a day on which 2 grams of penicillamine were administered. Since the daily dietary intake of copper is generally about 3 to 4 milligrams, the copper balance was negative. Quantitative copper analyses of serial biopsy samples of liver show that hepatic copper concentrations fall as treatment progresses (Table II). Visible evidence of the removal of tissue copper is shown in Plate 3* where the circumferential brownish deposit of copper at the periphery of the cornea - the Kayser-Fleischer ring - has disappeared after 9 years of treatment with penicillamine (14).

Discussion and Conclusions

Although copper is as inherently toxic as many other heavy metals and is present in a number of tissues in concentrations at which many metals exhibit toxicity, copper toxicosis in human beings occurs only in about five individuals in a million. Except where suicidal attempts involve the ingestion of thousands of times the normal daily intake of copper, or where copper enters the body by other than the gasto-intestinal tract, copper toxicosis of man is seen solely in those individuals who inherit a pair of abnormal autosomal genes, the effect of which appears to be to impair a normal excretory route for copper absorbed in excess of what is metabolically essential. Copper toxicosis eventuates not merely because there is too much in the body, but because the excess is not incorporated into specific copper proteins which would

* Color plates are located in the Appendix.

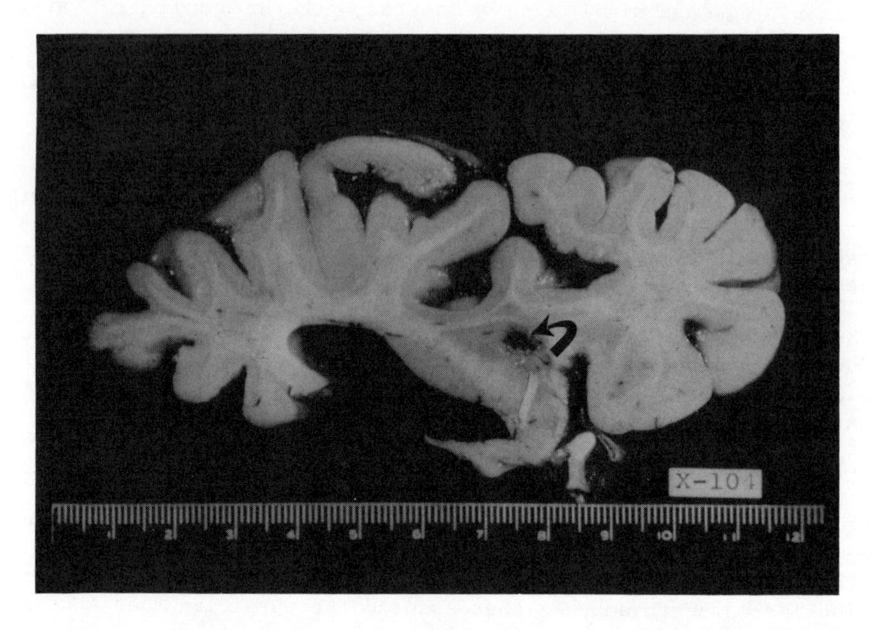

Figure 1. Cavity produced by toxicity of Cu in the putamen of a patient dead of Wilson's disease, indicated by an arrow in this sagittal section of the brain

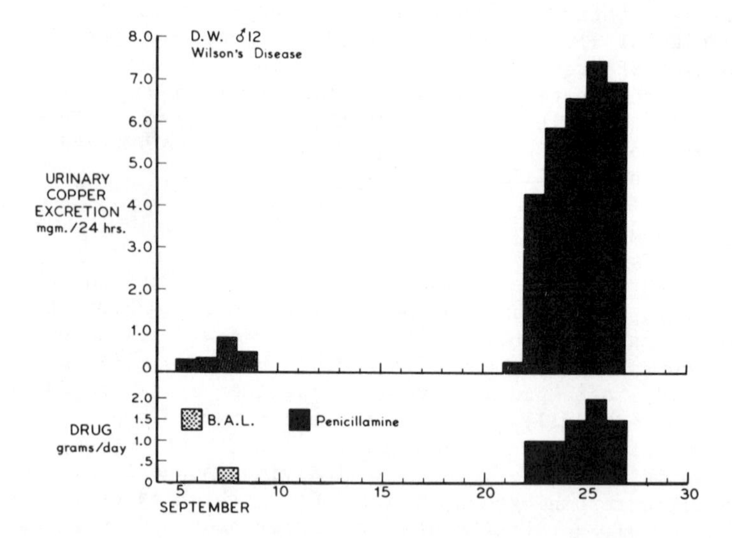

Figure 2. Increase of urinary Cu excretion of a patient with Wilson's disease, induced by the intramuscular administration of B.A.L. and the oral administration of penicillamine

TABLE II

EFFECT OF PENICILLAMINE THERAPY ON CONCENTRATIONS OF HEPATIC COPPER IN TEN PATIENTS WITH WILSON'S DISEASE

Hepatic Copper Concentration (μg/gm of dry liver)

Years of Treatment					
0	1	2	3	4	5
1092	944	-	-	696	-
971	-	-	146	-	-
1561	-	586	-	-	-
463	-	170	-	-	-
926	-	756	-	-	-
821	-	-	109	-	-
1004	-	-	-	945	-
866	-	-	-	737	-
1123	-	-	-	239	-
832	-	-	-	-	453

have had the effect of reducing the activity of the copper to zero.

The explanation of why almost all non-essential heavy metals are toxic, and thereby contrast sharply with the non-toxicity of essential iron, copper, and zinc is biological, not chemical. When the evolutionary process selected a heavy metal to carry out one or more essential biological functions it also apparently developed homeostatic mechanisms to rid the organism of inadvertently ingested excesses of the metal (and, incidentally, to conserve the element in times of dietary scarcity). Moreover, evolution was also required to reduce the inherent chemical toxicity of the free ions of an essential metal by providing a protective macromolecular structure (usually a protein) that would envelop the metal and permit it to play the physiologic role for which it was selected while reducing its thermodynamic activity to zero.

Literature Cited

1. Bruckmann, G.; Zondek, S. G. Iron, Copper and Manganese in Human Organs at Various Ages, Biochem. J., 1939, 33, 1845-1857.
2. Scheinberg, I.H.; Morell, A.G. Ceruloplasmin: in "Inorganic Biochemistry", Ed. Eichhorn, G.L., Elsevier Scientific Publishing Co., Ansterdam, 1973, Vol.1, pp.306-319.
3. Gregoriadis, G.; Morell, A.G.; Sternlieb, I. Catabolism of Desialyted Ceruloplasmin in the Liver, J. Biol. Chem., 1970, 245, 5833-3837.
4. Sternlieb, I.; van den Hamer, C.J.A.; Morell, A.G.; Alpert, S.; Gregoriadis, G.; Scheinberg, I.H. Lysosomal Defect of Hepatic Copper Excretion in Wilson's Disease (Hepatolenticular Degeneration), Gastroenterology, 1973, 64, 99-105.
5. Sternlieb, I.; Scheinberg, I.H. Prevention of Wilson's Disease in Asymptomatic Patients, N. Engl. J. Med., 1968, 278, 352-359.
6. Sternlieb, I.; Scheinberg, I.H. Wilson's Disease: in "Liver and Biliary Disease", Ed. Wright, R.; Alberti, K.G.M.M.; Karran, S.; Millward-Sadler, G.H., W.H. Saunders and Co., Philadelphia, 1979, pp.774-787.
7. Deiss, A.; Lee, G.R.; Cartwright, G.E. Hemolytic Anemia in Wilson's Disease, Ann. Intern. Med., 1970, 73, 413-418.
8. Wilson, S.A.K. Progressive Lenticular Degeneration: A Familial Nervous Disease Associated with Cirrhosis of the Liver, Brain, 1912, 34, 295-309.
9. Committee on Medical and Biological Effects of Environmental Pollutants, Copper. National Academy of Sciences, Washington, D.C., 1977, Publication 0-309-02536-2.
10. Cumings, J.N. The Effects of B.A.L. in Hepatolenticular Degeneration, Brain, 1951, 74, 10-22.
11. Denny-Brown, D.; Porter, H. The Effect of BAL (2,3-dimercaptopropanol) on Hepatolenticular Degeneration (Wilson's Disease), New Eng. J. Med., 1951, 245, 917-925,
12. Walshe, J.M. Wilson's Disease. New Oral Therapy, Lancet, 1956, 1, 25-26.
13. Sternlieb, I. Present Status of Diagnosis and Prophylaxis of Asymptomatic Patients with Wilson's Disease in: "Progress in Liver and Biliary Tract Diseases", Ed. Levey, C.M., S. Karger, New York, 1976, pp.137-142.
14. Sussman, W.; Scheinberg, I.H. Disappearance of Kayser-Fleischer Rings. Effects of Penicillamine, Arch. Opthal., 1969, 82, 738-741.

RECEIVED May 22, 1980.

22

Computer-Predicted Drugs for Urolithiasis Therapy

M. RUBIN and R. GOHIL Department of Biochemistry, Georgetown University Medical Center, Washington, DC 20007

A. E. MARTELL and R. J. MOTEKAITIS Department of Chemistry, Texas A&M University, College Station, TX 77843

J. C. PENHOS and P. WEISS Department of Physiology and Biophysics, Georgetown University Medical Center, Washington, DC 20007

To attain the goal of developing therapeutic agents on a predictive basis challenges the state of knowledge of physiology, metabolism and the relation of chemical structure and biologic activity. The fact that new drug development still depends upon extensive - and expensive - screening programs for initial leads of biological activity is a marker of the long path ahead. However, in the case of alkaline earth urolithiasis, which makes up approximately eighty-five percent of renal stones, it seems possible to predict potentially useful therapeutic compounds and subject the predictions to experimetnal verification. With the known composition of stones, the quantitative stability data now available for metal ligand interactions, and the current knowledge of the composition of biologic compartments in health and disease, there is now sufficient information for computer calculation of metal speciation in certain biological fluids. Also, it appears that proposed regulatory mechanisms for calcium and magnesium metabolism correspond to *in vivo* responses, and enough pharmacological data is at hand to allow a test of this approach to drug development. This paper describes the background of the clinical problem, the rationale, methods, results, and conclusions drawn from the computer simulation studies, and their successful experimental *in vivo* verification demonstrating the possibility of dissolving calcium hydrogen phosphate (brushite) bladder and magnesium ammonium phosphate stones or inhibiting their formation by treatment with an orally administered chelating agent.

The Problem of Urolithiasis

In a nationwide study based on hospital admission and discharge diagnosis (1) it has been reported that for adult patients an average of 9.47 persons per 10,000 population were admitted with a diagnosis of urinary calculi. Evidence of geographical variation was obtained with rates of 19.25 and 18.42 for South Carolina and Georgia at the highest side of the scale and 5.81 and 4.31 for Wyoming and Missouri at the low end. In general,

0-8412-0588-4/80/47-140-381$07.00/0

the southeastern and northeastern area center-states around Massachusetts and Maine had the highest incidence with the mid and southwestern states such as Wyoming, Arizona and Colorado at a lower level. Appropriate reservations were expressed concerning the scope and adequacy of the sample but nonetheless the data fulfilled basic requirements of statistical validity. "Stone belts" where the incidence of renal calculi is unusually high have also been well documented in other parts of the world. For the large majority of these patients a complex diagnostic workup to exclude a definitive etiologic basis for the formation of urinary calculi is neither feasible medically or on economic grounds. Thus, in the present absence of a generally useful medical treatment modality, surgical intervention is required when urinary obstruction takes place.

Alkaline earth renal lithiasis occurs when the ion solubility product is exceeded, a crystal nucleation site is present, and endogenous inhibitors of crystallization are decreased or absent (2,3,4,5). Mucoproteins (6), collagen (7), brushite (8,9,10), apatite (11), and monosodium urate (12), have been implicated as possible nucleation sites of alkaline earth stones (13). A high urinary concentration of magnesium (14) and pyrophosphate (15) on the other hand, have been cited as possible inhibitors of crystallization and crystal growth. Urinary citrate, by virtue of its capability to form unionized soluble calcium chelates, may serve to decrease the ion concentration of calcium and thus assist in prevention of calcium stone formation (16).

Therapeutic attempts to inhibit stone formation derive, as would be expected, from current understanding of the mechanisms responsible for their development (17). Numerous efforts have been made to bring about a reduction in the urinary concentration of potential precipitating ions. A large increase in fluid intake results in a general decrease in the ion activity of the consequent diluted urine. This form of therapy is difficult to sustain for a lengthy period. Various approaches have been used to bring about a reduction in urinary calcium, by far the most frequent of stone forming cations. Occasionally the presence of hypercalcuria can be traced to an underlying pathophysiologic process such as hyperparathyroidism, malignancy, hyperthyroidism or sarcoidosis. Effective therapy of the disease process can sometimes mitigate the extension of the renal lithiasis. Increased urinary excretion of calcium may also occur as a direct consequence of its increased intestinal absorption or as a result of a defect of its renal absorption. Intestional absorption can be modulated to some extent by dietary restriction but more effectively by oral administration of non-absorbable cellulose phosphate. This compound inhibits calcium absorption in the gut by complex formation (18). Correction of a renal tubular deficency in calcium reabsorption has been achieved by administration of thiazides (19). Although untoward side effects of protracted thiazide therapy are well established, their beneficial response for a selected group of patients appears unequivocal.

Oxalate and phosphate are the anions associated with the alkaline earth calculi, calcium oxalate, calcium hydrogen phosphate and magnesium ammonium phosphate. Hyperoxaluria may reflect increased oxalate synthesis or intestinal absorption. Although a procedure for decreasing oxalate synthesis is presently available, it has also been possible to modify the increased oxalate absorption often associated with various types of enteric malfunction. Restriction of oral oxalate (20), administration of calcium carbonate (21), cholesterylamine (22) and aluminum hydroxide antiacid preparations (23) can reduce urinary oxalate output by reducing its gastrointestinal absorption. Although a decrease in urinary phosphate can be attained by inducing systemic phosphate depletion through decrease of intestinal phosphate absorption using oral aluminum hydroxide (24), the process is associated with a significant number of side effects (25).

The view that nucleating sites are an essential induction factor in stone formation has led to a number of efforts to decrease their presence. Thus, oral cellulose phosphate has been utilized to decrease the brushite ($CaHPO_4 2H_2O$) activity in urine. Allopurinol (26), administered apparently with some success, reduces the hyperuricosuria observed concomitantly with some oxalate stone formation.

Efforts have also been made to diminish crystal formation and growth by increasing the urinary concentration of inhibitors of these processes. The side effects of oral magnesium therapy limit the possibility of attaining the high concentrations in urine needed to provide effective crystal formation and growth inhibition. Methylene blue, reputed to serve the same purpose, does not appear to be a useful agent (27). The positive effects obtained by the administration of oral phosphate have been ascribed to the resulting increase in the urinary excretion of the crystallization inhibitor, pyrophosphate (28). The most effective agent of this type developed to date is ethane-1-hydroxy-1,1-diphosphonate (29). This carbon analogue of pyrophosphate appears to function by serving as a significant inhibitor of phosphate and oxalate precipitation as well as slowing the aggregation of calcium oxalate crystals. Its use, however, results in hyperphosphatemia by increasing the tubular reabsorption of phosphate increasing bone resorption and affecting the renal metabolic formation of 1,25-dihydroxyvitamin D_3 (30-36).

The formation of magnesium ammonium phosphate (struvite) stones frequenly occurs in an alkaline urine resulting from urinary tract infection by urea cleaving gram negative organisms. Appropriate antibiotic therapy, or the utilization of acetohydroxamic acid to inhibit the activity of urease, coupled with management of phosphate excretion appear to be the treatment method of choice (37).

The capability of ethylenediaminetetraacetic acid EDTA to solubilize otherwise insoluble calcium precipitates led to its application in the attempted dissolution of alkaline earth urolithiasis (38,39,40). There are a number of disadvantages of this approach to therapy. EDTA is reported to be poorly absorbed by mouth (41), is potentially nephrotoxic (42), is not selective for calcium chelation in the presence of magnesium, and as we and others have observed (loc. cit.) causes intense pain and hematuria when administered as a retrograde irrigating solution. Despite intensive efforts at modification of the irrigating solution. Despite intensive efforts at modification of the irrigating solution, we were unable to solve this latter problem. Other attempts to develop litholytic solutions (43-51) have had only modest and inconsistent success.

There would be obvious advantages in the use of a calcium chelating ligand which could be given by mouth and end up in the urine in a form which would combine with urinary calcium to reduce calcium ion activity below the solubility product level. This possibility was examined in studies with ethylene bis-(β-aminoethylether)-N,N,N',N'-tetraacetic acid, EGTA (52). This compound appears promising on the basis of its low acute oral toxicity (3.96 ± 0.50 g/kg), lack of effect on serum calcium, absence of toxicity on oral feeding, and the evidence that 30-35% of the compound could be recovered in the urine after a single oral dose. It was observed, however, that when the compound was excreted into the urine it carried with it an equivalent of calcium derived from its combination with serum calcium. The net result consequently was an increase in the total urine calcium excretion with no decrease in the ionic calcium fraction.

Physico-Chemical Considerations

The Urinary Compartment. Recognition of the fact that the stone forming salts in urine are usually present in solution at a concentration exceeding their solubility products led to studies of the physico-chemical aspects of the factors involved in solid phase formation (53-63). For calcium oxalate it could be demonstrated that pH variation in the physiologic range was not significant. On the other hand, the "salting in" effect of urinary electrolytes and other constituents were potent positive solubilization factors. Sodium, potassium, magnesium, chloride, bicarbonate, phosphate, sulfate, citrate ion, and urea increased the supersolubility of calcium oxalate in urine. These early pragmatic studies were given a theoretical basis when computer programs provided a more facile solution of the numerous simultaneous equations governing the equilibria in urine. Using this approach it was possible to test the potential effects of various present modes of therapy on the solubility consequences for precipitating salts. Urine dilution, as would be anticipated *a priori*, decreases precipitability. Acidification of urine was likewise calculated to have a positive effect for calcium hydrogen phosphate, but in the physiologic range this is of little consequence for the solubility of calcium oxalate. The latter result agrees with earlier studies with rats. The computer calculations also suggested that a high urinary magnesium ion concentration in an acidified urine would increase calcium oxalate solubility, a finding also in keeping with animal studies. Therapeutic procedures, such as the administration of oral aluminum hydroxide gels to reduce urinary phosphate, were calculated to be of increasing potential effectiveness with a urinary pH of 5 and above. However, the lack of precise quantitative values for the stability constants of urinary species placed some limitations on the applicability of this approach to the problem of ionic equilibria in urine.

Improvement in the computer simulation became possible with the development of an iterative calculation process, utilizing better values for the stability constant of calcium oxalate. Good agreement was obtained between measured and calculated solubility and formation products of calcium oxalate in the urine of normal individuals and stone formers. In continued studies, application of this physico-chemical background together with laboratory measurements of the concentration of urinary constituents have further extended our understanding of the possibility of the formation and growth of alkaline earth renal stones in the clinical setting.

The Plasma Compartment. Computer calculations have also been utilized to provide data on the probable distribution of metal ions among the many potential binding ligands in plasma (64-69). The low relative concentrations of the plasma metal ions enhances the probability that the calculated results approximate the situation. Two major factors presently limit the potential applicability of such simulation efforts. The first is the exisitng uncertainty of the accuracy of some of the metal-ligand equilibria in this compartment. The second is the modulating influence of kinetic physiologic regulatory systems upon shifts in ion equilibria predicted by computer calculation based upon a static compartment model. However, it should be noted that the computer calculations of ionized calcium in plasma ultrafiltrate are in good agreement with measured values by spectrophotometry and potentiometry.

Rationale of Study

First consideration suggests that it would be unreasonable to expect that potent alkaline earth binding ligands would be able to move through calcium and magnesium laden body compartments to arrive in the urine in a form capable of additional alkaline earth combination and solubilization. Nonetheless, a more searching examination of the problem indicates that meeting the physicochemical requirements may be possible within the limits of change imposed by physiologic tolerance. The plasma compartment is illustrative of this situation. Calcium bound to protein represents almost half of the total calcium, low molecular weight complexes account for a small percentage, and the balance is ionic calcium. The entry of a chelating ligand such as EDTA upsets this carefully regulated homeostatic buffer with decrease in ionic calcium as a function of the rapidity and amount of EDTA introduced (70), and the formation of the physiologically unavailable calcium of the calcium chelate. The fact that plasma is a calcium buffer imples, however, that a calcium binding ligand can be selected which will provide calcium buffering activity in the same range of ionic calcium as exists normally at the plasma pH of 7.4. Such chelating ligands would exist in plasma partly free of calcium and partly as calcium chelates. In contrast to calcium proteinate, a large molecule, they would be freely filtrable at the kidney glomerulus for passage into the urinary tract. Transit of the glomerular filtrate to the urine is by way of the nephron (Fig. 1) which is uniquely controlled and selective in its capability to adjust the composition of the passing fluid. With the rejection of the high molecular weight calcium proteinate at the glomerulus and the reabsorption of almost all of the ionic calcium in the proximal and distal tubular regions, the equilibria will readjust to reflect the new ionic milieu. It could be anticipated that the chelating ligand would consequently have an enhanced capacity for calcium binding in the urinary compartment. A second factor may also serve to increase its effectiveness in solubilizing urinary calcium. In the passage of filtered plasma from the glomerulus to the urinary bladder sodium ion is reabsorbed with replacement by hydrogen and ammonium ions. To the extent that hydrogen ion acidification precedes and is dissociated from calcium ion reabsorption in the nephron the result could be the cleavage of filtered calcium chelate in an acidified environment, followed by reabsorption of liberated calcium ion. The ratio of calcium free ligand to calcium ion in urine would increase to provide additional calcium buffering capacity.

The gastrointestinal compartment is less complicated. Calcium binding ligands administered in the fasting state would be present in the absoprtive area primarily in the unchelated form. The local pH and the presence of endogenous cations and metal binding ligands would determine the extent of calcium chelation.

In the presence of food, however, one could anticipate that the food polyvalent cation composition, the competitive metal binding ligands such as amino acids and the local pH would establish the ratio of chelate and metal free ligand. It is probable that in for maximal effectiveness the potential therapeutic agent should reach the plasma compartment in metal free form. Thus consideration must be given to the possible interaction of all the metal binding ligands that would be present in the intestinal absorption region.

Selection of Test Compounds

Nitrilotriacetic acid was selected as a test compound to explore the applicability of the considerations that have been enumerated. At the plasma pH of 7.4 it buffers calcium ion at concentrations between 10^{-3} and 10^{-4} M (71). This value is close enough to the normal concentration of calcium ion in plasma, 1.2 x 10^{-3} M, to suggest that an animal could accommodate easily to an anticipated minor hypocalcemia. Fortuitously, because of its application in detergents as a polyphosphate substitute, an extensive literature has accumulated on its pharmacology and toxicology. It is known that the compound is readily absorbed from the gastrointestinal tract of rats and dogs, is non-metabolized and is almost quantitatively excreted in the urine. Its acute, subacute and chronic toxicity is low, with a no effect level of 0.5 percent in the diet of rats and dogs (72-75).

Materials and Methods

Nitrilotriacetic acid was supplied by the Hampshire Chemical Co., Division of W. R. Grace Co., Nashua, New Hampshire. For feeding studies it was neutralized with 1 M sodium hydroxide to provide a 1% NTA solution. Ethyl nitrilotriacetate was obtained as a clear colorless viscous oil by refluxing 30g of nitrilotriacetic acid in 1 liter of absolute alcohol in the presence of 3g of toluenesulphonic acid for forty-eight hours. The alcohol was removed by distillation from a water bath in vacuo, the residue mixed with 100 ml of 0.5 percent sodium bicarbonate solution and extracted with one liter of benzene. The benzene solution was extracted twice more with bicarbonate solution, washed with water and concentrated from a water bath in vacuo. The clear colorless residual oil was dissolved by shaking with an aqueous solution of 1.5 M citric acid and the volume adjusted by addition of distilled water to provide a final solution of 1% NTA ethyl ester as citric acid salt.

Animal Studies

Mixed brushite/struvite bladder stones in rats were produced according to the method of Vermeulen et al. (76). Pieces of zinc sheet, 20-30 mg., were surgically implanted in the bladders of 40 Sprague-Dawley male rats. Three control groups of ten animals were maintained on an ad lib diet of Purina rat chow for eight weeks. The fourth group of ten animals was given the same diet to which a neutral sodium NTA solution had been added to provide 5000 ppm (0.5%) of NTA. At the end of the study period one group of control animals and the group of NTA treated animals were sacrificed and examined for gross evidence of urolithiasis.

One of the two remaining groups was continued as a control on the regular diet while the other was now fed the treatment

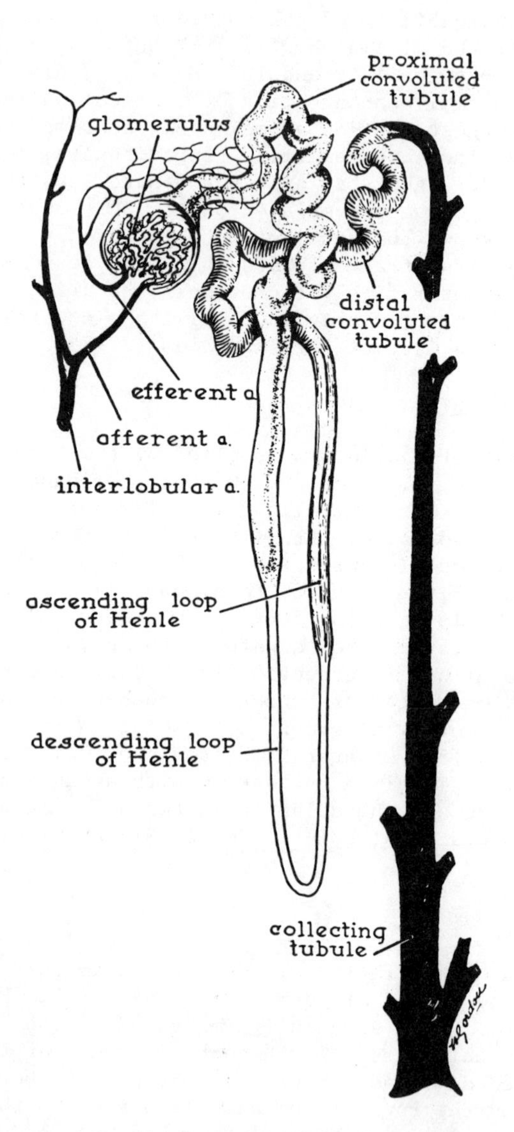

Figure 1. Diagram of the blood supply and functional anatomy of the kidney nephron

diet containing 0.5% NTA. Both groups were sacrificed after an additional period of four weeks and examined for the presence of renal stones.

Studies with ethylnitrilotriacetate were conducted in the same manner. For the therapeutic trial in a dog the neutral NTA solution was added to the regular diet to provide an NTA concentration of 0.5%. Feeding was continued for a period of two weeks. The animal was a four-year old female Scottish terrier. X-ray examination was done before and after the treatment period.

Computer Studies

The program utilized for this study provides the capability of establishing solution and solid phase equilibria for a multicomponent system of up to thirty five components and 500 solution and solid species. Using equilibrium constant data for metals, inorganic anions, and organic ligands (77) the iterative calculation provides for the determination of metal speciation as a function of pH and component concentration. Table I and II provide calculated constituent concentrations for the Purina rat chow used in the animal feeding studies for the basic urine composition.

Results of Computer Studies

The Plasma Compartment. A computer calculation of the speciation of plasma calcium and other ions in the presence of the chelating agent nitrilotriacetic acid (NTA) was performed on a typical plasma composition involving the ligands listed in Table III. The appropriate stability constants have been selected by restricting consideration to those that undergo significant interactions with Ca^{2+}, Mg^{2+}, Cu^{2+}, and Zn^{2+} ions in solution. Since nearly all amino acids (the monoamino monocarboxylic acids) have similar affinities for metal ions, an "average" stability constant and total concentration were employed. This procedure eliminated the need for consideration of a large number of unknown mixed-ligand stability constants for ternary "mixed-ligand" complexes. Data in the literature have been adjusted where possible (78), to 1.6 M ionic strength and 37°C. Typical data for plasma metal ion speciation in the presence of 1×10^{-3} M NTA is provided in Table IV. It may be noted that although the concentration of ionic calcium has decreased only somewhat (from a normal range of about 49-52% to a level of 45%) there has been a redistribution of the non-ionic calcium from the plasma proteins to NTA. It can be anticipated for other chelates that the extent of this redistribution will be dependent upon the stability constants of the calcium chelates at phasma pH. Data for the effect of up to 1.5 M NTA on the ionic calcium level of plasma is provided in Figure 2. If the plasma compartment were static then up to 1.5 M NTA could be present before ionic clacium levels were reduced to a non-physiological level. In fact, as we have reported (79-82), the

TABLE I

INTESTINAL CONCENTRATION OF FOOD COMPONENTS

Component	Concentration
Calcium	3×10^{-1} M
Phosphorus	3×10^{-1} M
Potassium	2×10^{-1} M
Magnesium	8×10^{-2} M
Sodium	1.3×10^{-1} M
Iron	6.4×10^{-3} M
Zinc	8×10^{-4} M
Manganese	3×10^{-3} M
Copper	1.7×10^{-4} M
Cobalt	1×10^{-5} M
Amino Acids	1.3 N

Calculated from the analyzed content of Purina rat chow in a wt/wt dilution with intestinal fluid.

TABLE II

SIMULATED URINE COMPOSITION

Component	Concentration
Phosphate	3.35×10^{-2} M
Sulfate	5×10^{-2} M
Citrate	2.3×10^{-2} M
Chloride	1.12×10^{-1} M
Bicarbonate	1.6×10^{-2} M
Sodium	9.7×10^{-2} M
Potassium	3.7×10^{-2} M
Calcium	3.2×10^{-2} M
Magnesium	7.7×10^{-3} M
Ammonia	3×10^{-2} M

Urea and trace concentrations of other constituents have been omitted.

TABLE III

METAL COMPLEXES IN BLOOD PLASMA
($t = 37°$, $\mu = 0.16$ M, pH = 7.4)

Complexing Ligand	Total Ligand Concentration (M x 10^3)	Ca^{2+}	Mg^{2+}	Cu^{2+}	Zn^{2+}	H^+
CO_3^{2-}	-	2.4	2.7	6.1	5	9.8
HCO_3^-	28.80	0.8	0.8	-	-	6.1
HPO_4^{2-}	1.13	1.3	1.8	3.3	2.4	11.6, 5.7
Alanine (etc.)	2.51	1.0	1.6	7.8, 14.3	4.5, 8.5	9.0, 2.3
Histidine	0.07	1.0	1.6	9.8, 17.5	6.3, 11.7	8.8, 5.8, 1.7
Proline	0.21	1.0	1.6	8.6, 15.9	5.1, 9.7	10.1, 1.9
Lactate	2.67	1.1	0.9	2.5	1.8	3.7
Citrate	0.11	3.2	3.2	5.9	4.7	(16.8), 5.7, 4.3, 2.8
Globulin (no SH)	1.5	2.2	2.2	3.7	2.9	5.9
Albumin (SH)	0.4	2.2	2.3	10.2	6.7	9.4

Total millimolar concentrations of metal ions: Ca^{2+} 2.5; Mg^{2+} 0.9; Zn^{2+} 0.05; Cu^{2+} 0.001.

Stability constants from Smith R. M.; Martell, A. E. (78).

TABLE IV

DISTRIBUTION OF METAL IONS IN BLOOD PLASMA

($[NTA]_t = 1.0 \times 10^{-3}$ M)

Ligand	Per Cent Metal Bound			
	Ca^{2+}	Mg^{2+}	Cu^{2+}	Zn^{2+}
CO_3^{2-}	1.9	1.3	-	-
HCO_3^-	7.6	10.2	-	-
HPO_4^{2-}	-	4.0	-	-
ALA et al.	-	-	9.0	-
Histidine	-	-	25.2	-
Proline	-	-	-	-
Lactate	1.5	1.3	-	-
Citrate	2.1	2.9	-	-
Globulin	8.2	11.0	-	-
Albumin		-	-	-
Uncomplexed	45.1	60.5	-	-
NTA	32.5	8.7	64.2	99.6
% Distribution of NTA	8.13	7.8	-	5.0

Calculated from Table III according to Morel, F.; Morgan, J. (77).

presence of calcium chelating ligands evokes the usual homeostatic regulatory responses which restore the ionic calcium concentration. In chelates given by mouth and absorbed from the gastrointestinal tract the final determinant of complexation will be the balance between the kinetics of ligand absorption, plasma calcium homeostasis, urinary excretion of the calcium chelate and possible renal dissociation of the compound with reabsorption of calcium and excretion of the chelating ligand. For NTA it appears that even at low plasma NTA concentrations some of the compound will be available for excretion into the urine in calcium free form.

The Urinary Compartment. The computer program permits a wide-ranging exploration of the potential effect of calcium binding ligands on the species distribution as a function of changes in the concentration of any constituent and changes in the pH of the medium. A number of these possibility have been explored for the presence of NTA in the urinary compartment. Variations in urine pH in the range of 4.8 to 8.0 at various concentrations of ammonia, oxalate, phosphate, sulfate, citrate, sodium, potassium, chloride, calcium, magnesium, and carbonate provide insight into the possible consequences for alkaline earth salt precipitation in the presence of NTA. Figure 3 demonstrates the sharp dependence upon pH, calcium and oxalate concentrations for concentrations of NTA that effectively inhibit precipitation when the total urinary calcium concentration is 1×10^{-3} M and the oxalate concentration is between 1×10^{-4} and 6×10^{-4} M. An increasingly high concentration of urinary NTA is required as the pH is decreased in the moderately acidic region.

The effect of variations in the calcium ion distribution in urine as a function of variations in NTA concentration has also been calculated for lower concentrations of NTA equivalent to those obtainable in feeding studies. Figure 4 provides data in terms of the change in the fraction of ionic calcium that would be present in solutions of varying pH from 5.0 to 8.0 with NTA concentrations of 3×10^{-3} and 6×10^{-3} M. Clearly, in these ranges of urinary NTA concentration, a major shift in the ionic concentration of calcium ion can occur. This takes place with a corresponding change in the fraction of free NTA ligand available for additional calcium chelation. At urine concentrations of 1×10^{-3} M for calcium, 6×10^{-3} M for citrate, and 6×10^{-3} M for NTA, and at pH 7.65 essentially all the calcium in urine is in the form of a non-ionized chelated species. Approximately 65% of the total NTA will be present as the unchelated calcium free ligand.

Related calculations for the solubilization of magnesium ammonium phosphate are presented in Figures 5, 6, and 7 at the stated concentrations of magnesium phosphate and ammonium ions. The high urinary concentrations of NTA required to inhibit solid phase formation at the alkaline pH associated with the clinical circumstances for this type of stone formation renders the potential applicability of NTA of uncertain value.

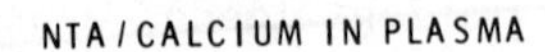

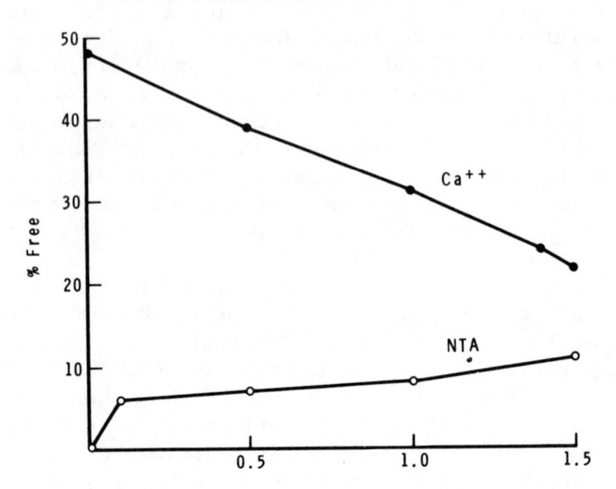

Figure 2. Computer-calculated change in the percentage of plasma calcium as Ca^{2+} and of the percentage of unchelated NTA as a function of increasing plasma NTA molar concentration

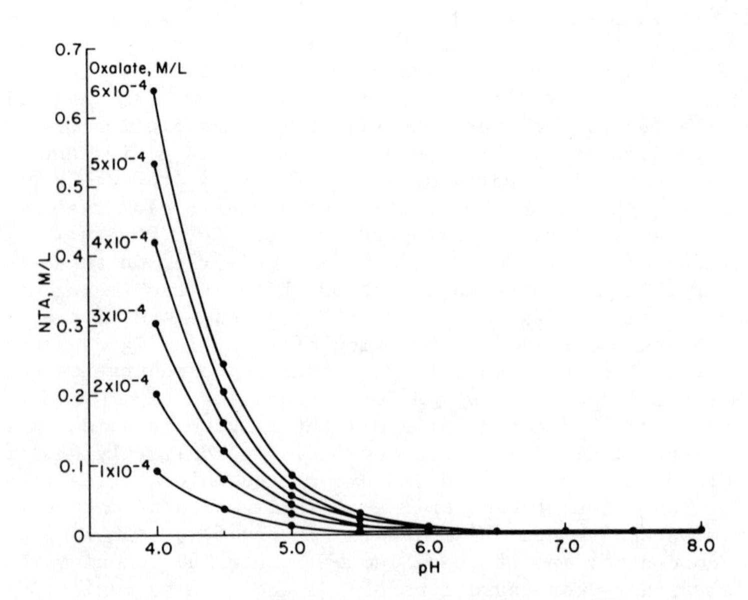

Figure 3. Computer-calculated concentration of NTA required to inhibit the precipitation of calcium oxalate at 1×10^{-3}M Ca and varied concentrations of oxalate in a simulated urine (Table III) as a function of urine pH

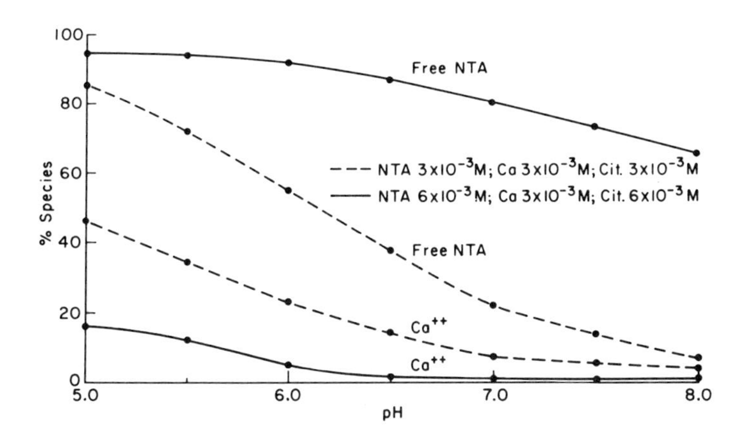

Figure 4. Computer-calculated percentage of NTA unchelated and Ca as Ca ion in simulated urine at two NTA and citrate concentrations as a function of urine pH

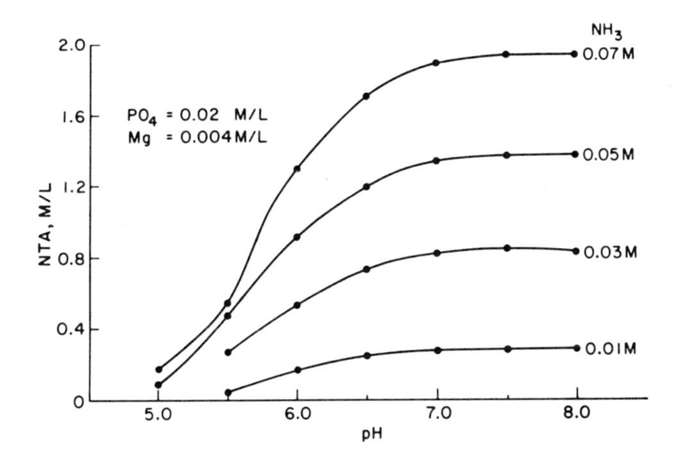

Figure 5. Computer-calculated concentration of NTA required to inhibit the precipitation of $MgNH_4PO_4$ in a simulated urine as a function of Mg concentration and urine pH

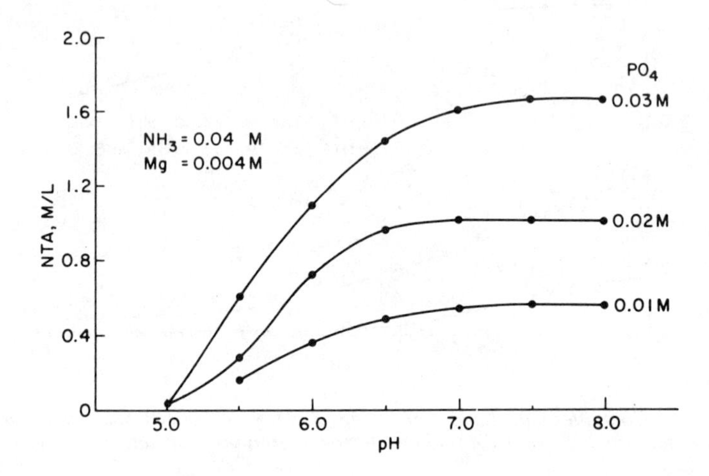

Figure 6. Computer-calculated concentration of NTA required to inhibit $MgNH_4$-PO_4 precipitation in a simulated urine as a function of phosphate concentration and urine pH

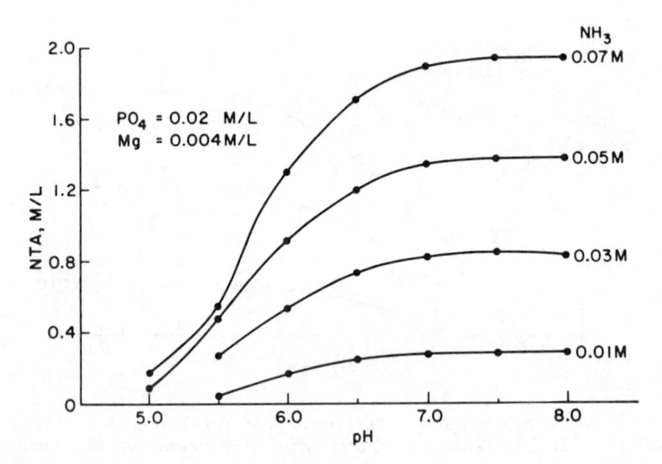

Figure 7. Computer-calculated concentration of NTA required to inhibit $MgNH_4$-PO_4 precipitation in a simulated urine as a function of ammonia concentration and urine pH

The situation appears to be quite different in the case of calcium hydrogen phosphate (brushite) stones. At a pH above 6.5 this species is converted, Figure 8, to that of the soluble calcium NTA. The important influence of the pH on the solubilizing role of NTA is clearly evident in the data plotted in Figure 9. As would be anticipated, an increased urine pH markedly reduces the concentration of the ligand required to form calcium NTA. The data suggest that NTA could effectively inhibit the formation or potentially cause the solubilization of calcium hydrogen phosphate stones in the urinary tract.

The Gastrointestinal Compartment. In the pH range of 6.0 to 8.0 which may occur in the absorptive area of the gastrointestinal tract the amount of free or unchelated NTA present upon ingestion of the usual rat diet is given in Figure 10. Thus, when NTA is added to the diet at 1.8 x 10^{-2} M approximately 45 to 80 percent of the compound will not be chelated to a polyvalent cation. Once again, the sharp pH dependence of chelation is evident. The fraction of unchelated NTA is directly related to the hydrogen ion concentration. Significantly a high percentage of NTA in a diet containing 5000 ppm NTA will be available for absorption in uncomplexed form even in the presence of the competitive calcium interaction of the dietary constituents.

Implications of the Computer Studies for Urolithiasis

The results of the computer calculations applied to simulated gastrointestinal, plasma and urine compartments suggest that the inclusion of a non-toxic concentration of 5000 ppm NTA in the diet of rats would:

a) provide a significant concentration of the calcium binding ligand available for absorption in its uncomplexed form;
b) result in plasma concentrations of NTA which would maintain plasma calcium ion in the physiologic range while allowing for a significant fraction of the compound to be excreted into the urine free of bound calcium;
c) convert essentially all of the urine calcium to the soluble calcium NTA chelate. Maintenance of the (calcium) chelated form would be difficult for NTA in the presence of increasing concentrations of oxalate at a urine pH below 6.0, and for increasing concentrations of magnesium, ammonium, and phosphate ions with increasing pH. However, NTA appears to have the potential for inhibiting the formation and inducing the solubilization of calcium hydrogen phosphate in urine.

Animal Studies

Rat Studies. The computer calculations were tested by inducing the experimental formation of brushite/struvite bladder stones in rats. The inhibition of stone formation by the feeding of 5000

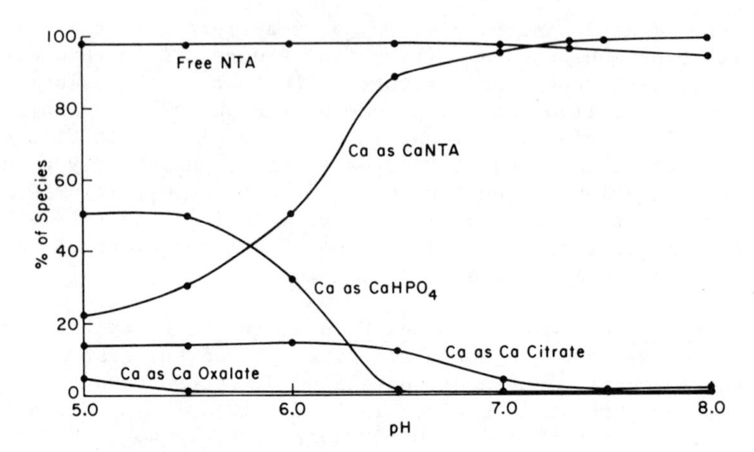

Figure 8. Computer-calculated Ca speciation in simulated urine in the presence of 0.2 M *NTA as a function of urine pH*

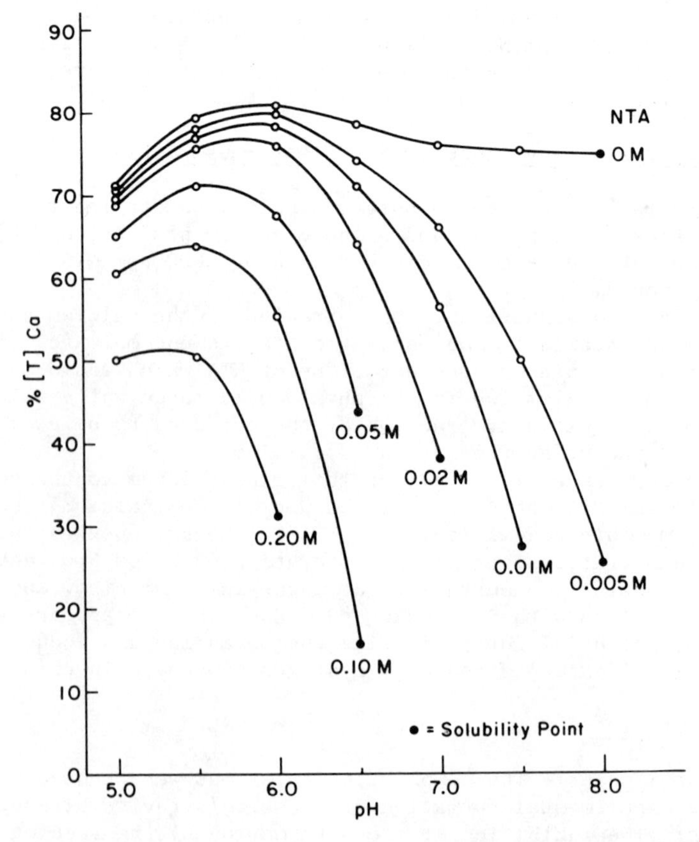

Figure 9. Computer-calculated percentage of the total Ca in a simulated urine present as solid-phase calcium hydrogen phosphate as a function of NTA concentration and urine pH

ppm of NTA in the diet of treated animals compared to a group of untreated controls is illustrated in Plate 9*. In a first study eight of ten animals in the untreated control group had single large stones or medium-sized stones with multiple small stones. The two remaining control animals had incipient stone formation as manifested by crystalline encrustation on the zinc implant. In the treatment group seven of the ten animals were free of bladder stones. In two animals, also without evidence of stone formation, we could not find the implanted zinc sheet. In the remaining animal there was a small bladder stone.

The study of inhibition of stone formation was repeated. One animal in the control group died during the four week period and was not autopsied. The nine control animals at the end of the study had a large stone or multiple stones. Eight of the ten treated animals had no evidence of stones. In three of these animals the implanted zinc pellet was not retrieved. The two remaining animals in the treated group yielded a pellet with a minor amount of zinc encrustation.

A third study was undertaken to determine whether the addition of NTA and its ethyl ester to the animal diet could cause the dissolution of existing calculi. Forty animals were implanted with zinc pellets. They were maintained on Purina rat chow for four weeks. At that time, ten animals were sacrificed and examined for bladder stones. Multiple stones were observed in 9/10 animals in this control group. The remaining thirty animals were divided into one control group of ten and two treatment groups of ten animals. One treatment group was given 0.5% of NTA in its diet. The second treatment group was given 0.5%, as NTA, of the NTA ethyl citrate. All animals were sacrificed at the end of four weeks. All of the control animals had developed bladder stones. Eight of the ten NTA treated animals were grossly free of stones. Two animals had small to moderate sized stones. The NTA ester citrate treated group showed some evidence of toxicity as indicated by dehydration and weight loss. Six animals were free of bladder stones. Four had medium sized stones.

Dog Study. A female four-year old Scottish terrier with clinical and X-ray evidence of phospaturic bladder stone formation, Figure 11a, was treated for two weeks by the incorporation of 0.5% of NTA in the diet in the form of the neutral 1% solution. The X-ray at the end of the treatment period, Figure 11b, showed no evidence of the presence of calculi. The animal was symptom free.

Discussion

The experimental verification of the computer generated predications provides some assurance of the soundness of the underlying assumptions. It is consequently of interest to explore possible limitations of this approach.

* Color plates are located in the Appendix.

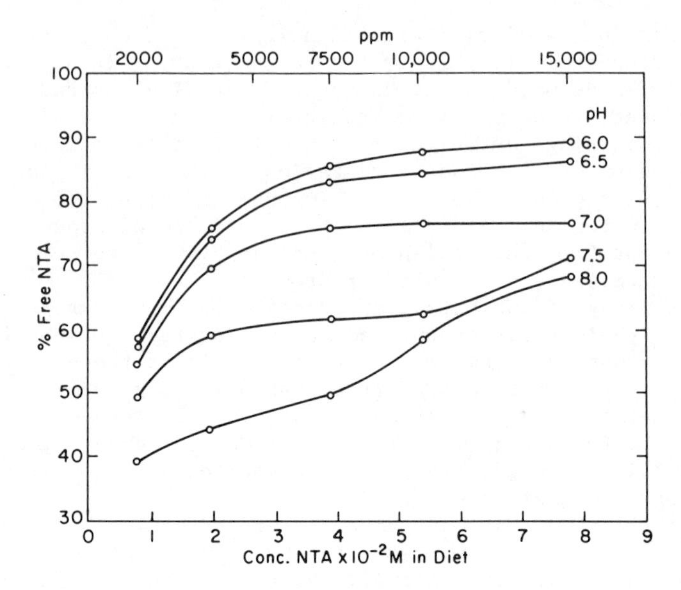

Figure 10. Computer-calculated percentage of unchelated ("free") NTA present in the simulated intestinal contents during digestion as a function of NTA concentration and intestinal pH

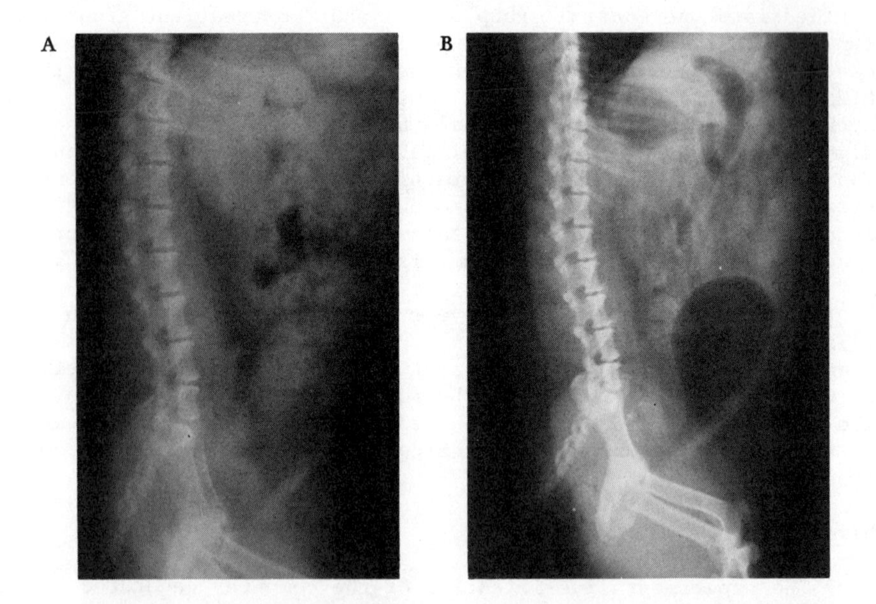

Figure 11. X-rays of female Scottish terrier (4 years) with clinical evidence of brushite bladder calculi. Treatment consisted of the incorporation of 0.5% Na_2-NTA wt/wt, in the diet for a period of two weeks.

The problem of the transport of metal-free ligands from gastrointestinal tract to plasma compartment is not a significant one. Administration of the therapeutic agent to the fasting patient, incorporation at dietary levels sufficient to provide some metal free chelating agent for absorption, encapsulation to bypass metabolic conversion and ensure absorption from the intestine, and structure modification to provide for post absorptive metabolic conversion to the active agent, all offer promising avenues.

A more difficult problem concerns the nature of the relations between the potential therapeutic agent and the plasma compartment constituents and homeostasis. A chelating agent potent enough to cause the solution of solid phase calcium oxalate in the urinary tract would first combine with ionic and protein bound calcium of the plasma. This follows since the stability constant of calcium proteinate at a plasma pH of 7.4 is several orders of magnitude lower than the stability constant of calcium oxalate. As we have shown, the resulting hypocalcemia will be a function of the rate of entry of the agent into the plasma compartment and the homeostatic regulation of ionic calcium (79-82). In any event, even if the hypocalcemia were to be maintained with physiologically tolerable levels, the therapeutic agent would have been calcium-saturated before it entered the urinary compartment. Whether it could still be of potential utility for the solution of urinary calcium oxalate calculi may depend upon some presently unresolved questions concerning the mechanisms of calcium metabolism in the nephron.

There is, however, an alternative approach to the problem. By the selection of potent calcium binding ligands with basic donor groups, advantage could be taken of the pH differential between the regulated 7.4 of plasma and the potential of increasing urine pH to a maximum of 8.0. Increased calcium binding capability would ensue in the passage of such a compound from plasma to urine. Potential limiations are less severe for efforts that would be directed toward inhibiting the formation of recurrent urinary calculi. In this case the need for reducing urinary calcium ion activity can be met by application of a less potent calicum binding chelating agent in combination with dietary restrictions of calcium, phosphate and oxalate.

The complexity of calcium metabolism in the nephron coupled as it apparently is with that of sodium, hydrogen, bicarbonate and ammonium ions, offers some unusual opportunities for control of the interaction of the chelating ligand with calcium and magnesium ions. The glomerular filtrate as it enters the proximal tubule is essentially protein free, isohydric and isoosmotic to the plasma. The calcium concentration of the fluid compared to plasma is decreased by approximately half due to the non-filtration of the plasma calcium proteinate. In its further passage along the proximal tubule the fluid loses sodium and calcium ion by their reabsorption across the cells lining the area. One can anticipate that some equilibrium readjustment could occur in the interaction of calcium and chelating ligand. Depending on the relative concentrations of the species, the result would be an increased ratio of calcium-free ligand to calcium chelate. This could increase the potential for decreasing calcium ion activity. Following water reabsorption in the loop of the nephron, a further set of ionic readjustments takes place in the distal tubular portion of the nephron. For purposes of present considerations, the most important changes are those in the concentrations of hydrogen, ammonium, sodium, bicarbonate, and calcium ions. The distal tubular reabsorption of sodium ions is associated with concomitant secretion of hydrogen and ammonium ions. The net result is an acidification of the resulting urine. Such an effect, which is variably controllable according to systemic homeostatic needs by drugs such as the diuretic carbonic anhydrase inhibitors and through dietary modification, could result in some pH dependent dissociation of the calcium chelate in transit. If, as seems possible, a further controlled reabsorption of liberated calcium ion were to occur, there would again be an enhanced calcium-free ligand calcium chelate ratio with further increase in the calcium sequestering capability of the therapeutic agent.

That these possibilities have potential reality may be inferred from some published studies. In a report on the long term influence of NTA on mineral metabolism it was pointed out that the only observed change was an increase in urinary zinc excretion (75). There was no change in calcium balance. The conclusion can be drawn from this work that the glomerular filtration of plasma calcium NTA was followed by dissociation of the calcium NTA and reabsorption of the calcium independently of the essentially quantitative urinary excretion of the ligand. The extensive use of ethylenediaminetetraacetic aicd, EDTA, and its calcium chelate in medicine has provided much information on the metabolism of the compounds in humans. Calcium EDTA is the treatment of choice in lead poisoning (83-86).

Spencer et al. demonstrated that administration of intravenous EDTA was followed by a variable and incomplete urinary excretion of calcium in relation to the molar equivalence of the administered ligand (87, 88). We subsequently reported that this effect was characteristic of experimetnal hypocalcemia and could be overcome by administration of carbonic anhydrase inhibitors (79). The conclusion may be drawn from these studies that renal dissociation and calcium reabsorption can occur even with a potent calcium binding chelating agent such as EDTA. It is unclear, however, that this process could be controlled for the purpose of dissolving renal calculi. Therapeutic trial of the compound by retrograde irrigation of the urinary tract has been unsuccessful in our hands, as well as in the other trials, due to local irritation by these irrigating solutions.

Abstract

A method has been developed to predict the potential utility of orally administered chelating ligands to inhibit the formation of alkaline earth urinary calculi or dissolve those present in animals. A computer program provides the calculated speciation of calcium and magnesium in simulated intestinal plasma and urinary compartments upon the addition of the ligand. Predicted levels of metal ion interaction with the added chelating agent provide suggested dosage levels and other parameters for biological testing. The addition of 0.5% of disodium nitrilotriacetate (Na_2NTA) to the diet of rats inhibited the formation of experimental mixed brushite ($CaHPO_4$) struvite ($MgNH_4PO_4$) bladder calculi formed by the implantation of zinc pellets. Established calculi were dissolved by this treatment. A brushite calculus in a dog was dissolved by Na_2NTA treatment as established by X-ray and clinical findings.

Literature Cited

1. Boyce, W. H.; Garney; F. K.; Strawcutter, H. E., J. Am. Med. Assn., 1956, 161, 1437-1442.
2. Steel, T. H. Ann. Rev. Pharmacol. Toxicol., 1977, 17, 11-25.
3. Pak, C.Y.C. J. Clin. Invest., 1969, 48, 1914-1922.
4. Robertson, W. G.; Peacock, M.; Marshall, R. W.; Marshall, D. H.; Nordin, B.E.C. N. Engl. J. Med., 1976, 294, 249-252.
5. Finlayson, B.; Both, R. Urology, 173, 1, 142-144.
6. Boyce, W. H. Am. J. Med., 1968, 45, 673-683.
7. Pak, C.Y.C.; Ruskin, B. J. Clin. Invest., 1970, 49, 2253-2261.
8. Pak, C.Y.C.; East, D. A.; Sanzenbacher, L. J.; Delea, C.; Bartter, F. C. J. Clin. Endocrinol. Metab., 1972, 35, 261-270.
9. Pak, C.Y.C.; Earnes, E. D.; Ruskin, B. Proc. Natl. Acad. Sci. USA, 1971, 68, 1456-1460.
10. Pak, C.Y.C.; Ciller, E. C.; Smith, G. W. II; Howe, E. S. Proc. Soc. Exp. Biol. Med., 1969, 130, 753-757.
11. Meyer, J. L.; Bergert, J. H.; Smith, L. H. Clin. Sci. Mol. Med., 1975, 49, 369-374.
12. Coe, F. L.; Lawton, R. L.; Goldstein, R. B.; Tembe, V. Proc. Soc. Exp. Biol. Med., 1975, 149, 926-929.
13. Pak, C.Y.C.; Arnold, L. H. Proc. Soc. Exp. Biol. Med., 1975, 149, 930-932.
14. Melnick, I.; Landes, R. R.; Hoffman, A. A.; Burch, J. F. J. Urol., 1971, 105, 119-122.
15. Russell, R.G.G.; Fleisch, H. Proc. Int. Symp. Renal Stone Res., Madrid, 1962, Ed. Cifuentes Dellata, L.; Rapado, A.; Hodkinson, A., Basel and New York, Karger, 307-312.
16. Harriosn, H. E.; Harrison, H. C. J. Clin. Invest., 1955, 34, 1622-1670.
17. Williams, H. E. "Metabolic Aspects of Renal Stone Disease. 1975-1976. The Year in Metabolism", Ed. Freinkel, N., Plenum

18. Pak, C.Y.C.; Delea, C. S.; Bartter, F. C. N. Eng. J. Med., 1974, 290, 175-180.
19. Yendt, E. R.; Guay, G. F.; Garcia, D. A. Can. Med. Assoc. J., 1970, 102, 612-620.
20. Marshall, R. W.; Cochran, M.; Hodgkinson, A. Clin. Sci., 1972 43, 91-99.
21. Smith, L. H.; Hofmann, A. E. Gastroenterology, 1976, 66, 1257-1261.
22. Smith, L. H.; Fromm, H.; Hofmann, A. F. New Engl. J. Med., 1972, 286, 1371-1375.
23. Earnest, D. L.; Gancher, G. R.; Admirand, W. H., Gastroenterology, 1976, 70, A23/881.
24. Larengood, R. W. Jr.; Marshall, V. F. J. Urol., 1972, 108, 368-371.
25. Bruin, W. J.; Baylink, D. J.; Wergedal, J. E. Endocrinology, 1975, 96, 394-299.
26. Coe, F. L. Clin. Res., 1976, 24, 396A (Abstract).
27. Smith, M.J.V. J. Urol., 1972, 107, 164-169.
28. Fleisch, H.; Bisaz, S.; Care, A. D. Lancet, 1964, 1, 1065-1067.
29. Fraser, D.; Russell, R.G.G.; Pohler, O.; Robertson, W. G.; Fleisch, H. Clin. Sci., 1972, 42, 197-207.
30. Ohata, M.; Pak, C.Y.C. Kidney Int., 1973, 4, 401-406.
31. Pak, C.Y.C.; Ohata, M.; Holt, K. Kidney Int., 1975, 7, 154-160.
32. Recker, R. R.; Hassing, G. S.; Lau, J. R.; Saville, P. D. J. Lab. Clin. Med., 1973, 81, 258-266.
33. Bonjour, J. P.; Treschsel, J.; Fleisch, H.; Schenk, R.; DeLuca, H. F.; Baxter, L. A. Am. J. Physiol., 1975, 229, 402-408.
34. Jowsey, J.; Riggs, R. L.; Kelly, P. J.; Hoffman, D. L.; Bordier, P. J. Lab. Clin. Med., 1971, 78, 575-584.
35. Jowsey, J.; Holley, K. E. Lab. Clin. Med., 1973, 82, 567-575.
36. Robertson, W. G.; Peacock, M.; Marshall, R. W.; Knowles, F. Clin. Sci. Mol. Med., 1974, 47, 13-22.
37. Musher, D.; Templeton, G.; Griffith, D. Clin. Res., 1976, 24, 26A (Abstract).
38. Rubin, M.; Kimbraugh, J. Georgetown University Medical Center and Walter Reed Army Hospital, 1950, unpublished.
39. Gehres, R. H.; Raymond, S. J. Urol., 1951, 65, 474-483.
40. Abeshouse, B. S.; Weinberg, T. J. Urol., 1951, 65, 316-331.
41. Foreman, H.; Trujillo, T. T. J. Lab. Clin. Med., 1954, 43, 566-571.
42. Foreman, H.; Finnegan, C.; Lushbaugh, C. C. J. Am. Med. Assn., 1956, 160, 1042-1046.
43. Kallistratos, G. Drug Design, 4, 1973, 237.
44. Timmermann, A.; Kallistratos, G. Isr. J. Med. Sci., 1971, 7, 689-695.
45. Kallistratos, G. Urol. Int., 1974, 29, 93-113.
46. Kallistratos, G. Eur. Urol., 1975, 1, 261.

47. Timmerman, A.; Kallistratos, G. J. Urol., 1966, 96, 469-475.
48. Heap, G. J.; Perrin, D. D.; Cliff, W. J. Med. J. Austral., 1976, 1, 714.
49. Brigman, J.; Finlayson, B. Invest. Urol., 1978, 15, 496-497.
50. Ziolkowski, F.; Perrin, D. D. Invest. Urol., 1977, 15, 208-211.
51. Riet, B. V.; O'Rear, C. E.; Smith, M.J.V. Invest. Urol., 1979, 16, 201-203.
52. Wynn, J. E.; Reit, B. V.; Borzelleca, J. F. Tox. Appl. Pharm., 1970, 16, 807-817.
53. Finlayson, B.; Miller, G. H. Jr. Invest. Urol., 1969, 6, 428-440.
54. Robertson, W. G. Clin. Chim. Acta., 1969, 24, 149-157.
55. Holzbach, R. T.; Pak, C.Y.C. Am. J. Med., 1974, 56, 141-143.
56. Robertson, W. G.; Peacock, M.; Nordin, B. E. Clin. Sci., 1968, 34, 253-260.
57. Marshall, R. W.; Robertson, W. G. Clin. Chim. Acta, 1976, 72, 253-2601.
58. Finlayson, B.; Smith, A.; DuBois, L. Invest. Urol., 1975, 13, 20-24.
59. Robertson, W. G.; Peacock, M.; Nordin, B. E. Clin. Sci., 1971, 40, 365-374.
60. Nordin, B.E.C. "Metabolic Bone and Stone Disease", Williams and Wilkins, 1973, p.309.
61. Pak, C.Y.C. J. Clin. Invest., 1969, 48, 1914-1922.
62. Prien, E. L. Jr. Ann. Rev. Med., 1975, 26, 173-179.
63. Finlayson, B. Kidney Int., 1978, 13, 344-360.
64. Perrin, D. D.; Sayce, I. G. Talanta, 1967, 14, 833-842.
65. Ingri, N.; Kakolowicz, W.; Sillen, L. G.; Warnquvst B. Talanta, 1967, 14, 1261-1286.
66. Nancollas, T.P.I.; Nancollas, G. H. Anal. Chem., 1972, 44, 1940-1950.
67. Lavender, A. R.; Pullman, T. N.; Goldman, D. J. Lab. Clin. Med., 1964, 63, 299-305.
68. Hallman, P. S.; Perrin, D. D.; Watt, A. E. Biochem. J., 1971, 121, 549-555.
69. Perrin, D. D.; Agarwal, R. P. "Metal Ions in Biological Systems" Vol.2, Ed. Sigel H., Marcel Dekker, N.Y., 1973, Ch.4.
70. Popovici, A.; Geschickter, C. F; Reinovsky, A.; Rubin, M. Proc. Soc. Exp. Biol. Med., 1950, 74, 415-417.
71. Chaberek S.; Martell, A. E. "Organic Sequestering Agents", John Wiley & Sons, N.Y., 1959, pp.587-588.
72. Budny, J. A.; Niewenhuis, R. J.; Buehler, E. V.; Goldenthal, E. I. Tox. Appl. Pharm., 1973, 26, 148-153.
73. Nixon, G. A.; Buehler, E. V.; Niewenhuid, R. J. Tox. Appl. Pharm., 1971, 21, 244-252.
74. Michael, W. R.; Wakim, J. M. Tox. Appl. Pharm., 1971, 18, 407-416.
75. Michael, W. R.; Wakim, J. M. Tox. Appl. Pharm., 1972, 22, 297.

76. Vermeulen, C. W.; Grove, W. J.; Goetz, R.; Ragins, H. D.; Correll, N. O. J. Urol., 1950, 65, 541-548.
77. Morel, F.; Morgan, J. Env. Sci. Tech., 1972, 6, 58-67.
78. Smith, R. M.; Martell, A. E. "Critical Stability Constants" Vols, 1-4, Plenum, N.Y., 1974, 1975, 1976, 1977.
79. Rubin, M.; Lindenblad, G. E. Ann. N.Y. Acad. Sci., 1956, 64, 337-342.
80. Rubin, M.; Alexander, R.; Lindenblad, G. Ann. N.Y. Acad. Sci., 1960, 88, 474-478.
81. Rubin, M. in "Transfer of Calcium and Strontium Across Biological Membranes", Ed. Wasserman, R. H., Academic Press, N.Y., 1963, pp.25-47.
82. Thomas, R. O.; Litovitz, T. A.; Rubin, M. I.; Geschickter, C. F. Am. J. Physiol., 1952, 169, 568-575.
83. Hardy, G. L.; Foreman, H.; Rubin, M.; Kissin, G.; Aub, J. C.; Butler, A. M.; Byers, R. K.; Harrison, H. E.; Shipman, T. L. AMA Arch. Indust. Med., 1953, 7, 137.
84. Rubin, M.; Gignac, S.; Bessman, S. P.; Belknap, E. L. Science, 1953, 117, 659-660.
85. Bessman, S. P.; Rubin, M.; Leikin, S. Pediatrics, 1954, 14, 201-208.
86. Bessman, S. P.; Reid, H.; Rubin, M. Ann. Med. Soc. D.C., 1952, 21, 312.
87. Spencer, H.; Vaninscott, V.; Lewin, I.; Laxzlo, D. J. Clin. Invest., 1952, 31, 1023-1027.
88. Spencer, H.; Greenberg, J.; Berger, E.; Perrone, M.; Laszlo, D. J. Clin. Med., 1956, 47, 29-41.

RECEIVED May 22, 1980.

APPENDIX

A

B

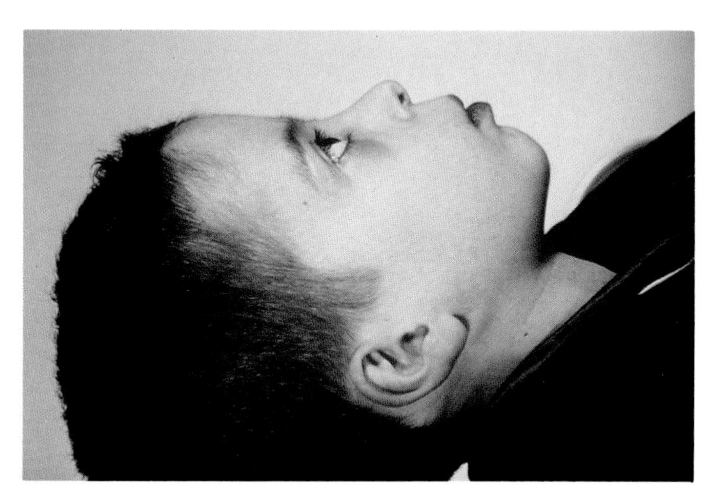

Plate 1. A patient with homozygous beta thalassemia, also known as thalassemia major or Cooley's anemia

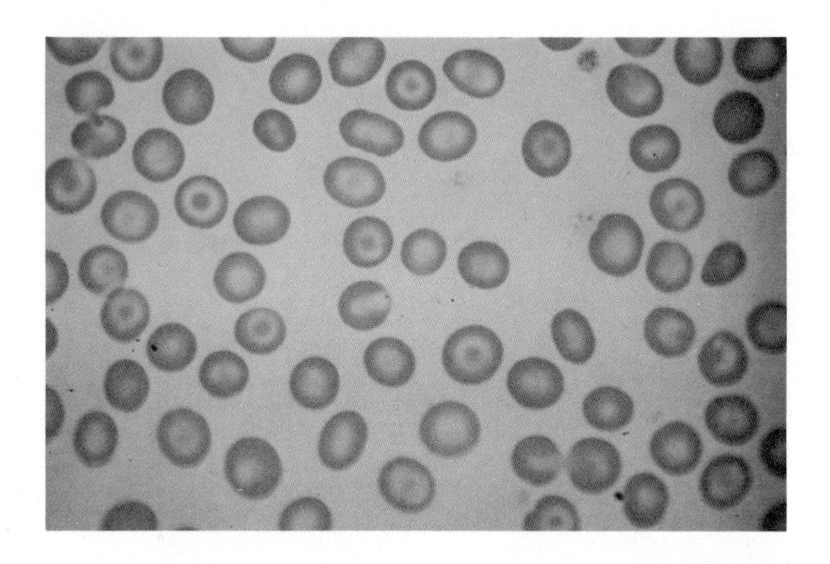

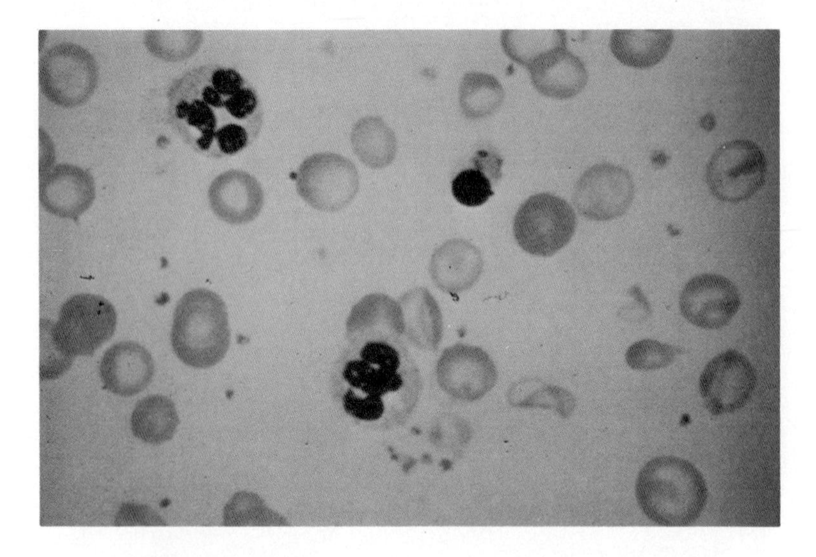

Plate 2. Normal blood smear (top); blood smear of a patient with Cooley's anemia (bottom)

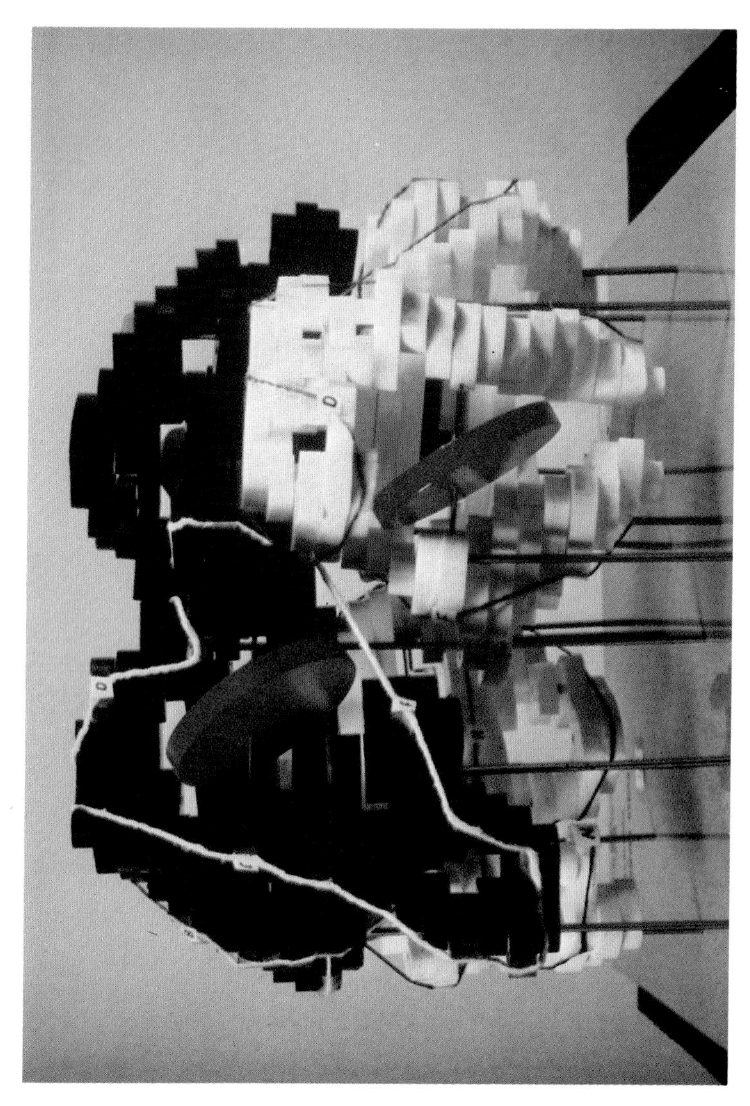

Plate 3. *Photograph of the Perutz 2-Å-to-1-cm model of hemoglobin. The string traces the primary sequence and the tags with letters labeling the helices.*

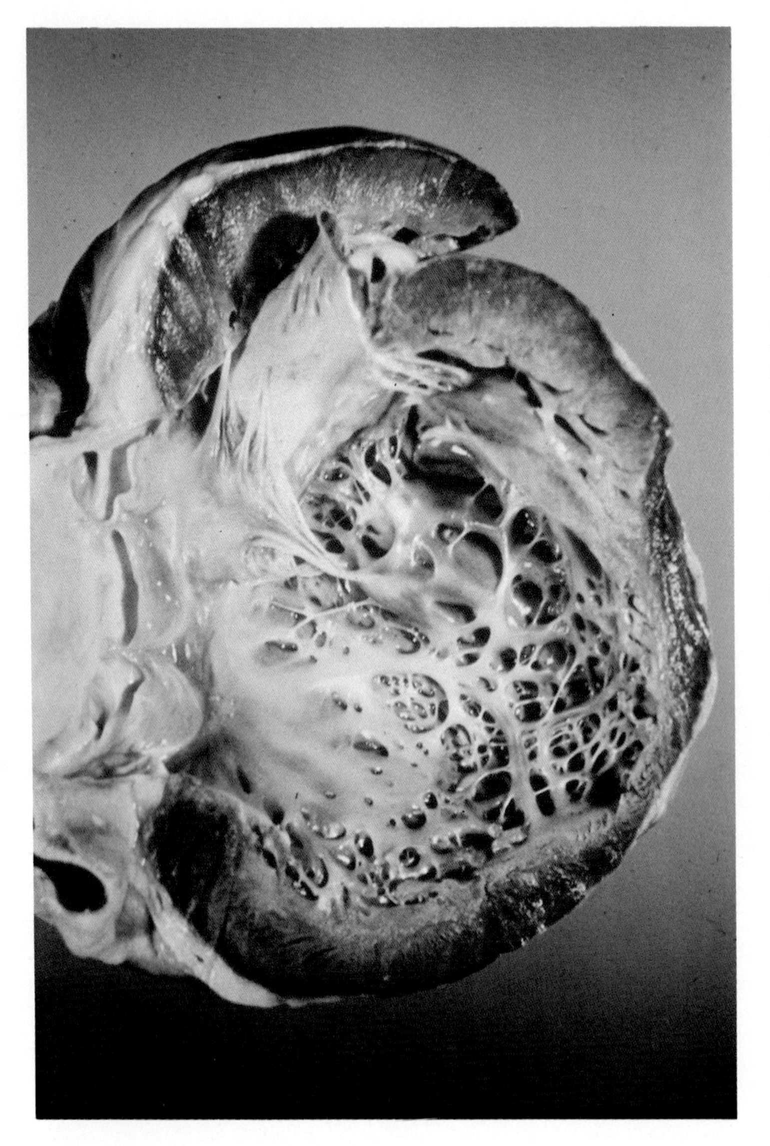

Plate 4. Cross section of the heart from a 23-year-old male who died from iron overload (transfusional hemosiderosis)

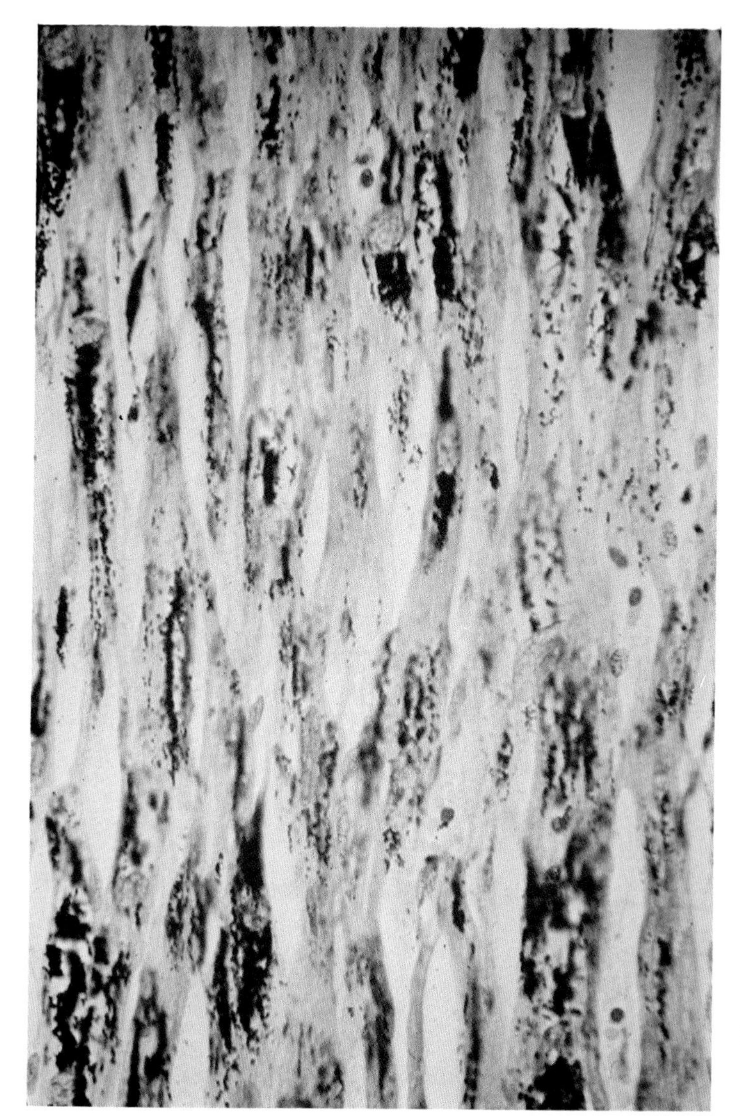

Plate 5. Iron stain of a histological section of cardiac tissue taken from the heart shown in Figure 7; $\times$ *136*

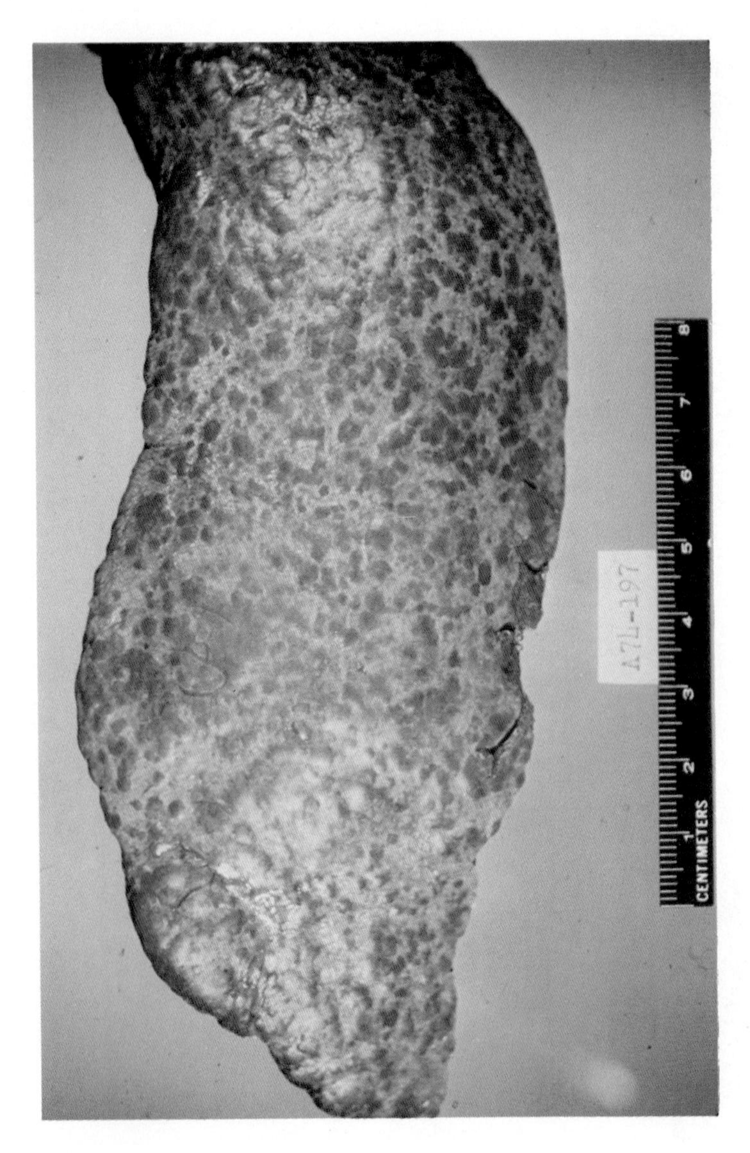

Plate 6. The liver from the same patient

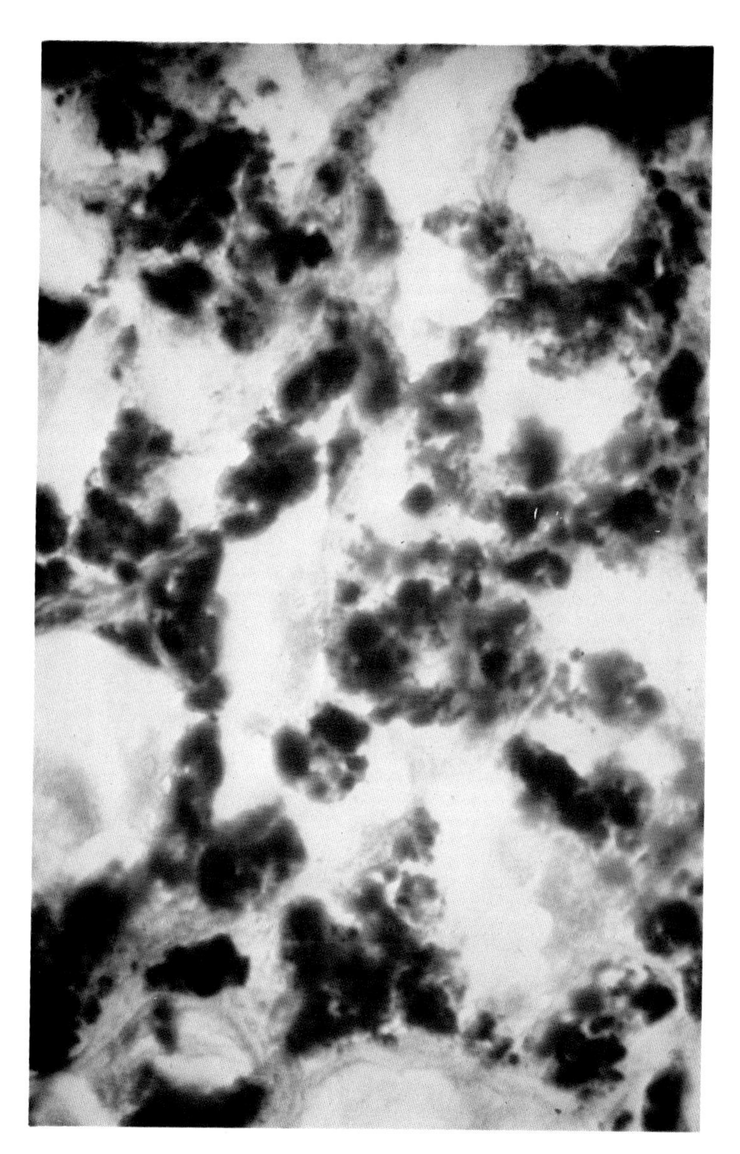

Plate 7. Iron stain of a section of pituitary gland from the same patient; × *136*

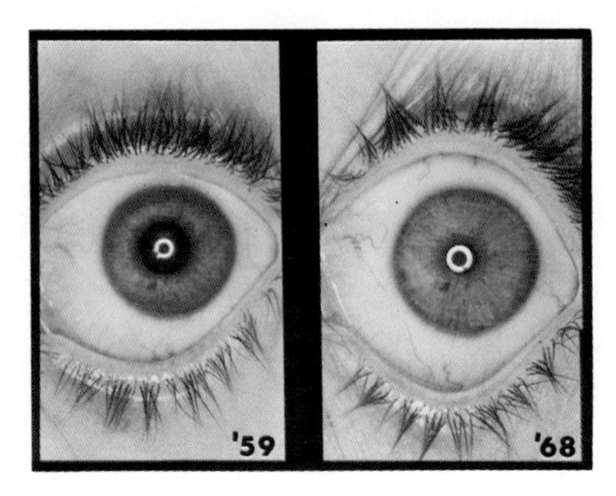

Plate 8. Virtual disappearance of corneal Cu deposits in the periphery of the cornea (Kayser–Fleischer ring) of a patient with Wilson's disease, following 9 years of treatment with penicillamine

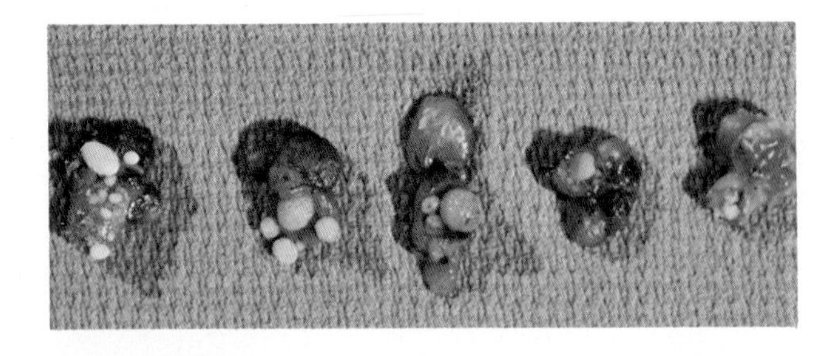

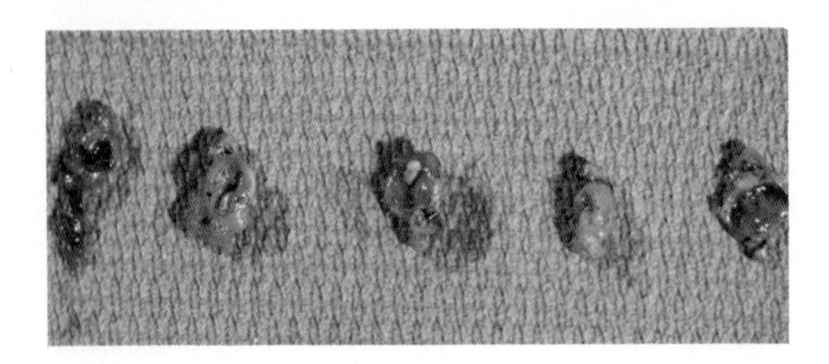

Plate 9. Mixed calcium hydrogen phosphate and magnesium ammonium phosphate bladder stones generated in rats four months after the implantation of zinc sheet (control group): (top) untreated controls, (bottom) NTA treated animals. The treated group were fed a diet containing 0.5% of Na_2NTA for the same period.

INDEX

INDEX

H

I

K

L

M

N

O

R

S